Introduction To
Social Statistics

Prentice-Hall Methods of Social Science Series

Editors

Herbert L. Costner

Neil Smelser

Introduction To Social Statistics

Vanderlyn R. Pine

State University of New York
College at New Paltz

Prentice-Hall, Inc., Englewood Cliffs, New Jersey 07632

Library of Congress Cataloging in Publication Data

PINE, VANDERLYN R.
Introduction to social statistics.

Includes bibliographical references and index.
1. Social sciences—Statistical methods.
I. Title.
HA29.P623   300′.1′82   76–52932
ISBN 0-13-496844-1

Printed in the United States of America

10 9 8 7 6 5 4 3 2 1

Prentice-Hall International, Inc., *London*
Prentice-Hall of Australia Pty. Limited, *Sydney*
Prentice-Hall of Canada, Ltd., *Toronto*
Prentice-Hall of India Private Limited, *New Delhi*
Prentice-Hall of Japan, Inc., *Tokyo*
Prentice-Hall of Southeast Asia Pte. Ltd., *Singapore*
Whitehall Books Limited, *Wellington, New Zealand*

Contents

v

Preface

This book is intended to provide an easy-to-understand approach to some commonly used procedures in social statistics. It is not intended to produce professional statisticians, and if you are inclined to become one, this book may be a good place to begin that career, but it certainly is not a good place to end it. If you are seriously interested in a career as a sociologist specializing in research, data analysis, mathematics, or social statistics, you should plan right now to take as many advanced statistics and quantitative analysis courses as you can.

We cover "some commonly used procedures in social statistics" in order to demonstrate the way they are used to analyze data gathered through social research. Today, research and analysis of data comprise an important aspect of sociology. One of our main purposes is to show how social statistics may be used as (1) a practical research device and (2) an analytical tool for examining quantitative (numerical) data.

Another aim of this book is to help teach you how sociologists carry out research. Thus, throughout the book we try to connect elements from the language of sociology to the language of statistics. Yes, statistics may be thought of as a language. As such, social statistics is used to help sociologists deal with numerical information about people. It is an essential aspect of the quantitative brand of sociology which examines such matters as population trends, unemployment, occupational inequality, delivery of health services, the extent of racial prejudice, demography, ecology, and many other concerns. It is difficult to analyze such matters when we are confronted with large aggregates of people and groups without the use of statistics.

Naturally, statistics can be manipulated, and sometimes we hear the comment, "The same statistics can be used to prove two completely opposite points of view." To some extent, such statistical manipulation is a philosophical problem, and it is not within the scope of this book to deal with the philosophical implications of statistical analysis. Instead, we leave such judgment up to you after you become familiar with the language and use of social statistics.

There are a number of reasons why this book is different from others which deal with the same material, and these differences help make it more useful for students of social statistics and more sensible for their teachers. The first reason comes in the form of a confession. I am neither a mathematician nor a statistician. Instead, I am a somewhat humanistically inclined sociologist with a background in English literature and creative writing. Thus, I must confess that social statistics was one of the most frightening courses I encountered as an undergraduate.

When I scanned the pages of my first statistics textbook and found unbelievably complex formulas, I was absolutely certain that I had made the wrong decision. At about that time, however, I carried out a research project in which I gathered some data and tried to parcel out the findings. As I did so, I found that using statistical analytical tools could be fun and illuminating as well as challenging.

Part of the fun came from realizing that it is possible to work with numbers in a simple and straightforward fashion, and that the underlying principles of many social statistics involve only *a few* basic models, theorems, and procedures. I also discovered that a knowledge of the *rationale* underlying a statistic and an awareness of *how it is derived* considerably simplify its interpretation and allow one more easily to assign meaning to the statistical characteristics of the data. This means that to understand this book, you need no more math background than simple arithmetic and a small amount of elementary algebra.

The second reason this book is different is because of the way the material is presented. We begin with some familiar concepts from everyday life, which we then translate into some familiar statistics, which we then integrate with some less familiar ones. These initial steps become the basis for a number of related statistical procedures which is what the remainder of the book is about. Thus, once you understand the initial material, you only need to see how the basic concepts are incorporated in each of the new statistical procedures. This enables us to cover those "commonly used social statistics" mentioned earlier.

In order to determine which statistical procedures are "commonly used," we examined the leading books and the research articles in the major sociology journals over the past few years. We selected those statistical procedures which are increasingly appearing in the most recent literature. This means that you do not have to struggle through statistical procedures which are of little or no help when it comes to reading present-day sociology journal articles or books. It also means that after completing this course, you should be able to read *and* understand those sociological studies which use numbers and frightening-looking statistical procedures.

The third reason why this book is different is because of the way we present the basic kinds of statistics. We have divided the book into six parts, each of which deals with an aspect of statistics that has unique characteristics and applications.

Students have found that the following five steps provide an excellent way to use this book. These steps result in considerably greater success, especially for those of you who are somewhat shy of statistics.

> First, carefully study the table of contents of the entire book. This provides an overview *in advance* of the material that will be covered. Then, examine the detailed table of contents for the specific chapter to be read. Familiarity with the outline seems to make a big difference.
>
> Second, as chapters are assigned, first read, but *do not study* one chapter at a time. Students have found that they tend to do much better and understand more easily after the modest familiarity that a "once-through-lightly reading" provides. This is especially true since "knowledge of the facts *before* the facts" seems to make them more accessible.
>
> Third, carefully read the text material and only glance at the tables or graphs cited.
>
> Fourth, go back and examine the tables and graphs in considerable detail without looking at the text.
>
> Fifth, read and study the chapter thoroughly, integrating text and tables.

This may sound like a drawn-out way of reading a chapter. However, step one is a good idea with any book, and steps two through four should take no more than three quarters of an hour. This 45 minutes spent in advance of studying the chapter has reduced the actual reading time and study of the chapter by as much as one half. Furthermore, students who follow this format generally do far better in terms of understanding the material covered.

A legion of fine people provided me with advice, assistance, admonishment, encouragement, suggestions, and helpful comments over the past five years through the creation of this book. I would like to thank them for their efforts. First, there are the students who have studied numerous draft versions in the course Social Statistics. Their class participation, careful reading, discovery of errors and oversights, and general willingness to serve as "experimental subjects" with a developing book has endeared them to me. Even though they may have suffered a certain lack of precision in newly emerging ideas, I hope that the benefits of necessarily careful reading and the fun of being part of creating something made it worthwhile for them. Thanks to all of you unnamed but appreciated former students.

Expert advice and suggestions were offered by many professional sociologists. I am especially grateful to Herbert L. Costner, James A. Davis, Edmund D. Meyers, Jr., Lynne Roberts, and Jerald G. Shutte, all of whom provided extensive reviews and detailed helpful suggestions. Thanks also to a number of anonymous reviewers whose comments added to various aspects of the book.

Encouragement and solid advice were provided by Prentice-Hallers Neale Sweet, Irene Fraga, and Kitty Woringer. Guidance above and beyond the call of duty generously was given by Sociology Editor Edward H. Stanford. His unfaltering support often was the major reason why I did not scrap the project altogether. A hearty thanks to all of you at Prentice-Hall.

Good editorial and statistical input is a precious item and often all too hard to come by. I have been fortunate enough to have received it from my wife Patricia P. Pine, and to her I express my great thanks. I have also received it from my associate and friend Kathy D. Williams. She has had the onerous task of reading every word, typing every version, studying the material, and helping in every way to get this book done. To her, my deepest appreciation is too little to offer, but "Thanks, Kathy" anyway.

Vanderlyn R. Pine

The Application of Statistics in Social Research

Statistics plays a very important role in sociology, even though it is often maligned, at times misunderstood, and more than occasionally misused. The purposes of this chapter are to illustrate the relevance of statistical analysis and to provide an orientation to statistics as it is presently used in sociology.

1.1 THE DEVELOPMENT OF SOCIAL STATISTICS

Social statistics has evolved from three seemingly unrelated fields of interest over a period of approximately three centuries and in stages of about one hundred years. These three fields were not welded together consciously; rather they merged as time passed and more and more attention was given to sociological thought.

The first field that contributed to the development of statistics was probability theory. It came into prominence about the middle of the seventeenth century, when a great deal of attention was given to gambling and games of chance. For example, members of the French royalty plotted the regularity with which the outcomes of certain gambling events, such as rolling dice, could be predicted, given specific circumstances. These calculations, combined

1

with a mathematical interest in chance, gave rise to the development of theories of probability.

The second field that contributed to the development of statistics involved the recording of government facts and figures. It came to bloom in the middle of the eighteenth century, when most advanced technological nations paid increasing attention to records of state concern. Interest in population change, the registration of births and deaths, the number of children per family, questions about fertility (birth), morbidity (illness), mortality (death), and many other social phenomena became matters of concern to most growing industrialized countries. Records of these events were called "statistics," a word combining the Latin *status* (or state) and *istic* (or concern with): thus, a concern with state matters.

Interest in numerical data continues in these and many other areas. For example, Table 1.1.1 shows the population of the United States from 1790 to 1970 in thousands of people. Records like this are published by the United States Department of Commerce and can be found in a number of sources available in most college libraries under such titles as *Historical Statistics, Colonial Times to 1957*; *Census Reports in Population, U.S. Census of Population*; and the *Statistical Abstract of the United States.*

A third field contributed to the development of *social* statistics, and it was

Table 1.1.1   U.S. population: 1790 to 1970
(in thousands)

1790	3,929
1800	5,308
1810	7,240
1820	9,639
1830	12,866
1840	17,070
1850	23,192
1860	31,365
1870	38,469
1880	49,984
1890	62,590
1900	75,643
1910	91,560
1920	105,284
1930	122,178
1940	131,081
1950	149,984
1960	177,315
1970	200,285

Source: Adapted from *Statistical Abstract of the United States, 1971*. U.S. Bureau of the Census, Washington, D.C., 1971 (92nd edition), page 24, No. 22

sociology itself. It arose around the middle of the nineteenth century when the French philosopher Auguste Comte (1798–1857) introduced the word *sociology*, meaning the study of society. It was clear to early sociologists that there was much to be learned from numerical analyses of society. The Belgian statistician Adolphe Quetelet (1796–1874), for example, analyzed the annual regularity of crimes, suggesting that predictions about criminal activity could be made based on the regularity of certain crimes. Later, French sociologist Emile Durkheim (1858–1917) analyzed the rates of suicide in Europe, delineating a typology of suicide that is still accepted.

Toward the end of the nineteenth century and during the early years of the twentieth century, statistical techniques from agriculture, biology, genetics, and mathematics were adapted to and incorporated in social statistics. Notably, the work of Englishman Sir Francis Galton (1822–1911) and his student Karl Pearson (1857–1936) focused on the application of statistical methods to biology. Today, the field of social statistics is expanding and developing on its own. As statistics is currently constituted, its general purpose is to provide order to large amounts of information. From this perspective, social statisticians seek to describe, estimate, explain, predict, and infer by using sociological data.

The way statisticians work is neither arcane nor arbitrary. As a matter of fact, there is a close similarity between statistics in social research and the way most people use available information in making everyday decisions. In a common sense way, all of us are statisticians in that we use our own system of logical steps in thinking through the way the world works. Most of us gather evidence by observing the way things happen around us; then this evidence is pulled together in our mind's eye, and we come to certain conclusions about what it all means. For example, before we cross a busy street, we generally examine the evidence by looking at the oncoming cars from both directions, and then we make a conclusion based on our best estimate about the traffic. "Sophisticated" social research works in a similar fashion, utilizing more precise but just as sensible statistical methods.

1.2 QUANTITATIVE SOCIAL RESEARCH

The relevance of social statistics is clearest when viewed in the framework of sociological methodology and *empirical* (based on observation or experiment) research. In general, social research is an attempt to provide evidence about hypotheses, to test existing theories, and to provide a basis for new ones. A fundamental goal of social research is to achieve accurate explanations, predictions, and descriptions of human behavior. Even so, it is important to be aware that answers are *sought* by research, but they are not necessarily found by it.

The word *hypothesis* can mean many things, but as many sociologists use it, it refers to descriptions, propositions, guesses, and speculations.[1] For example, if we suspect that academic performance is somehow related to a student's motivation, we have formed an hypothesis. Often hypotheses arise from our own observations, which are, after all, sensory impressions or "messages" about things and events that are seen, heard, or experienced. When we report on such observations, we are attempting to describe "reality." In this framework, hypotheses ought to be "provable" because of evidence that exists in the form of observations that "fit" or "agree with" the hypotheses. Social research is the gathering of this kind of evidence, and in this sense it tests generalizations about "reality".

Social research is interested primarily in the relationships between two or more characteristics or attributes called variables. This is largely because social theories are seldom limited to one variable. For example, it would be unusual for us to be concerned with academic performance unless we assumed that grades influence or reflect other aspects of a student's life—admission to graduate school, attitude toward education, and the like. Furthermore, multivariate relationships (those involving several variables) are necessary for explanations of social behavior and for causal analysis.

1.3 STATISTICAL TECHNIQUES AND SUBSTANTIVE QUESTIONS

The issues and problems that we have been discussing have been incorporated in the actual research projects of many sociologists. In order to illustrate the wide variety of ways in which social statistics can be used in research, we will examine and consider some projects that use statistical procedures. In all of them, statistical techniques are used to help arrive at answers to substantive questions about such things as voting behavior, attitudes toward communism, the distribution of wealth in American society, the formation and maintenance of small groups, the attitudes of high school and college students, occupational mobility and social structure, demographic studies, and many others. In this chapter we merely mention the statistical method, but we do not define the terms. Later on, we will explain method and techniques in greater detail. From this early discussion, you will get an idea of the various terms employed and the uses of statistical methods and techniques.

Politics and the formation of opinion during a presidential campaign are the subjects of a study conducted by Bernard Berelson, Paul Lazarsfeld, and

[1] William N. Stephens, *Hypotheses and Evidence* (New York: Thomas Y. Crowell Company, 1968).

William McPhee. Their findings are reported in their book *Voting*.[2] This research was conducted by studying residents from a small town during the presidential campaigns of Harry S. Truman and Thomas E. Dewey in 1948. The data were collected through several repeated interviews with the same voters. This technique is called a *panel study*, and it enables the researcher to examine trends over time. Many variables were investigated in this study, including such things as voters' perception of politics, reactions to issues, attention to the mass media, the influence on political preferences of social class, religion, and institutional leadership of the local community. By using cross-sectional surveys, trend tabulations on changes in attitudes and behavior from one interview to another, panel tabulations examining the turnover in variables at two or more points in time for the same people, percentage tables, histograms, line graphs, but "no formal statistical techniques,"[3] the authors present a number of important, substantive findings. They found that discussions about politics occur mainly within certain groups of people and that such discussions are likely to take place among friendship groups, religious groups, and civic groups. They also found that the changing attitudes of the apathetic segment of society tend "to hold politics together" because of "swinging" to one candidate or another. Finally, they found that people tend to become more conservative as the election draws nearer.

In the early 1950s, a great deal of public attention was directed at governmental repression of American communism. To study *Communism, Conformity and Civil Liberties*,[4] Samuel A. Stouffer investigated (1) the reaction of American society to the communist conspiracy outside and inside the United States and (2) whether or not people, in their zeal to halt the communist conspiracy, might be willing to sacrifice some of the very civil liberties the conspiracy supposedly would destroy. A large-scale national sample of six thousand men and women was selected for interviews by probability sampling methods in the summer of 1954. In addition, a special sample of civic leaders was also drawn. A major purpose of the study was to determine what Americans thought about the problem of communism during the height of the nationally televised Senator Joseph McCarthy hearings. The variables analyzed include age, education, region of the country, sex, and religion. Through the use of percentage tables, histograms, and cross-tabulation tables, Stouffer demonstrates that rising level of education, decline in authoritarian childraising, geographic mobility, and exposure to television give rise to increased independent thinking and respect for people with different ideas.

[2]Bernard Berelson, Paul Lazarsfeld, and William McPhee, *Voting* (Chicago: The University of Chicago Press, 1954).

[3]*Ibid.*, p. xii.

[4]Samuel A. Stouffer, *Communism, Conformity and Civil Liberties* (New York: John Wiley and Sons, Inc., 1955).

Sociologists have long been fascinated with the question of why some groups are maintained while others die. In *Great Books and Small Groups*,[5] James A. Davis examines one aspect of this question, analyzing whether or not exposure to the Great Books program during 1957 led to changes in group members' knowledge, their reading habits, attitudes, and civic participation. Davis studied over nineteen hundred individuals who were sampled from discussion groups, and he measured the loss, retention, or growth of membership over a one-year period. The variables with which he dealt were preparation for the program, marital status, religion, political preference, education, occupation, subjective role classification, and the dropout rate from the program. By the use of line graphs, percentage tables, and Yule's Q values, Davis found that group discussion and group relationships are important factors in program retention, that discussion activity is the keystone for the success of the program, that small groups are affected considerably by the social structure both within and from outside the group itself, that most individuals play a large number of roles, and that their behavior in the group is a rather complex combination of their present situation and residues and habits from earlier patterns outside the group.

One of the pressing concerns in American society is the impact of education on the social life of students. In *Adolescent Society*,[6] James S. Coleman is interested in comparing the consequences on the students within two status systems: one rewarding achievement in only one particular activity, and the other rewarding achievement equally in many activities. His concern is with those activities that have one social meaning in certain settings and different social meanings in other settings. Coleman studied students from ten high schools located throughout northern Illinois during the academic year 1957–58, examining all the students, their teachers, and their parents. The variables examined by Coleman are the interests and attitudes of the teenagers as measured by leisure activity, drinking and/or smoking, watching TV, homework, sex, values as measured by the importance placed on such things as popularity and good grades, and the effect of the social system on all of these. Using such statistical techniques as percentage tables, stratified percentage tables, line graphs, frequency distributions, averages, scales, standard errors, chi-square, proportions, probability, and standard deviations, Coleman concludes that adolescents have their own culture, which is considerably influenced by peer pressures and standards. Furthermore, he concludes that these values are formed by the student microcosm and not by society at large, although, he points out, society might influence the formation or selection of certain values.

Occupational mobility in American society is examined in great detail

[5]James A. Davis, *Great Books and Small Groups* (New York: The Free Press of Glencoe, Inc., 1961).

[6]James S. Coleman, *Adolescent Society* (New York: The Free Press, 1961).

by Peter M. Blau and Otis Dudley Duncan in their study of the *American Occupational Structure.*[7] The issue of social stratification and the legacies of sociologists such as Karl Marx, Max Weber, and William Lloyd Warner form the backdrop of their study, which is a sophisticated, multivariate statistical analysis of the factors that influence social mobility. Their data are drawn from a sample of over twenty thousand American men in the labor force between the ages of twenty and sixty-four. Many statistical techniques are utilized throughout the study, including ranking, percentage tables, means, correlations, regression coefficients, partial regressions, and path analysis. The variables utilized are size of residence, ethnic background, family type, geographic mobility, sibling patterns, region, color, marital status, and education of father, siblings, and respondent. One of the important findings is that the dynamics of the stratification system in the United States are to a great extent dependent upon the occupational structure. Notably, greater or lesser mobility is dependent upon the complex interrelationships between education, first job, number of siblings, and occupation.

Otis Dudley Duncan, David Featherman, and Beverly Duncan seek to elaborate upon the complexities of how and to what extent social and economic status is intergenerationally transmitted in their study *Socioeconomic Background and Achievement.*[8] They analyze the relationships among education, occupation, and income from a major survey of social mobility based on a large-scale national sample. Duncan, Featherman, and Duncan delineate the factors that can be identified as influencing the achieved status of a given occupation. Analyzing most of the variables covered by Blau and Duncan, this study uses such statistical techniques as mean scores, correlations, regression, partial regression, and path analysis to measure the relationships among the variables.

Throughout this book, we will use examples from the above-mentioned studies as well as from the *Statistical Abstract of the United States.* The point we will be making is that social statistics are used to analyze substantive questions, the associations among variables, and causal assertions.

1.4 FLAWS IN THE USE OF STATISTICS

It is important to be aware that overly simple statistical analyses may lead to inappropriate conclusions. In the first place, the size of a sample and the way it is selected may distort the findings considerably. Suppose, for example,

[7]Peter M. Blau and Otis Dudley Duncan, *American Occupational Structure* (New York: John Wiley and Sons, Inc., 1967).

[8]Otis Dudley Duncan, David Featherman, and Beverly Duncan, *Socioeconomic Background and Achievement* (New York: Seminar Press, 1972).

you read that in a current survey of college students, 80 percent of the respondents reported smoking more than two packs of cigarettes a day, compared to only 10 percent in a similar study conducted last year, suggesting that the amount of smoking is on the increase. If, however, you learn that the current sample was of five students in a smoke shop and the earlier sample was of five students in a gym, you would be wise not to trust such a conclusion. Both studies are examples of *sample size distortion* and *selection distortion*, i.e., the samples are (1) too small and (2) not necessarily representative of college students in general.

In a fascinating little book entitled *How to Lie With Statistics*,[9] Darrell Huff points out some of the ways in which statistics are commonly misused. One of his examples demonstrates how a change in measurement procedures when studying trends can seriously distort conclusions that are derived from statistics. Figure 1.4.1 presents Huff's "gee-whiz graph" and his explanation, both demonstrating the danger of such a distortion.

Another example of how statistics may be misleading occurs in a situation in which there is a spurious (apparent) relationship between variables but not a substantive one. For example, it could be argued that a correlation between air pollution and long life means that air pollution brings about long life. We can weed out what is causing such a spurious relationship by analyzing the relationship controlling for a third variable. In this case, we might find that there also are high correlations between the level of technological advancement and both air pollution and long life. If so, we could argue that this third variable causes both, and therefore the impression that air pollution and long life are causally related is false. Thus, additional statistical techniques can be used to clarify actual relationships.

Another flaw that arises in overly simple statistics is often referred to as the *ecological fallacy*. This fallacy occurs when we conclude that because there is a correlation between two variables in a community, there is also the same correlation for individuals in the community. For example, in a community with three candidates running for one office, a vote that is evenly divided may reflect community indecision. However, it would be an ecological fallacy to conclude that individual voters were indecisive about the candidates. Without any complex calculations, we can analyze the vote by looking at the proportion of votes cast for each candidate (33.3 percent–33.3 percent–33.3 percent). The split shows that the indecision is not on an individual level but rather on the community level.

Throughout this book, we will explain some of the intricacies of many statistical techniques. You should be mindful that these techniques can be misused, and you should develop an abiding conscience about their applica-

[9]Darrell Huff, *How to Lie With Statistics* (New York: W. W. Norton & Company, Inc., 1954).

We'll let our graph show how national income increased ten per cent in a year.

Begin with paper ruled into squares. Name the months along the bottom. Indicate billions of dollars up the side. Plot your points and draw your line, and your graph will look like this:

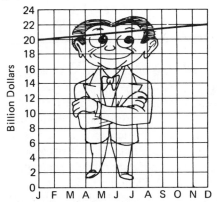

Now that's clear enough. It shows what happened during the year and it shows it month by month. He who runs may see and understand, because the whole graph is in proportion and there is a zero line at the bottom for comparison. Your ten per cent *looks* like ten per cent – an upward trend that is substantial but perhaps not overwhelming.

That is very well if all you want to do is convey information. But suppose you wish to win an argument, shock a reader, move him into action, sell him something. For

that, this chart lacks schmaltz . . .

(A further trick) will make your modest rise of ten per cent look livelier than one hundred per cent is entitled to look. Simply change the proportion between the ordinate and the abscissa. There's no rule against it, and it does give your graph a

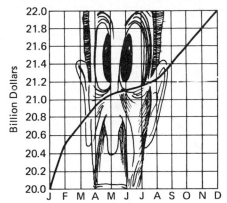

prettier shape. All you have to do is let each mark up the side stand for only one-tenth as many dollars as before.

That *is* impressive, isn't it? Anyone looking at it can just feel prosperity throbbing in the arteries of the country. It is a subtler equivalent of editing "National income rose ten per cent" into " . . . climbed a whopping ten per cent." It is vastly more effective, however, because it contains no adjectives or adverbs to spoil the illusion of objectivity. There's nothing anyone can pin on you.

Figure 1.4.1. Huff's gee-whiz graph and his explanation of it. (*Darrell Huff,* How to Lie with Statistics. *New York: W.W. Norton & Company, 1954, pp. 61–63.*)

tion. Knowing the detailed use of statistics will help in understanding them in the long run. Huff puts it this way:

> The secret language of statistics, so appealing in a fact-minded culture, is employed to sensationalize, inflate, confuse, and over simplify. Statistical methods and statistical terms are necessary in reporting the mass data of social and economic trends, business conditions, "opinion" polls, and the census. But without writers who use the words with honesty and understanding and readers who know what they mean, the result can only be semantic nonsense.[10]

[10]*Ibid.*, p. 8.

1.5 QUANTITATIVELY DESCRIBING RELATIONSHIPS AND ANALYZING SOCIAL SCIENCE DATA

As you become familiar with some of the computations used to come up with statistical answers, do not lose sight of the purpose of social statistics. Let us state the basic idea of the last few pages: The primary interest in social statistics is in the association between theoretically and substantively important variables.

The mathematics of statistics is relatively straightforward and easy to understand. However, using this math at all is not very sensible unless there is some interesting theoretical or substantive question under consideration. Interest in the variables in the studies cited in Section 1.3 exists because the investigators believed that attributes and characteristics such as socioeconomic status, attitudes towards communism, political interest, distribution of wealth, subjective role classification, peer group pressure, religion, employment, and so forth, all influenced certain other variables.

It is with two-variable relationships that our interest begins. Sometimes we look at a two-variable relationship at various constant states of the third variable, or we view it as if we adjusted for the association of the third variable.

When we examine the simple relationship between two variables and do not control for any others, we call it a *zero-order relationship*. When the relationship between two variables is examined holding constant or controlling for one other test variable, we call it a *first-order relationship*. For each control or test variable, the relationship is described as *nth-order*, depending upon how many there are. The third (or more) variable might clarify the original relationship or it might have no effect. Later, our interest will be in measuring quantitatively what happens to associations between variables when others are held constant.

The research process is not nearly as precise as the following description suggests; however, this or a similar sequence of steps indicates a common approach to research. We start with a sociological problem that we want to investigate. Then we formulate a general research problem and develop a theoretical framework. We move from the general research problem to specific research questions, often called *research hypotheses*. The research problem can be thought of as a set of questions that can be answered by the research rather than as a set of hypotheses about the problem. From the general problem there should be a logical extension to the specific research questions. Put differently, the general problem should grapple with an issue in a somewhat abstract fashion, whereas the specific research questions should deal more concretely with the crucial questions that later become hypotheses.

A *unit of observation* is the most elemental item (unit) observed in one's research. For example, we may use individuals enrolled in the school we are studying as the basic units of observation, or we may use the school itself as the basic unit of observation. After the problem has been determined, an evidence-gathering device or measuring instrument is designed and developed for the gathering of the data. Once data have been gathered, they are processed for analysis. This is where social statistical methods become useful analytical tools.

Statistics enables us to reduce a large volume of numeric values to a much smaller volume. This reduction enables us to summarize distributions, to measure relations among variables, to assess population values on the basis of sample values, and other similar steps. By reducing the quantity of information by statistical methods, we increase our ability to interpret the findings.

EXERCISES

1. In your own words, define, describe, or discuss the following terms and give an hypothetical example of each:

   empirical research
   hypothesis
   multivariate relationships
   spurious relationship
   zero-order relationship
   first-order relationship
   unit of observation

2. There are many definitions of the term "statistics." Check some of these definitions in reference sources for fields other than sociology. In what ways do they differ, and in what ways are they similar?

3. Suppose someone you know is taking a course in business math. She asks you how it differs from social statistics, claiming that both work with numbers and both are used to analyze everyday occurrences. In your own words, provide a convincing answer.

4. List at least 6 uses of statistics in sociology which are already familiar to you.

5. Statistics are important in American society. What effect do they have on such things as social policy, planning, government funding, and other related matters?

6. Note the number of times newspaper or magazine advertisements use statistical data and if the ads appear to distort the data. Describe and discuss the problem of "lying with statistics."

7. In your own words, briefly describe the development of present-day social statistics and the interconnection between it and substantive sociological issues.

8. Find 5 recent issues of the *American Sociological Review* and the *American Journal of Sociology*.

   A. compile the following information:

NAME OF JOURNAL	VOLUME NUMBER	NUMBER OF ARTICLES	NUMBER OF ARTICLES CONTAINING STATISTICS

   B. Do a cursory analysis of these articles. How do the statistics used differ from those you discussed in questions #4 and #6?

9. In your own words, describe the role of statistics in the research process.

10. Using any one of the sociological research projects described in Section 1.3, in a short essay describe the substantive sociological issues treated and the kinds of statistics employed.

part I

Classificatory Descriptive Statistics

2

Classification and Measurement in Social Research

In this chapter we describe the nature of quantitative data used in social statistics, focusing on the variety of observable units (cases) common in social research. These observable units include individuals, families, groups, organizations, societies, and so forth. We will see how sociologists deal with the attributes and characteristics that are possessed by individuals or groups and how sociologists classify people and organizations according to these attributes.

Classification is a common-sense procedure with which most of us are quite familiar. What we do in social statistics is formalize the classification procedures somewhat and deal with such things as levels of measurement and the way people are distributed over these levels of measurement.

2.1 CASES AND VARIABLES

In social research, we are generally interested in assigning variables to cases.[1] A *variable* is an attribute, characteristic, or trait that is commonly possessed

[1]For a more complete treatment of this and the following discussion, see William N. Stephens, *Hypotheses and Evidence* (New York: Thomas Y. Crowell Company, 1968), esp. pp. 1–20 and 77–113.

in differing levels, amounts, or categories by a number of cases. Cases are the objects of our inquiry and may be individuals, groups, organizations, events, and so forth. A case is the basic unit we observe, and we can say that a case is that which the hypothesis talks about, while a variable is that which the hypothesis imputes to the case. For example, a student (the case) has some *level* of intelligence, some *amount* of popularity, and some *amount* of energy (all variables). Or the sex (variable) of a person (the case) is either male or female (both categories). Quite simply, cases are the objects whose characteristics vary, and variables are the characteristics on which the cases differ.

If we were going to study 200 college students, each one would be a case, and each would possess such characteristics as age, sex, a level of education, attitudes, preferences, etc. If, instead, we were going to study 50 colleges, the enrollments of which total 250,000 students, each college would be a case, and each would possess such characteristics as an overall grade point average, a total number of students, a certain number of faculty members and administrators, a dropout rate, etc. If, instead, we were going to study 10 societies, the number of colleges in which total 15,000 and the enrollments of which total 4,500,000 students, each society would be a case, and each would possess such characteristics as a rate of college enrollment, a suicide rate, an unemployment rate, a fertility rate, a distribution of races, a distribution of ages, a form of government, an ethnic composition, a gross national product, a consumer price index, etc. Clearly, there is much grist for the social statistician's mill in terms of cases and variables.

In addition to its name, for statistical purposes each variable possesses a *verbal definition*, its own categories or classes, and an *operational* (working) *definition*.[2] A verbal definition is the words used to define a variable. An operational definition is a procedure (operation) for carrying out the sortings of cases into the categories. Clearly, some variables can be measured by sorting the cases into classes along a continuum, and others can be measured by sorting the cases into two or more categories that are mutually exclusive (no case in more than one category) and totally inclusive (includes every case). Thus, each individual or group (case) studied possesses a specific category, amount, or level of a particular measurable variable. For example, in the study *Voting* the authors are interested in the concept of political preference, which they treat as a variable. Its verbal definition is the political party with which a person is affiliated. Its categories are Republican, Democrat, and Other. The operational definition is the answer of each respondent to the question, "In which political party are you registered?"

Importantly, operational definitions are measurement-sorting procedures (such as actual questions) that attempt to connect an abstract concept (poli-

---

[2]This section blends some of the ideas developed by Stephens, *Hypotheses and Evidence* and those of James A. Davis, *Elementary Survey Analysis* (Englewood Cliffs, New Jersey: Prentice-Hall, Inc., 1971), esp. pp. 9–62.

tical preference) to a single, observable variable (party membership). They are intended to answer specific research questions, which are then connected to the theoretical formulations or general hypotheses of the research. It is important to keep clear the distinctions between theories (or theoretical definitions), research hypotheses (or questions about variables), and procedures (or operational definitions). The three are different.

Measurement is a process, and what is being measured is a unit of observation. Thus, measurement differs from that which is being measured. We first have a theory, say, that there are different "quality" students; next, we have a measure, for example, grade point average; and finally, we have a unit, an individual student. Importantly, we actually *measure* things with operational definitions of variables. If we talk about "good students and poor students," we are treating "quality" in a theoretical way. On the other hand, if we measure the quality of given students, we operationally define quality in terms of, say, GPA. Thus, measurement refers to giving cases numerical scale scores with respect to some variable. In general, social research measures with verbal instruments such as questionnaires and scales.

Variables may be classified according to whether they are discrete or continuous. *Discrete variables* are ones in which not all values are possible. For example, it is impossible to have 1 and 3/4 living children; thus, the number of children for a given family will be a discrete number, even though the *average* number of children per family may be a fraction. *Continuous variables* exist when *all* values are possible. This means that we could subdivide into smaller and smaller levels that blend until any value may be designated. For example, weeks may be divided into days, hours, minutes, seconds, and so forth. Such divisions of variables and data are necessarily a compromise with reality, but in order to count units and summarize and analyze them statistically, we must accept such a compromise.

In addition to being discrete or continuous, variables also may be classified according to whether they are observed or inferred.[3] *Observed variables* generally can be directly measured with a specific measuring instrument. *Inferred variables* are presumed characteristics or attributes that cannot be measured directly. For instance, the concepts of political preference or academic performance are inferred variables, while someone's party membership or a student's GPA are observed variables. It is possible to connect the two types by using observed variables to represent (stand for) inferred ones.

As the name implies, we can measure observed variables directly by observing the specific characteristics or attributes in which we are interested,—sex, age, and so forth. With inferred variables, however, we must choose or develop indexes that represent the characteristics or attributes that are inferred— social class, attitudes toward politics, etc. Moreover, an observed variable

[3]For a more complete treatment of this topic see Stephens, *Hypotheses and Evidence*, esp. Chap. 5.

can be measured by observing one characteristic, whereas an inferred variable can be measured by many different indexes. In sociology, we are most interested in inferred variables, and even though we may treat some observed variables as background characteristics, the hypotheses we analyze are primarily about the relationships between inferred variables.

When we use an index, it amounts to constructing a conception of something intangible, and such abstractions are called *constructs*.[4] For example, an index of social class characteristics is a construct, and this index may be thought of as a devised conception of something intangible, that is, the inferred variable social class. Dealing with constructs gives rise to at least three problems. First, many constructs of inferred variables are not clearly (if at all) defined. Second, often the inferred variables themselves have vague or confusing definitions. Third, we are seldom certain whether or not there is a connection between a constructed index and the inferred variable it is supposed to be measuring. Thus, we might have an index that we believe measures an inferred variable, but the index and the variable really may not go together. This final problem raises the two serious issues of validity and reliability.[5]

Our concern with *validity* is whether or not we are justified in linking an inferred variable to the index that we use to measure it. The ideal situation is to have good links between our theoretical definition, our verbal definition, and our operational definition. The question is, therefore, do the differences in our indicators (operational definitions) reflect true differences in the variable? Put differently, does an index bear close correspondence to the underlying variable that it is supposed to represent?

Our concern with *reliability* is (1) whether or not repetitions of our measure will give similar results or (2) whether or not different measuring instruments give the same results. Put differently, we are concerned with the extent to which variation on a given variable is due to measurement inconsistencies. For instance, to measure inches consistently we would want a ruler that is not made of flexible rubber. Similarly, each time a sociological measuring instrument is repeated, we want the results to be reliable (consistent). A reliable measure may be either valid or not; however, an unreliable measure can never be valid. Even though a balance scale (the kind in your doctor's office) is a very reliable instrument, it would not be valid for measuring political preference. Thus it makes sense that a highly reliable instrument may be either valid or invalid. What about the second part of our statement—that a highly

[4]For a fuller discussion see Claire Selltiz, Marie Jahoda, Morton Deutsch, and Stuart W. Cook, *Research Methods in Social Relations* (New York: Holt, Rinehart and Winston, 1965), esp. pp. 41–42.

[5]Many of the validity questions are discussed in more detail by Derek Phillips, *Knowledge From What?* (Chicago: Rand McNally and Company, 1971), esp. pp. 16–19; Stephens, *Hypotheses and Evidence;* Selltiz, et. al., *Research Methods in Social Relations;* and Davis, *Elementary Survey Analysis.*

unreliable measure can never be valid? It would not be reliable to measure academic performance according to grades that were determined by drawing cards from a well-shuffled deck of cards. Such a measure is not reliable; thus it cannot be valid.

What we have said so far sets the stage for much of what comes later. Before we move on, however, let us determine how social statistics plays a practical part in the research process and touch on the ways in which sociologists actually carry out classification.

2.2 CLASSIFICATION AND STATISTICS

Classification is such a common practice that the familiar idea of it needs very little explanation. Just consider everyday language in which nouns and adjectives provide means for classifying things as variables and putting things into categories. For example, when we say "look at the big crowd," we have classified an observation about the size of a collection of people. In this instance, "size" is the variable name, "crowd" describes the nature of the case, and "big" is the category. At this level, it is clear that not only is classification common, but also it may be straightforward and easy.

Sociological classification grows out of the social importance of such things as sex, race, social class, age, education, occupation, residence, region of the country, academic performance, attitudes toward communism, and so forth. Each of these variables has important sociological implications, especially when examined in terms of "cause" and "effect." For example, Stouffer points out that one's attitude toward communism is influenced by one's level of education; that is, education is said to "cause" attitudes.[6] We accept the oversimplification that this statement contains because we are concerned with the *concept* of causation. Our interest in the causal connection between two variables makes classification by variables and categories an integral ingredient in social statistics.

In order to carry out statistical classification of sociological variables, we have said that the categories or classes must be mutually exclusive and totally inclusive. This means that to sort cases into separate categories of a variable, each case must fit in one and only one category (be mutually exclusive), and that there must be a unique category for every case (be totally inclusive). In this way, it is possible to classify individuals precisely according to their possession of one category or another of each variable.

Since it is possible to classify individuals according to their possession of a certain category of a variable, it is simply an extension to classify them

[6]Samuel A. Stouffer, *Communism, Conformity, and Civil Liberties* (New York: John Wiley & Sons, Inc., 1955), p. 90.

according to their possession of more than one variable at a time. This conforms to the common-sense notion of the way in which we think of the world in everyday terms. For example, we might say, "There were several blacks in the big crowd," thereby classifying by size and race simultaneously. We could call this crossing two classifications or, more simply, *cross-classification*.

This is analogous to the procedures of the social statistician who examines such things as high quality versus low quality students, depending upon their year in college. In this case, the variables academic performance and year in school would be cross-classified. Similarly, if we were interested in differences between upper class versus lower class, whites compared to blacks, we would cross-classify the variables social class and race.

Stouffer was interested in the influence of education on tolerance. Table 2.2.1 shows the percentage of cases which scored high on his scale of tolerance. The data show that "the more schooling, the more tolerance."[7]

Table 2.2.1.   Percent more tolerant by education

	More Tolerant	Number of Cases
College Graduates	66%	308
Some College	53	319
High School Graduates	42	768
Some High School	29	576
Grade School	16	792

Source: Adapted from Samuel A. Stouffer, *Communism, Conformity, and Civil Liberties* (New York: John Wiley & Sons, Inc.), 1955 p. 90.

Later on, we will deal with cross-classification of discrete variables in greater detail. We also will deal with that larger portion of statistics that is concerned with the combining of observations on continuous variables. It is enough to say at this point, however, that the procedures, no matter how complex they become, are actually extensions of common sense and everyday procedures.

2.3 LEVELS OF MEASUREMENT

Measurement involves a set of procedures that is familiar in everyday life, especially if you think of it as a special kind of classification. Consider when we measure level of education in terms of years of school completed. In reality we are classifying individuals according to their possession of a certain number of years of education. Since this is the same thing as sorting cases into

[7]Stouffer, *Communism, Conformity, and Civil Liberties*, p. 90.

quantitatively varying categories or classes, measurement can be seen as a special example of classification.

Categories are generally the classes of variables that have qualitative attributes, such as sex (male versus female), religion (Roman Catholic, Protestant, and Jew), and so forth. One of the most obvious and common ways to categorize cases is with *dichotomous measurement*. Dichotomy means "divided into two parts," and even though a dichotomy may also be and often is treated as a nominal *or* an ordinal scale, we isolate it as a separate type of measurement. Dichotomy refers to any variable that has just two states. For example, sex is a dichotomous variable, the categories being "male" versus "female." With all dichotomies, there are two classes: "either"/"or," "yes"/ "no," "plus"/"minus," "high"/"low," "the possession of"/"the absence of" some attribute, and so forth.

A more general way to handle various kinds of classes is to distinguish scales according to their *level of measurement*. Levels of measurement focus on the measured attributes of a variable that a case possesses, and we use levels of measurement to classify people according to their location on a variable. Four levels of measurement are important for us:

1. Nominal measurement
2. Ordinal measurement
3. Interval measurement
4. Ratio measurement

The first level is *nominal measurement*. Nominal means "by name only," and a nominal scale is one in which there are two or more category classifications dividing a variable according to name only. Religion is a nominal variable. One may be Roman Catholic, Protestant, Jew, Greek Orthodox, and so forth. These are names that distinguish people according to their possession of a category of the variable religion. Nominal scales, then, are ones whose categories and any numbers assigned to them serve only as names, just as the player numbers of a football team have no inherent mathematical value other than identification. Thus, nominal scales enable us to *classify* the cases but not to arrange them in any special order nor to measure relative or absolute distances between them. Some important nominal variables are region of the country, political party membership, and ethnic background.

The second level is *ordinal measurement*. Ordinal means "order by rank or succession," and an ordinal scale is one in which there are two or more classes divided according to rank.[8] With ordinal measurement of the classes, we can measure the rank order of individuals according to the category in which they fall. We can rank the order of candidates for a political office according to their "popularity"; we can rank the order of colleges according to their

---

[8]All variables can be made into dichotomies and treated as if they were ordinal scales, but for our purposes such a step is not necessary.

"place" in an athletic league; but we cannot measure the distance between the categories. For example, "social class" may be an ordinal measure in that we can order people according to the levels lower, working, lower middle, upper middle, and upper. In all instances, we can *move up or down* from one level to the next. Even though there is a rank order in, say, "social class" location, we cannot say that a working class person is "*twice* as much as" a lower class person. The most important characteristic of ordinal scales is that they possess transitivity, that is, if X_1 is larger than X_2 and X_2 is larger than X_3, then X_1 must be larger than X_3. Occupational prestige, amount of prejudice, and social class are important ordinal variables. Others are easily identifiable because they are stated in such terms as "less than" or "greater than."

The third level is *interval measurement*. Interval means "order by rank or succession and distance," and interval scales are those that are divided into two or more classes that are ordered into measurable distances apart. How much the intervals are apart is an aspect of the measurement procedure itself; thus interval scales do not have an absolute zero point. In other words, the zero point of the measuring scale for an interval scale is located arbitrarily.

There are very few social science scales with arbitrary zero points and established, countable intervals; however, an example of such measurement using an arbitrary zero point is the elevation of a land above "sea level." Generally, the zero point of sea level is fixed at the mean level of the oceans midway between high and low tide, and the height of land masses is determined by their stable elevation in feet above this point rather than above an actual "level" of the water surface. Thus, when the tide comes in, the height of land above the actual sea level is less than when the tide is out. Obviously, this also changes the geographical dimensions of a land mass, and the number of square feet of a body of land may change considerably depending upon when one measures the surface around its perimeter. As can be seen in Figure 2.3.1, an island with sloping shores would "lose" more of its land mass with a rising tide than would an island with steep shores. Furthermore, the ratio of the comparable elevations of two land masses would change with the tide.

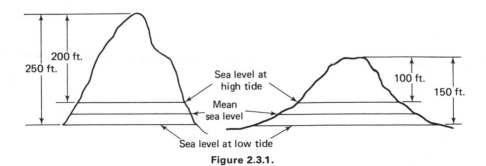

Figure 2.3.1.

Specifically, if the tidal range is a maximum of 50 feet, as it is in parts of the Bay of Fundy, the elevation of two land masses in the diagram would change from 100 and 200 feet to 150 and 250 feet. Clearly, the heights above tidal sea level do not possess a constant ratio, for at low tide it is 2 to 1 or 0.5, and at high tide it is 25 to 15 or 0.6.

Another example of an interval scale involves the measurement of the temperature of some object, that is, the intensity of its heat. Depending upon our purpose, we would probably use any one of several standardized scales, called thermometers, to determine an object's relative "hotness" or "coldness." The most common thermometers are the Celsius (formerly called centigrade), Fahrenhcit, and Kelvin temperature scales. All three are intended to measure the same thing, but the fixed points for the freezing and boiling points of water and the intervals dividing them are calibrated differently. Specifically, on Celsius or Fahrenheit thermometers, the zero points *do not* indicate a nonarbitrary, fixed, or absolute value that indicates the total absence of heat. The Celsius scale sets the zero point equal to the freezing point of water. The Fahrenheit scale sets the zero point approximately equal to the temperature produced by combining equal quantities of snow and common salt. On the other hand, the Kelvin scale sets the zero point equal to the total absence of heat, which is equal to "absolute zero." Figure 2.3.2. indicates that

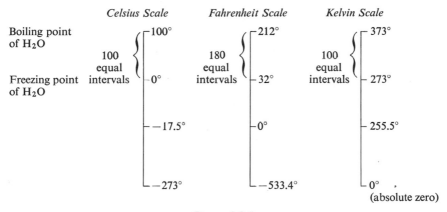

Figure 2.3.2.

if the intervals between the two fixed points for the freezing and boiling points of water are set equal, then the location of zero is arbitrary and not fixed. Thus, Celsius scales with 100 degrees between freezing and boiling and an arbitrary zero, and Fahrenheit scales with 180 degrees between freezing and boiling and an arbitrary zero are true interval scales.

Even though such scales are seldom available for sociological research, the importance of interval scale measurement is that many statistics require that we assume interval measurement, even though the variables may be measured with the next level of measurement, that is, ratio measurement. In general, if it is possible to establish the size of the intervals and to qualify the scale as an interval one, we can use the familiar arithmetic operations of addition, subtraction, multiplication, division, square roots, powers, and logarithms.

The fourth level is *ratio measurement*. Ratio means "order by rank or succession and distance from a fixed zero." Ratio scales possess all of the characteristics of the ones already mentioned; in addition, they have an "absolute zero." Thus, we may speak about the "ratio" between scores. From a computational viewpoint, we need not be too concerned with the distinction between interval and ratio measurement, because for sociological purposes, we often analyze ratio scales with statistics that require the assumption of interval measurement. Thus, even though such things as age and GPA are ratio scales, we generally handle them as interval scales. Commonly used ratio variables include such things as the birth rate, the death rate, rates of various crimes, county size in square miles, percentage of blacks and whites, various indexes of state characteristics, size of community, population in thousands, and so forth.

Table 2.3.1 shows that the level of measurement can be assessed according to the property of the scale used to measure it. Importantly, the properties of the scales are cumulative, and they can be arranged in an unambiguous rank order. Thus, the scales themselves form an ordinal scale; that is, each higher level scale possesses all the properties of the lesser level scales as well as its own property. It turns out that the higher the level of measurement possessed by a scale, the more high-powered the statistics that can be used for analytical purposes.

Table 2.3.1. Property of scale

Level of Measurement	Classi- fication	Rank Order	Relative Distance	Absolute Distance
Nominal	+	0	0	0
Ordinal	+	+	0	0
Interval	+	+	+	0
Ratio	+	+	+	+

This emphasizes a compromising situation. In the first place, the higher the level of measurement, the more high-powered the statistics that can be used, but the variables must *meet all the assumptions* required for that level. On the other hand, the lower the level of measurement, the less high-powered the statistics that can be used, but the easier it is to meet the assumptions for each lower level. Some sociologists compromise by making risky assumptions in order to use the most sophisticated statistical methods. Other sociologists compromise by making safe assumptions, but in doing so they must use simpler statistical methods.

2.4 THE DISTRIBUTION OF CASES

So far, we have seen how individuals may vary on a given attribute, characteristic, or trait called a variable. We also have seen that such variation is a central interest in social statistics. Since cases may vary on a particular characteristic, we can think of them as being "distributed over" that variable. When we do this, it is common to summarize the results of our research by listing the frequencies for each category of each variable. This is carried out by counting all of the cases studied and sorting them according to the level of each variable. Thus, when we speak about *frequencies*, we refer to the number of actual units or cases observed for the categories of each variable.

For example, when we distinguish between people according to the variable sex, some are male and some are female. We can say that the cases, in this instance, people, are distributed over the two categories of the variable. Suppose that we are interested in 112 of the social science students at our college. Of these 112 students, 60 are female and 52 are male. The *frequency distribution* of the 112 students for the variable "sex" will be as follows:

Sex	Frequency
Male	52
Female	60
Total	112

Similarly, when we divide the population of the United States into various age categories, we may think of the individuals as being distributed over the variable age, as shown in Table 2.4.1.

This table is fairly easy to construct. We start by locating each individual according to the categories (intervals) of the variable age and then assigning them to the appropriate category shown in the table. Once we have classified all the cases into the categories of the variables, we merely count the number (frequency) in each. This leads to a frequency distribution.

Our interest in measurement depends upon cases being *spread out* or *distributed over* a given variable for what may seem to be an obvious reason but one which is of great importance. Namely, if everyone were, say, the same age, there would be no sense in measuring it. In other words, for measurement

Table 2.4.1.   Frequency distribution of the population of the United States by age in 1975 (projected)

Age	(In Thousands)
Under 5 years	19,968
5–9 years	17,851
10–14 years	20,714
15–19 years	20,806
20–24 years	19,205
25–34 years	31,320
35–44 years	22,608
45–54 years	23,671
55–64 years	19,912
65 and over	21,503
Total	217,557

Source: *Statistical Abstract of the United States, 1973*. Washington, D.C.: U.S. Bureau of the Census, p. 6.

purposes it is essential for variables actually to vary. Later on, we will see that the association between variables assumes variation on the variables. This is because if there is not individual variation, there cannot be covariation, and it is the way variables covary or vary together that is the hub of association.

2.5 PERCENTAGES, MARGINALS, AND GRAPHIC PRESENTATIONS

In Section 2.4, we saw that a frequency distribution portrays the way in which the subjects are distributed among the categories or intervals of a given variable according to the way the data are grouped. Often we are interested in more than the frequencies alone, and it is common in social research to use percentages to help describe frequency distributions. *Percentage* means the number of cases *per 100* for any given category of a variable.

Let us start with the idea of proportions. If 41 of the 112 social science majors are sociology majors, in order to calculate the proportion of sociology students, we divide $41/112 = 0.366 =$ the proportion. The percentage is calculated by multiplying the proportion by 100. In our example,

$$\text{the percentage} = \frac{41}{112} \cdot 100 = 36.6 \text{ percent.}$$

Thus, the formula for proportions is N_1/N, and the formula for percentages is $(N_1/N) \cdot 100$, when N_1 refers to the number of cases in the first category, and N refers to the total number of cases on that variable.

When variables are summarized in tables, the frequencies or percentages for each category appear in the margins of the table. When the frequencies or percentages are displayed in this manner, they are referred to as the *marginals*. For example, suppose that we ask each of the 112 students the following question:

What is your major in the division of social sciences? (Check one.)

[ ] Sociology
[ ] Psychology
[ ] Anthropology
[ ] Government
[ ] Economics

The answers to this question may be summarized in the marginals for the question by listing the frequencies and percentages. We can see that there are 41 (36.6 percent) sociology majors, 26 (23.2 percent) psychology majors, and so forth. After the study is done, the marginals are usually written in a code book which resembles a blank questionnaire and looks like this:

What is your major in the division of social sciences? (Check one.)

	N	Percentage
[ ] Sociology	41	36.6
[ ] Psychology	26	23.2
[ ] Anthropology	15	13.4
[ ] Government	21	18.8
[ ] Economics	9	8.0
	112	100.0%

In addition to presenting data in tabular form, sociologists commonly present the same material graphically. The aim of both tables and graphs is to organize a body of data into sensible configurations in a way that will help interpret its meaning. Our concern now is with converting tabular data into pictorial or graphic form so that they will be more easily comparable or interpretable, thereby assisting statistical analysis visually.

For nominal and ordinal variables, it is customary to present the data in *bar graphs*. In a bar graph, a vertical bar is drawn for each category of the variable; the height of the bar represents the number of cases in that category. The width of each bar is arbitrary, but it is common and sensible to set the width equal to one unit in height. When this is done, the area of each bar represents the frequency of the appropriate category, and the total area of all the bars is equal to the total number of cases under consideration.

Figure 2.5.1 is a bar graph that displays information about the Federal budget and compares the proportionate expenditure for defense outlays

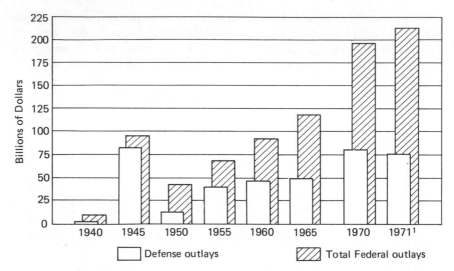

Figure 2.5.1. National defense and total budget outlays: 1940 to 1971. (*From U.S. Bureau of the Census,* Statistical Abstract of the United States: 1971 (*92nd edition.*) *Washington D.C., 1971, p. 241, Fig. XIV.*)

versus total outlays. A comparative bar graph such as this helps clarify the relationship between one category of a variable compared to its total distribution.

Another way to present a bar graph is to compare the categories of a variable. Figure 2.5.2 on page 30 shows the percentage of Republicans compared to Democrats (categories) for each selected personality test item (variable) from *Voting*. It shows relatively small differences between the parties on these key questions from the study.

For dichotomies and nominal variables such as sex, religion, academic major, and so forth, the ordering of the intervals (bars) is arbitrary; however, with ordinal, interval, and ratio variables, the proper ordering gives a useful graphic representation of the frequency distribution.

A straightforward method of presenting ordinal and interval data in order to enable comparison of the differences among the frequencies is to draw a figure called a *histogram*. A histogram is comprised of connected vertical bars, the heights of which are proportional to the frequencies in the class and the widths of which are proportional to the size of the intervals of the variable. A histogram may show the absolute class frequencies as well as the relative size of each frequency compared to the others.

The construction of a histogram can be accomplished on ordinary graph paper. It is good practice to construct a histogram with equal frequency intervals and to set the length of the horizontal axis equal to the length of the

28

vertical axis. It is possible to establish unequal axes, but the extension of either the vertical or the horizontal axis may distort the data. When unequal class intervals are used, it is important for comparative purposes to maintain the spatial ratio of the intervals. Thus, with unequal intervals, it is common to reduce the height of the bar so that the *area* is equivalent to that in the equal intervals. The histogram shown in Figure 2.5.3 represents the distribution of 112 students according to the variable grade point average. The horizontal axis (width) of each section equals the size of the interval for grade point average, and the vertical axis (height) equals the frequencies in the intervals, that is, the number of students. Assume that we ask the question:

What is your cumulative grade point average? (Check one).

	N	Percentage
[ ] 0.0–0.5	1	1.1
[ ] 0.6–1.0	6	5.3
[ ] 1.1–1.5	11	9.8
[ ] 1.6–2.0	18	16.1
[ ] 2.1–2.5	23	20.5
[ ] 2.6–3.0	26	23.2
[ ] 3.1–3.5	17	15.1
[ ] 3.6–4.0	10	8.9
	112	100.0%

The histogram graphically shows that the frequency distribution of students over the variable grade point average tapers somewhat gradually from the lowest to the highest, and then drops off more sharply. The same substantive conclusions about which intervals contain how many students may be made by examining the distribution according to the number in the interval (*N*), the percentage of cases in the interval (%), or the histogram.

It is a straightforward procedure to convert a histogram to a *frequency polygon*. The purpose of a frequency polygon is to approximate a curve that displays the frequency distribution. A frequency polygon can be constructed either by superimposing it over a histogram or by connecting with straight lines the dots that mark the midpoint of the top of the bars that would have resulted from a histogram. It is common practice to utilize the histogram to display discrete distributions and to use the frequency polygon to display continuous distributions or ones in which continuity is assumed. It is important to observe that the vertically inverted triangles created by the frequency polygon lines bisecting the bars are of equal size. Thus a frequency polygon and a histogram of a given frequency distribution contain the same area. The frequency polygon resulting from connecting the midpoints of each interval at the height of the frequencies is shown in Figure 2.5.4.

Percentage Agreeing with Each Statement

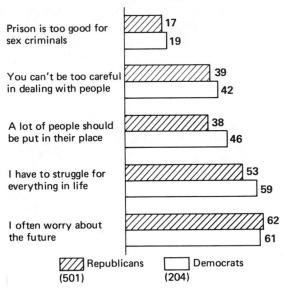

Figure 2.5.2. Republicans and democrats do not differ in selected personality test items. (*From Bernard R. Berelson, Paul F. Lazarsfeld, and William N. McPhee,* Voting. *Chicago: The University of Chicago Press, 1954, p. 193.*)

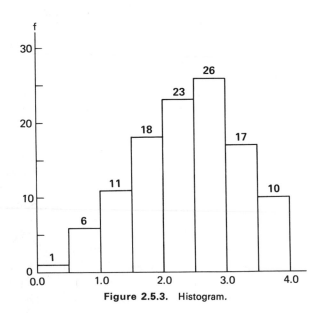

Figure 2.5.3. Histogram.

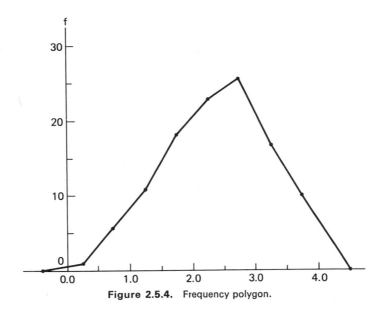

Figure 2.5.4. Frequency polygon.

It is possible to display percentages in a similar fashion in a *percentage polygon*. Figure 2.5.5 is a percentage polygon showing the relationship between informal social relations among printers and interest in union politics, indicating that those active in the occupational community are also involved and active in the union.

Another effective form of graphic presentation displays the percentage distribution of cases for all the categories of a variable in the form of a *pie chart*. A pie chart is so named because it is a circular graph that is divided into slices. The area of each slice is drawn to be proportionate to the frequency of the category or interval. In order to divide the pie, we set the circumference of 360° equal to 100 percent. Then the percentage frequencies are measured around the circumference of the pie chart with a protractor. The points along the circumference are then connected to the center of the circle to form the radii, which are the boundaries for the proportional slices of the pie. Commonly, pie charts are shaded to emphasize the proportionate sizes of the various segments. Figure 2.5.6 indicates the change between 1950 and 1970 in the pattern of naturalization in the United States, showing that there has been a major change in the means by which immigrants become citizens. Namely, in 1950, 61 percent became citizens by marrying U.S. citizens, while in 1970, 72 percent became citizens under general naturalization procedures.

It is possible to design graphs that may be used to display additional one-variable distribution information. A good example of one such graph is the

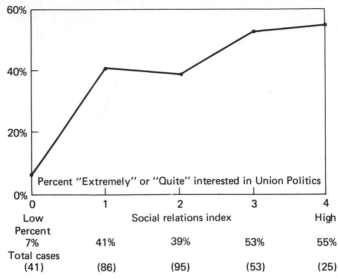

Figure 2.5.5. Relationship between informal social relations with other printers and interest in union politics.* (*From Seymour Martin Lipset, Martin Trow, and James Coleman,* Union Democracy. *Garden City, New York: Anchor Books, Doubleday & Company, Inc., 1956, p. 80.*)

triangular graph presented by Coleman in *Adolescent Society.*[9] Table 2.5.1 presents the percentage distribution of respondents over three categories of the qualitative variable *self-image in school.*

We can construct a triangular graph to display the relative choices of boys and girls for the categories brilliant student, athletic star for boys, or

Table 2.5.1. How boys and girls want to be remembered in school
(fall term, percentages)

	Boys	Girls
Brilliant Student	31.3%	28.8%
Athletic Star for boys	43.6%	
Leader in Activities for Girls		36.1%
Most Popular	25.0%	35.2%
N	(3,696)	(3,955)

Source: James S. Coleman, *The Adolescent Society.* New York: The Free Press, Inc., 1961, p. 30.

[9]James S. Coleman, *The Adolescent Society* (New York: The Free Press, 1961), p. 29.

32

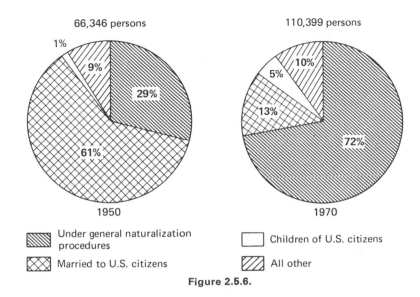

66,346 persons   110,399 persons

1%

9%

29%

61%

1950

10%

5%

13%

72%

1970

| Under general naturalization procedures | | Children of U.S. citizens |
| Married to U.S. citizens | | All other |

Figure 2.5.6.

leader in activities for girls, and most popular according to the question "How do you want to be remembered in school?" In the triangle graph in Figure 2.5.7, a specific group is plotted by a point within an equilateral triangle. Each corner may be thought of as the top point of a pyramid and is equal to 100 percent response. Thus, in Coleman's study, at the top is 100 percent brilliant leader; at the left is 100 percent athletic star or activities leader; and at the right is 100 percent most popular. The more respondents who give a particular response, the closer the point will be to that corner. Thus points along one edge of the triangle indicate a 0 percent response to the category of the opposite point, and the exact center of the triangle indicates 33.3 percent responses on the three categories. From the triangle graph in Figure 2.5.7, we can conclude that for boys the image of the athletic star is the most important, and for girls the images of activities leader and most popular are about equally more important than the image of a brilliant student.

So far, we have learned about ways of summarizing and describing data and of examining one or more variables for analytical purposes. It is important to point out that we have been dealing with various statistics *as measures*. It may seem obvious, but it is crucial to distinguish between the use of the word "statistics" when it refers to measures and the use of the word "statistics" when it refers to the subject matter of this book.

Sometimes the use of the word "statistics" when referring to measures is combined with the number of variables with which we are concerned. In these

33

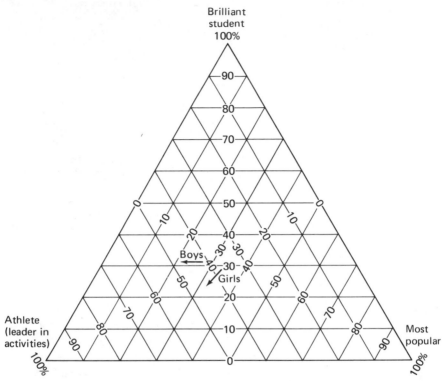

Figure 2.5.7. Relative choice of image of athletic star, (leader in activities for girls) brilliant student, and most popular, in fall and spring. (*From* The Adolescent Society *by James S. Coleman. New York: The Free Press, 1961, p. 29.*)

situations, we speak of univariate, bivariate, and multivariate statistics. *Univariate statistics* measure one variable at a time. They are useful for descriptive purposes and for facilitating comparison with other populations or with the same population at different points in time. *Bivariate statistics* deal with two variables at a time and focus on measuring relationships between variables. We have now seen how it is possible to look at the association between variables with bivariate means of summarizing. Later on, we will explore additional ways to measure associations between variables, going beyond merely describing them with two-way frequency distributions. *Multivariate statistics* involve three or more variables and are used to explore the important features of bivariate relationships of particular substantive interest. Later on, we will examine three- and four-variable relationships.

34

EXERCISES

1. In your own words define the following terms:
   validity
   reliability
   variable
   verbal definition
   operational definition
   levels of measurement
   frequencies
   marginals

2. The following variables are often used in sociological research. For each variable below indicate whether it is a dichotomous measure (D), a nominal measure (N), an ordinal measure (O), an interval measure (I), or a ratio measure (R). In answering this question, use the highest possible level of measurement for each variable.
   A. Region of the country
   B. Father's education (Years of school completed)
   C. County size (number of square miles)
   D. Occupation (Professional, white collar, blue collar)
   E. Race (White, Black, other)
   F. Age (in years)
   G. Occupational prestige (high, middle, low)
   H. Political orientation (conservative vs. liberal)
   I. Sex (male vs. female)
   J. Racial discrimination (percentage of Blacks present)

3. Assume that you have gathered the following information about the wages of a sample of 52 work-study students in your college.

$25	$40	$23	$26
$32	$27	$30	$29
$28	$31	$17	$39
$21	$24	$28	$27
$23	$38	$20	$30
$37	$35	$19	$22
$16	$18	$22	$34
$25	$21	$24	$29
$33	$28	$36	$26
$27	$30	$28	$28
$27	$26	$29	$25
$31	$29	$32	$33
$34	$31	$35	$28

A. Construct a one-way frequency distribution using intervals of $1.00.

B. Calculate the percentage distribution and report it with the frequency distribution of Part A.

C. Draw a histogram on graph paper.

4. Collapsing a detailed table to a fourfold table frequently better demonstrates the relationship between two variables. Using the figures from the following table, construct a fourfold table showing the relationship between level of education and level of interest in issues. (Note: choose a cut-off point that you can support for "high" and "low" levels of education.)

What kind of people are most likely to be communists?
(national cross-section)

		Percentage who say that U.S. Communists are likely to be:		
	Number of Cases	White-Collar, Better Educated, etc.	Working Class, Less Educated, etc.	Other or Nonspecific
Total Sample	4908	18%	35%	47%
Respondents in Cross-Section by Interest and Education:				
More Interested in Issues				
College graduates	308	31%	59%	10%
Some college	319	27	57	16
High school graduates	768	22	47	31
Some high school	576	19	37	44
Grade school	792	15	31	55
Less Interested in Issues				
College graduates	68	22%	58%	20%
Some college	141	22	49	29
High school graduates	440	16	38	46
Some high school	446	14	26	60
Grade school	1050	12	14	74

Source: From Samuel A. Stouffer, *Communism, Conformity, and Civil Liberties.* New York: John Wiley & Sons, Inc., 1955, page 173, Table 4.

5. Look at the variables employed in two research articles in a recent issue of the *American Sociological Review* or the *American Journal of Sociology.* Determine their operational definitions and whether they are *observed* variables or *inferred* variables.

6. Construct a histogram to depict the changes in employment of males from 1955–1970 in agriculture and non-agricultural industry. Interpret the histogram.

Class of worker of employed persons, by sex: 1955 to 1970
[in thousands of persons 14 years old and over through 1965;
16 and over thereafter.]

Class of Worker	1955[1] Male	1955[1] Fe-male	1960 Male	1960 Fe-male	1965 Male	1965 Fe-male	1970 Male	1970 Fe-male
Employed in agriculture ..	5,487	1,243	4,678	1,045	3,729	856	2,861	601
Wage and salary workers ..	1,415	285	1,558	308	1,243	249	979	174
Self-employed workers	3,582	149	2,687	116	2,170	137	1,722	88
Unpaid family workers	489	810	433	621	316	470	160	339
Employed in nonagricul-								
tural industries	37,803	18,661	39,807	21,151	43,304	24,289	46,099	29,066
Wage and salary workers ..	32,934	17,119	34,689	19,287	38,434	22,331	42,116	27,330
In private households	254	1,962	288	2,201	334	2,214	189	1,565
Government workers	4,234	2,604	4,788	3,155	5,685	3,938	6,781	5,643
Other wage and salary								
workers	28,446	12,554	29,613	13,931	32,415	16,179	35,145	20,122
Self-employed workers	4,809	1,077	5,027	1,340	4,794	1,419	3,929	1,288
Unpaid family workers	60	465	91	524	77	540	53	449

[1]Data not adjusted for change in definition of employment adopted in 1957. For adjusted totals, see table 327.

Source: U.S. Bureau of the Census, *Statistical Abstract of the United States:* 1971. (92nd edition.) Washington, D.C., 1971, p. 222, No. 349.

7. Using graphs, show the information presented in the following table. (Note: collapse the table, if you wish.)

Average number of choices received on all elite criteria by athlete-scholars, athletes, scholars, ladies' men, and all boys, in the five small schools*

	Farmdale	Marketville	Elmtown	Maple Grove	Green Junction
Athlete-scholar	18.0	17.0	27.5	28.1	33.6
	(1)	(3)	(4)	(7)	(5)
Athlete	12.3	22.1	18.2	18.7	23.3
	(13)	(15)	(18)	(15)	(16)
Scholar	7.5	14.1	8.7	9.0	8.9
	(10)	(11)	(17)	(10)	(22)
Ladies' man	13.2	17.8	9.1	16.2	16.6
	(5)	(5)	(7)	(5)	(13)
All boys	7.5	6.4	6.2	6.4	6.6
	(70)	(198)	(266)	(217)	(281)

*The number of cases is given in parentheses below each relevant category.

Source: From James S. Coleman, *The Adolescent Society*. New York: The Free Press, 1961, p. 153, Table 27.

8. Discuss the differences between the terms "reliability" and "validity." Give examples of each.

9. Nominal variables such as race, religion, and ethnicity, often cause social differentiation within a society. However, ordinal, ratio, and interval variables, such as income, education, and occupation, frequently cause social stratification within a society. Discuss the differences between social differentiation and social stratification and how these differences relate to the differences between nominal and ordinal variables.

10. The number of children in families of 35 children in a day-care center are as follows. Construct a frequency distribution of the number of children in the 35 families.

2	3	3	1	6	10	4
6	4	2	2	3	1	3
7	5	4	1	2	1	5
10	6	8	7	5	11	12
9	4	2	3	6	2	1

3

Measurement and Cross-Classification

Our interest in the association between variables leads us to examine such things as the influence of socioeconomic status on the way people vote, or the influence of region and type of community on the willingness to tolerate nonconformity. We have seen the ways in which these variables are associated in some of the tables and graphs already presented; however, we have not explored all the possibilities. In this chapter, we expand our discussion of measurement and incorporate it with the idea of cross-classifying more than one variable. To do this, we start with the empirical indicators we use to measure variables relevant to our research questions. First we compare two separate classifications, and then we cross-classify two distributions. It is a natural and relatively straightforward step to introduce a third variable; however, before tackling the three-variable situation, let us take up where we left off in Chapter 2 and move from one variable to two variables.

3.1 EMPIRICAL INDICATORS AND STATISTICAL NOTATION

In all social research, the verbal and operational definitions of the variables are of the utmost importance. For example, the variable *socioeconomic status* may be verbally defined as "one's location in the status structure of the community." The variable *two-party vote* may be verbally defined as "one's location in the political composition of the community in terms of political affiliation and identification." The variable *where Americans live* may be verbally defined as "one's geographical location." Finally, the variable *willingness to tolerate nonconformity* may be verbally defined as "one's tolerance for nonconforming secular and religious behavior."

All of these variables also may be defined operationally; that is, they may be measured. In the *Voting* study we discussed earlier, *socioeconomic status* is measured by an index that uses answers concerning the breadwinner's occupation and education and the interviewer's rating of the breadwinner. The variable *two-party vote* is measured by the question "How did you vote in the last election?" The operational definition of *where Americans live* is made according to answers people give to questions about residence in the four major geographical regions of the U. S. Bureau of the Census. The variable *willingness to tolerate nonconformity* is measured by use of an index of several questions regarding nonconforming situations and each respondent's answers to those questions.

The importance of the connection between the verbal definition and the operational definition cannot be emphasized too strongly, for the central feature of social statistics ultimately involves an attempt to connect a theoretical position with a set of findings. Thus, concepts such as social class, tolerance, political preference, and so forth, must be grappled with according to the sorting procedures (operational definitions) we use in classifying the cases under investigation. In other words, operational definitions form a code system that enables us to classify units of observation according to their possession of a variable.

So far we have been talking about variable names, verbal definitions, and operational definitions. Now it is necessary to introduce a bit more detail about the abbreviations we give to variables and to explain how these abbreviations fit into the idea of a code system.

Often in statistics we use a single capital letter, such as U, V, W, X, Y, Z, etc., to designate variables. Instead of using different designations for a person's score, we commonly use one of these single capital letters together with a lower-case subscript (i, j, k, m, etc.). For example, we would use the symbol Y_i for the generalized person i on variable Y. The subscript ranges $1 \leq i \leq N$, and Y_i has the numeric value of the category. Thus, the first person's score could be designated Y_1, the second person's score Y_2, the third person's score Y_3, and so forth. The first case in the study *Voting*, for example, could be assigned the code name "1," the second case "2," the third case "3," and so forth. Then, we could assign the variable *two-party vote* the code name "Y." If case "1" voted Republican, we would give it the following designation, $Y_1 = $ Republican. Naturally, we would assign each party a number and then indicate each case's "score" as a number. For example, if we designate Democrats as being category 1, Republicans as being category 2, and other parties as being category 3, then individual "1"'s score on variable Y would be 2, thus $Y_1 = 2$. Before considering how this helps us, let us quickly give some thought to the notion of handling things quantitatively with mathematics.

One of the most commonly used statistical operations is summation (addition). Summation is designated by the Greek capital letter sigma $\sum$ It is used to tell us that we are to add together all of the elements to the right of the symbol. $\sum$ is called the "summation sign." Usually, there are notations above and below sigma. These are used to indicate that the subscript of the variable in question takes on successive values from the number shown below the sign through the number shown above the sign. Thus, the general sign $\sum_{i=1}^{N} Y_i$ means that we are to go from the first person ($i=1$) to the last person (N), adding up (summing) the individual scores for the variable Y. Since each individual or case possesses some level or category of variable Y, and since each individual must be counted separately, we use Y_i to represent the generalized individual. Thus,

$$\sum_{i=1}^{N} Y_i = Y_1 + Y_2 + Y_3 + Y_4 \cdots + Y_N$$

Consider the way the summation sign is actually used. The notation $\sum_{i=3}^{10} Y_i$ is read "sum the Y_i's when i goes from 3 to 10," or "add the scores of the third through the tenth cases." This is carried out like this:

$$\sum_{i=3}^{10} Y_i = Y_3 + Y_4 + Y_5 + Y_6 + Y_7 + Y_8 + Y_9 + Y_{10}$$

Now, consider the practical way this notation works, remembering that its main purpose is to *simplify* statistical procedures. Suppose that we are interested in the total number of hours spent studying statistics each week by four outstanding students, Tom, Nancy, Alan, and Martha. Let us designate the variable "hours of study" $= X$, and the students Tom $= 1$, Nancy $= 2$, Alan $= 3$, and Martha $= 4$. We can summarize these steps as follows:

Name	Student Number	Symbol	Hours of Study (X)
Tom	1	X_1	16
Nancy	2	X_2	15
Alan	3	X_3	12
Martha	4	X_4	10

As we carry out the summation, remember that notation is nothing more than a code system designating information about the cases in which we are interested in a concise, numerical fashion.

$$\sum_{i=1}^{4} X_i = X_1 + X_2 + X_3 + X_4$$

$$\sum_{i=1}^{4} X_i = 16 + 15 + 12 + 10$$

$$\sum_{i=1}^{4} X_i = 53$$

It is very important to understand a few basic rules about notation and the way to use mathematical symbols such as $\cdot$ or $(\ )(\ )$, $\div$ or $/$, $+$, and $-$. There is an order of priority in algebraic operations, and the correct order is (1) parentheses, (2) exponentiation, (3) multiplication and/or division, and (4) addition and/or subtraction. Where ambiguity exists, operations are performed from left to right.

Let us state and carry out several algebraic operations for variables X and Y.

Symbol	Algebraic Operation	
$\sum_{i=1}^{N} X_i$	$= X_1 + X_2 + X_3 \ldots + X_N$	$=$ summation
$\sum_{i=1}^{N} Y_i$	$= Y_1 + Y_2 + Y_3 \ldots + Y_N$	$=$ summation
$\sum_{i=1}^{N} X_i^2$	$= (X_1 \cdot X_1) + (X_2 \cdot X_2) + (X_3 \cdot X_3) \ldots + (X_N \cdot X_N)$	$=$ square and sum
$\sum_{i=1}^{N} Y_i^2$	$= (Y_1 \cdot Y_1) + (Y_2 \cdot Y_2) + (Y_3 \cdot Y_3) \ldots + (Y_N \cdot Y_N)$	$=$ square and sum
$\left(\sum_{i=1}^{N} X_i\right)^2$	$= (X_1 + X_2 + X_3 \ldots + X_N) \cdot (X_1 + X_2 + X_3 \ldots + X_N)$	$=$ sum and square
$\left(\sum_{i=1}^{N} Y_i\right)^2$	$= (Y_1 + Y_2 + Y_3 \ldots + Y_N) \cdot (Y_1 + Y_2 + Y_3 \ldots + Y_N)$	$=$ sum and square

It is often necessary to locate individual cases according to two variables simultaneously. To do so, we use the appropriate letters, so that each *variable* is designated by its own capital letter, such as X and Y, and each case is designated by a unique subscript. Then we may state algebraic operations for two variables just as we did for one. Notice that in each instance we examine person 1, 2, 3, etc., on *both* variables X and Y. This means that we use 1's case scores on two variables in combination with each other.

Symbol	Algebraic Operation	Name
$\sum\limits_{i=1}^{N} X_i \cdot Y_i$	$= (X_1 \cdot Y_1) + (X_2 \cdot Y_2)$ $+ (X_3 \cdot Y_3) \ldots + (X_N \cdot Y_N)$	$=$ multiply and sum, commonly called "the sum of the products."
$\left(\sum\limits_{i=1}^{N} X_i\right) \cdot \left(\sum\limits_{i=1}^{N} Y_i\right)$	$= (X_1 + X_2 + X_3 \ldots + X_N)$ $\cdot(Y_1 + Y_2 + Y_3 \ldots + Y_N)$	$=$ sum and multiply, commonly called "the product of the sums."

3.2 COMPARISONS OF TWO CLASSIFICATIONS

As we have seen, it is analytically useful to examine the way in which cases are distributed over a single variable. By using two frequency distributions (classifications) at least one of which is a nominal variable, we can compare the relative distributions of the second variable for each of the nominal classes. For example, suppose that we are interested in analyzing the associations between education and sex. Using the *Current Population Survey* of persons 25 years old and over in 1970, 109,310,000 persons 25 years and over are classified according to *the number of years of school completed*.[1] The frequencies themselves are useful, but for comparative purposes it is a good idea to utilize percentages as well, as can be seen in Table 3.2.1. The same persons also are classified according to *sex*, which is also shown in Table 3.2.2 in raw frequencies as well as percentages.

Since our interest is in the association between sex and education, we compare percentage distributions of the two categories for the variable *sex* over the classes of the variable *years of school completed* as is shown in Table 3.2.3. One does not have to be a statistical wizard to appreciate the distinctions that emerge from comparing the percentage distributions for the *years of school completed* over the categories of the variable *sex*. For example, we see that there are only minor (1 or 2 percent) differences between the percentages of males and females who have completed various years of elementary school and the first three years of high school. However, we see that proportionately more females than males (8 percent) have completed four years of high school. There is a minor (1 percent) difference between the percentage of males and females who have completed one to three years of college. Finally, propor-

[1]U.S. Bureau of the Census, *Statistical Abstract of the United States:* 1971. (92nd edition.) Washington, D.C., 1971, p. 108, No. 162 and p. 109, No. 163.

Table 3.2.1 **Current population survey of 1970 persons 25 years old and over years of school completed (in thousands)**

		Frequency	Percentage
Elementary	0–4 years	5,747	5%
School	5–7 years	9,924	9
	8 years	14,595	14
High	1–3 years	18,682	17
School	4 years	37,134	34
College	1–3 years	11,164	10
	4 or more	12,063	11
Total (*N*)		109,310	100%

Source: Adapted from *Statistical Abstract of the United States, 1971.* U.S. Bureau of the Census, Washington, D.C., 1971 (92nd edition), page 109, No. 163.

Table 3.2.2 **Sex (in thousands)**

	Frequency	Percentage
Male	51,784	47%
Female	57,526	53
Total (*N*)	109,310	100%

Source: Adapted from *Statistical Abstract of the United States, 1971.* U.S. Bureau of the Census, Washington, D.C., 1971 (92nd edition), page 108, No. 162.

tionately more males than females (6 percent) have completed four years of college or more.

In the study *Voting*, an examination of political support within the electorate indicates that there is a basic difference between minorities and "the majority" when one controls for the nominal variable *ethnic and/or religious group*.[2] The utility of examining two distributions in the above-described fashion also is apparent if displayed graphically. Figure 3.2.1, displaying the percentage voting Republican on the variable *two-party vote*, shows the extent to which the minorities supported the Republican party compared to white, native-born Protestants. We have seen that even at an elementary level, it makes good sense to consider the comparative distributions of one variable

[2]Bernard R. Berelson, Paul F. Lazarsfeld, and William N. McPhee, *Voting*. (Chicago: The University of Chicago Press, 1954).

Table 3.2.3   Years of school completed by sex
(in thousands)

		Male		Female	
		f	%	*f*	%
Elementary	0–4 years	3,031	6%	2,716	5%
School	5–7 years	4,884	9	5,041	9
	8 years	7,041	14	7,554	13
High	1–3 years	8,355	16	10,327	18
School	4 years	15,571	30	21,653	38
College	1–3 years	5,580	11	5,584	10
	4 years or more	7,321	14	4,743	8
	Total (*N*)	51,784	100%	57,526	100%

Source: Adapted from U.S. Bureau of the Census. *Statistical Abstract of the United States: 1971*. (92nd edition.) Washington, D.C., 1971, p. 109, No. 163.

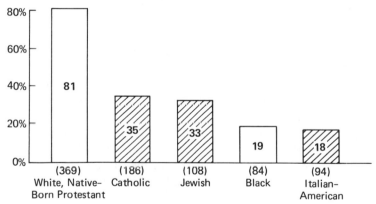

Figure 3.2.1. The minorities and the "majority" vote differently. (*Bernard F. Lazarsfeld, and William N. McPhee,* Voting. *Chicago: The University of Chicago Press, 1954, p. 62.*)

considering another. When we do so we are examining the association between two variables.

3.3 CROSS-CLASSIFICATION OF TWO DISTRIBUTIONS

Cross-classification is a comparison of the distribution over row categories between columns, or a comparison of the distribution of column categories between rows. For example, Table 3.3.1 shows the frequency distribution and

**Table 3.3.1 Current population report of 1970 on
employed persons 16 years of age and over**

Occupation Group	N		Percentage
White-Collar Workers			
Professional and technical workers	11,140		14%
Managers, officials, and proprietors	8,289		11
Clerical workers	13,714		17
Salesworkers	4,854		6
Total		37,997	48
Blue-Collar Workers			
Craftsmen and foremen	10,158		13
Operatives	13,909		18
Nonfarm laborers	3,724		5
Total		27,791	36
Service Workers		9,712	12
Farmworkers		3,126	4
Total		78,627	100%

Source: From U.S. Bureau of the Census, *Statistical Abstract of the United States*: 1971. (92nd edition.) Washington, D.C., 1971, p. 222, No. 347.

the percentage distribution of employed persons 16 years of age and over according to occupation group.

Now let us cross-classify the variable "occupation group" with the variable sex. Table 3.3.2 indicates the frequencies and percentages of males and females in given occupation groups.

To determine these percentages, we divide the number within each occupation group by the total number for all occupation groups. Thus, to determine the percent of male white-collar workers in the category "Professional and technical workers," we divide and multiply, $(6,842/20,054) \cdot 100 = 34.1$ percent.

It also is possible to compute the percentages in the other direction and to examine what percentage of each occupation group is male and what percentage is female. The steps for such a computation are straightforward. If we were interested in the percentage of professional and technical workers which is male, we would divide and multiply, $(6,842/11,140) \cdot 100 = 61.4$ percent. The calculation of percentages is not as arbitrary as it may appear; it depends on such matters as our theoretical assumptions, the subject of our research, and so forth. At the end of this section more atttention is given to the *direction* of such calculations. For now, however, it is important to realize that our research interests influence such decisions.

Cross-classifications also can be displayed graphically. In *Communism, Conformity, and Civil Liberties*, for example, the evidence shows that the

Table 3.3.2   Occupation group and sex

	Male	Percent-age	Female	Percent-age
White-Collar Workers				
Professional and technical workers	6,842	14%	4,298	14%
Managers, officials, and proprietors	6,968	14	1,321	4
Clerical workers	3,481	7	10,233	34
Salesworkers	2,763	6	2,091	7
Total	20,054	41	17,943	60
Blue-Collar Workers				
Craftsmen and foremen	9,826	20	332	1
Operatives	9,605	20	4,303	14.5
Nonfarm laborers	3,589	7	136	.5
Total	23,020	47	4,771	16
Service Workers	3,285	7	6,427	22
Farmworkers	2,601	5	525	2
Total	48,960	100%	29,667	100%

Source: *Adapted from Statistical Abstract of the United States, 1971.* U.S. Bureau of the Census, Washington, D.C., 1971 (92nd edition), page 222, No. 347.

region of the country in which one lives has an important influence on one's tolerance.[3] To measure this association between the variables *region* and *tolerance*, Stouffer presented the sliding bar graph shown in Figure 3.3.1. It is apparent from a visual examination that the greatest percentage of respondents were more tolerant in the West, the next greatest in the East, the next greatest in the Middle West, and the least in the South.

The tables presented thus far have already been constructed and require little detailed explanation as to how they were developed. Our knowledge of

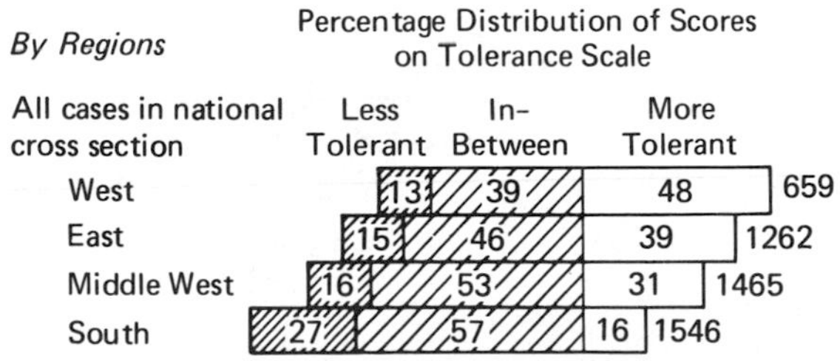

Figure 3.3.1. Where Americans live as related to their willingness to tolerate nonconformity. (*From Samuel A. Stouffer,* Communism, Conformity, and Civil Liberties. *New York: John Wiley and Sons, Inc., 1955, p. 112.*)

[3]See Samuel A. Stouffer, *Communism, Conformity, and Civil Liberties.* (New York: John Wiley & Sons, Inc., 1955). Chapter 5.

statistical notation tells us that it ought to be possible to devise a code system of variable abbreviations and case numbers to construct such tables. As you probably suspected, statisticians use a standard system of notation in which individual cases may be located according to the possession of two or more variables simultaneously. Assume, for example, that we wish to cross-classify cases according to the two variables *education* and *age*. We already have information about the variable *education*. It was operationally defined as *years of school completed* and was reported in Table 3.2.1, which lists the frequencies and percentages for each level of education. Table 3.3.3 reports the frequency and percentage distributions of persons 25 and over for the *same* 109,310 people on the variable age.

Table 3.3.3   Age (in thousands)

	Frequency	Percentage
25–34	24,865	23
35–44	23,021	21
45–54	23,398	21
55 and over	38,026	35
Total	109,310	100%

Source: Adapted from *Statistical Abstract of the United States, 1971*. U.S. Bureau of the Census, Washington, D.C., 1971 (92nd edition), page 110, No. 165.

Now let us examine the code system mentioned above, which helps us cross-classify these two variables. First we can subdivide the variable education into years of school completed using the following abbreviations:

Variable	Y	
Name	Education	
Verbal Definition	The amount of formal schooling	
Operational Definition	The answer to the question "How many years of school have you completed?"	
Interval	Years of School	Code Number
Elementary School	0–4 years	1
	5–7 years	2
	8 years	3
High School	1–3 years	4
	4 years	5
College	1–3 years	6
	4 and more	7

Now let us subdivide the variable age. Sorting the cases according to age emphasizes an important feature of sociological analysis, namely, there is a tendency to group scores into unambiguously defined categories and/or convenient divisions of the variable. This procedure enables us to examine relationships more coherently because the data are collapsed into interpretable classes. In our example, the variable age is collapsed into 10-year intervals leaving the last class open and using the following abbreviations:

Variable	X	
Name	Age	
Verbal Definition	The age attained by one's last birthday	
Operational Definition	The answer to the question "On your last birthday, what was your age?"	
Interval	Age	Code Number
	25–34	1
	35–44	2
	45–54	3
	55 and over	4

With this code system it is easy to see how individual cases may be located in the appropriate category of the cross-classified variables. For instance, a 37-year-old high school graduate could be designated as 52. A 29-year-old college graduate could be designated 71. A 60-year-old who left high school after two years could be designated 44. By assigning all cases the appropriate code numbers, it is possible to assign them according to their joint possession of any two variables. Table 3.3.4 shows how each cell in the table has a unique code number and how the marginal frequencies are distributed.

Cross-classification is the process of assigning each person to his or her unique location in the table according to the possession of a specific age and a specific amount of education. Generally this step is accomplished by some kind of electronic sorting machine, and then the total number of cases in each cell is listed. Table 3.3.5 reports the cross-classified frequency distribution for the variables *years of school completed* and *age*.

Earlier we mentioned that it is possible to calculate percentages in more than one direction. In Table 3.3.5, for example, let us look at one cell and determine its three possible percentage values. There are 10,916 high school graduates in the age level 25–34 years. First, if we divide this number by the overall N of 109,310 and multiply by 100, we find that about 10.0 percent of the respondents are high school graduates, ages 25–34. Second, if we divide the same number by the frequency for all high school graduates of 37,134 and multiply by 100, we find that about 29.4 percent of the high school graduates are ages 25–34. Third, if we divide the same number by

Table 3.3.4   Code numbers for the cross-
classification of years of school completed and age

Years of School		Code No.	25–34 years 1	35–44 years 2	45–54 years 3	55 years and over 4	Marginal Fre-quencies
Elementary School	0–4 years	1	11	12	13	14	5,747
	5–7 years	2	21	22	23	24	9,924
	8 years	3	31	32	33	34	14,596
High School	1–3 years	4	41	42	43	44	18,682
	4 years	5	51	52	53	54	37,134
College	1–3 years	6	61	62	63	64	11,164
	4 years or more	7	71	72	73	74	12,063
Marginal Frequencies			24,865	23,021	23,398	38,026	109,310

Source: Adapted from *Statistical Abstract of the United States, 1971*. U.S. Bureau of the Census, Wqshington, D.C. 1971 (92nd edition) pages 109 & 110.

the frequency for all persons, ages 25–34, of 24,865 and multiply by 100, we find that about 43.9 percent of the people, ages 25–34, are high school graduates. Thus,

$$\frac{10,916}{109,310}\cdot 100 = 10.0 \text{ percent} \qquad \frac{10,916}{37,134}\cdot 100 = 29.4 \text{ percent}$$

$$\frac{10,916}{24,865}\cdot 100 = 43.9 \text{ percent}$$

When we use either the second or the third type of percentage, we refer to having calculated it in one *direction* or the other. In this case, the 29.4 percent for all high school graduates is in the "direction" of the variable education, and the 43.9 percent for all persons ages 25–34 is in the "direction" of the variable age. The direction we select is determined according to which variable we feel has a causal influence on the other.

We have been assuming without saying as much that there is a *causal* ordering between the variables we study. Thus we assumed that sex influences education, that age influences education, that sex influences occupation, that majority/minority status influences political activities, and so forth. This assumption of causal influence makes good sense and conforms to the every-day observations that we make about the way the world works. However, it will not always be so easy to assume causal direction. Later on, we will

**Table 3.3.5   Years of school completed by age
(in thousands) frequencies**

Years of School Completed		Age				
		25–34 years	35–44 years	45–54 years	55 years and over	Total
Elementary School	0–4 years	322	621	866	3,938	5,747
	5–7 years	796	1,404	1,802	5,922	9,924
	8 years	1,144	1,934	2,808	8,710	14,595
High School	1–3 years	4,252	4,259	4,305	5,866	18,682
	4 years	10,916	9,324	8,915	7,979	37,134
College	1–3 years	3,506	2,532	2,363	2,763	11,164
	4 years or more	3,929	2,947	2,339	2,848	12,063
Total		24,865	23,021	23,398	38,026	109,310

Source: Adapted from: U.S. Bureau of the Census. *Statistical Abstract of the United States: 1971.* (92nd edition.) Washington, D.C., 1971, page 110, No. 165.

consider the hairy question of two-way causation. At this point, however, it seems reasonable to recommend common sense in the matter of the influence of one variable on another.

One of the simplest means by which causal influence can be determined is by considering events according to the order in which they occur, called *temporal sequencing.* Again, common sense suggests that when one event precedes another and is associated with it, the first event influences or in one sense "causes" the second event. It is through this perspective that we treat such things as parents' education "causing" respondent's education or as one's education "causing" one's income.

Back to our example. Since we assume that age influences the amount of schooling people complete, we "percentage" in the direction of the causal variable *age.* In general, it is customary to percentage the dependent variable in the direction of the independent variable, since we assume that the latter "causes" the former.

Table 3.3.6 shows the percentages that result from such calculations. A cursory examination of the percentages indicates that as *age* increases the percentage with more *years of school completed* decreases. Once again, statistics can facilitate our examination of the association between variables.

There are useful conventions that should guide the construction of percentage tables. Here are six:[4]

[4]The following rules are adapted from James A. Davis, *Elementary Survey Analysis.* (Englewood Cliffs, New Jersey: Prentice-Hall, Inc., 1971), p. 65.

Table 3.3.6 Years of school completed by age
(in thousands) percentages
$N = 109{,}310$

		25–34	35–44	45–54	55 and over	Total
Elementary	0–4 years	1.3%	2.7%	3.6%	10.4%	5.3%
School	5–7 years	3.2	6.1	7.7	15.5	9.1
	8 years	4.6	8.4	12.0	22.9	13.4
High	1–3 years	17.1	18.5	18.4	15.4	17.1
School	4 years	43.9	40.5	38.1	21.0	34.0
College	1–3 years	14.1	11.0	10.1	7.3	10.2
	4 or more	15.8	12.8	10.0	7.5	11.0
	Total	100.0% (24,865)	100.0% (23,021)	100.0% (23,298)	100.0% (38,126)	100.0% (109,310)

Source: Adapted from U.S. Bureau of the Census, *Statistical Abstract of the United States, 1971.* (92nd edition.) Washington, D.C., 1971, p. 110.

Rule 1: Always present your major findings in percentage tables as well as in more sophisticated and complex tabular form.

Rule 2: Always present the number of cases (case base = N) locating them below and in parentheses and, when possible, to the right of the percentages. This enables readers to compare percentages without case frequencies interfering.

Rule 3: In general, report percentages rounded to whole numbers. The use of decimals with percentages suggests scientific precision which, except in large samples, is not warranted. When decimals are presented, round off to one decimal place.

Rule 4: Always place percentage signs in both the table and the marginals locating them next to the percentages to make clear the distinction between frequencies and percentages.

Rule 5: When presenting elaborate multivariate percentage tabulations, the raw frequencies may be reported in an appendix if including them in a percentage table would be cumbersome.

Rule 6: Always exclude "no answers" from the case base reported, and account for them by reporting the number of "no answer" or "not applicable" cases.

3.4 INTRODUCING A TEST VARIABLE

Earlier we mentioned that one of the guiding assumptions of social statistics is that sociological variables are interrelated and that several may operate simultaneously. This is referred to as *multivariate influence*, and it occurs when a number of independent variables influence or "cause" a dependent variable. For analytical purposes, we single out a set of variables and study them as a system. Let us start with the idea of a three variable system.

Often we are interested in examining the association between two variables *holding constant* a third one. For example, we might be interested in the association between the number of years of school completed and sex holding race constant. Table 3.4.1 shows the percentage distribution of this three-variable relationship.

Table 3.4.1   Number of years of school completed by sex and race (1970)
(percentage, *f* in thousands)

		White		Black		Total—Both Races and Both Sexes
		Male	Female	Male	Female	
Elementary School	0–4 years	4.5%	3.9%	18.6%	12.1%	5.3%
	5–7 years	8.8	7.8	16.0	17.3	9.1
	8 years	13.9	13.4	11.1	11.3	13.4
High School	1–3 years	15.6	17.3	21.9	24.5	17.1
	4 years	30.9	39.0	22.2	24.4	34.0
College	1–3 years	11.3	10.1	5.7	6.0	10.2
	4 years or more	15.0	8.6	4.6	4.4	11.0
Total		100.0% (46,606)	100.0% (51,506)	100.0% (4,619)	100.0% (5,470)	100.0% (109,310)

Source: U.S. Bureau of the Census. *Statistical Abstract of the United States, 1971.* (92nd edition.) Washington, D.C., 1971, p. 108.

As with many aspects of statistical analysis, three-variable relationships can be portrayed graphically. For example, we can examine population characteristics such as the distribution of the population by age and sex, as seen in Figure 3.4.1, which is a horizontal percentage histogram.

Our interest in such a table or graph is in the three-variable relationship in which we have one dependent variable, abbreviated Y, one independent variable, abbreviated X, and one test (control) variable, abbreviated T. The table indicates the percentage distribution of Y given X for *constant* categories of T. Thus we can look at the number of years of school completed by sex holding race constant. Put differently, in Table 3.4.1, we can examine the percentage value of Y (years of school completed) given X (sex) holding

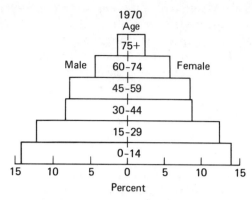

Figure 3.4.1. Percent distribution of the population by age and sex: 1970. (*Adapted from: U.S. Bureau of the Census,* Statistical Abstract of the United States: 1971 (*92nd edition.*) *Washington, D.C., 1971, p. 23, No. 21.*)

T (race) constant. Our interest is in the two-variable relationship between education and sex for (1) all whites and for (2) all blacks; that is, race is held constant.

Our interest in classifying people according to the possession of three variables simultaneously means that we need a code system to assign people appropriately. By using the letter abbreviations, a code system, and statistical notation, we can summarize information concisely. Suppose that we are interested in examining the association between extracurricular activity and academic performance holding constant the year in college. After gathering some data, we assemble the information reported in Table 3.4.2 about 24 outstanding students who actively participate in extracurricular organizations.

It is apparent that this alphabetical listing does not provide us with an easy means of testing the influence of the independent variable, number of extracurricular organizations (X), in terms of the dependent variable, grade point average (Y). An alternative arrangement would be to separate the students according to their year in college and the number of extracurricular organizations to which they belong. In other words, we can cross-classify variable T and variable X. Rather than tallying the number of students who fit into the cells of this cross-classification, we can list their scores on variable Y, as can be seen in Table 3.4.3 on page 53.

Since we can categorize each individual's score on variable Y given X and T, we use the code notation of Y_{XT}. A first-year student who is active in one extracurricular organization would be designated according to his or her GPA as Y_{11}. Looking over the list, we find that Bruce, Ernest, and Tom fit the description $X = 1$, $T = 1$, and we record their scores on variable Y in the table. Similarly, Philip and Walt are third-year students active in three

54

Table 3.4.2 List of students, year in college, number of extracurricular organizations, and grade point average

Name	Year in College (T)	Number of Extracurricular Organizations (X)	Grade Point Average (Y)
Alan	3	1	4.0
Barbara	1	3	3.3
Betty	2	3	3.5
Bob	4	2	3.9
Bruce	1	1	3.5
Ernest	1	1	3.6
Fred	2	2	3.7
Gail	1	2	3.1
Gary	2	1	3.7
George	2	2	3.4
Ilene	2	3	3.2
Joan	1	2	3.6
John	4	3	3.8
Jonathan	3	2	3.6
Lisa	3	2	3.7
Martha	4	1	4.0
Nancy	2	1	3.9
Peggy	3	2	3.8
Philip	3	3	3.3
Sandy	4	2	3.7
Terri	4	3	3.5
Tom	1	1	3.9
Walt	3	3	3.6
Wayne	4	3	3.5

extracurricular organizations; thus they fit into location Y_{33}. Each student's grade point score can be assigned in this fashion, as can be seen in Table 3.4.3.

We can now set up a general notation table to summarize how we locate people on three variables at once. The operations are merely an extension of those already stated, and the notation is a shorthand means of expressing them. At first glance, the general notation format as shown in Table 3.4.4 may appear complex, but it really is not.

The vertical (up and down) listings are called columns and the horizontal (side to side) listings are called rows. The individual case scores are represented by the symbols Y_{11}, Y_{21}, Y_{31}. . . Y_{nk}. The symbol Y_{XT} is the general case representing the score on variable Y in the Xth row and the Tth column. n indicates the number of categories of X for each category on T, and k indicates the number of categories of T for each category on X. The variable classes or levels may be noted as descending from top to bottom or ascending from bottom to top depending primarily upon one's personal preference.

The symbols in the body of the table, i.e., Y_{11} through Y_{nk}, are merely shorthand designations or code names for the scores of N individuals on the

Table 3.4.3   Grade point average, year in college, and number of extracurricular organizations

X = Number of Extracurricular Organizations	T = year in college				
	1	2	3	4	
1	3.9 3.5 3.6	3.9 3.7	4.0	4.0	26.6 = the sum of row 1. 7 students
2	3.6 3.1	3.7 3.4	3.8 3.6 3.7	3.9 3.7	32.5 = the sum of row 2. 9 students
3	3.3	3.5 3.2	3.6 3.3	3.8 3.5 3.5	27.7 = the sum of row 3. 8 students
	21.0 = the sum of column 1. 6 students	21.4 = the sum of column 2. 6 students	22.0 = the sum of column 3. 6 students	22.4 = the sum of column 4. 6 students	86.8 = the sum of the sums of columns 1–4. 24 students

variable Y. The single summation signs $\sum_{T=1}^{k_1} Y_{1T}$ through $\sum_{T=1}^{k_n} Y_{nT}$ that appear in the right-hand margin of the table tell us to sum the scores of the cases across the categories of X, from the first through the kth cases on variable Y. For example, $\sum_{T=1}^{k_1} Y_{1T}$ indicates that all the scores in the first row are to be summed. Similarly, the single summation signs $\sum_{X=1}^{n_1} Y_{X1}$ through $\sum_{X=1}^{n_k} Y_{XK}$ in the bottom margin tell us to sum the scores of the nth cases on variable Y. For example, $\sum_{X=1}^{n_1} Y_{X1}$ indicates that the scores in the first column are to be summed.

In the bottom right-hand corner of the table, the complex-appearing double summation sign $\sum_{X=1}^{n} \sum_{T=1}^{k} Y_{XT}$ is read from right to left and means that we first sum the Y cases in each of the T columns, and we then sum the sums of the T columns across all the X rows. In other words, we add together the individual column sums across the bottom of the table to get the overall sum. The essence of the double summation sign is that we add together all of the individual scores on variable Y, but we use the sums already calculated in each column instead of summing all the individual scores over again.

The simplicity of this imposing format becomes clearer if we reconsider how the students mentioned above fit into this notational scheme of things.

Table 3.4.4   General notation format

X	1	2	3	$\cdots$	k	
			T			
1	Y_{11}	Y_{12}	Y_{13}	$\cdots$	Y_{1k}	$\displaystyle\sum_{T=1}^{k_1} Y_{1T}$ k_1
2	Y_{21}	Y_{22}	Y_{23}	$\cdots$	Y_{2k}	$\displaystyle\sum_{T=1}^{k_2} Y_{2T}$ k_2
3	Y_{31}	Y_{32}	Y_{33}	$\cdots$	Y_{3k}	$\displaystyle\sum_{T=1}^{k_3} Y_{3T}$ k_3
$\vdots$	$\vdots$	$\vdots$	$\vdots$	$\vdots$	$\vdots$	$\vdots$
n	Y_{n1}	Y_{n2}	Y_{n3}	$\cdots$	Y_{nk}	$\displaystyle\sum_{T=1}^{k_n} Y_{nT}$ k_4
	$\displaystyle\sum_{X=1}^{n_1} Y_{X1}$ n_1	$\displaystyle\sum_{X=1}^{n_2} Y_{X2}$ n_2	$\displaystyle\sum_{X=1}^{n_3} Y_{X3}$ n_3	$\cdots$	$\displaystyle\sum_{X=1}^{n_k} Y_{Xk}$ n_k	$\displaystyle\sum_{X=1}^{n}\sum_{T=1}^{k} Y_{XT}$ N

Looking at those students in the second year of college helps further clarify the generalized single summation sign $\displaystyle\sum_{X=1}^{n} Y_{XT}$. Since we use n to indicate the number of categories on X, we read the sum sign in the second column $\displaystyle\sum_{X=1}^{n_2} Y_{X2}$ as "add together the scores of the students in the second category of T from the first ($X=1$) through the last ($n_2=6$) cases."

Thus,

$$3.9 + 3.7 + 3.7 + 3.4 + 3.5 + 3.2 = 21.4$$

as is seen in Table 3.4.3, or

$$Y_{12} + Y_{12} + Y_{22} + Y_{22} + Y_{32} + Y_{32} = \sum_{X=1}^{n_2} Y_{X2}$$

as is seen in Table 3.4.4.

The same procedure may be carried out for the rows, except that since we use k to indicate the number of categories of T, we read the sum sign

$\sum_{T=1}^{k_1} Y_{1T}$ as "add together the scores of the students in the first category of
X from the first (1) through the last ($k = 7$) cases." Thus we look at all the
students who are active in one extracurricular organization and add across
the table.

Now let us look at the double sigma in the lower right-hand corner. The
double summation sign $\sum_{X=1}^{n} \sum_{T=1}^{k} Y_{XT}$ is read as "add together the scores of
the students in each category of T from the first to the last (n) cases and add
together all the sums of the T columns for all the categories of X from the
first through the last (n_k) cases."

Thus,

$$\sum_{X=1}^{n} \sum_{T=1}^{k} Y_{XT} = \quad (3.9 + 3.5 + 3.6 + 3.6 + 3.1 + 3.3)$$
$$+ (3.9 + 3.7 + 3.7 + 3.4 + 3.5 + 3.2)$$
$$+ (4.0 + 3.8 + 3.6 + 3.7 + 3.6 + 3.3)$$
$$+ (4.0 + 3.9 + 3.7 + 3.8 + 3.5 + 3.5)$$
$$= \sum_{X=1}^{n} (21.0 + 21.4 + 22.0 + 22.4) = 86.8$$
$$= 86.8$$

as seen in Table 3.5.3.

Much of the upcoming materials are actually extensions of many things
that you already know about everyday life and the matters we have covered
in the first three chapters. If you have a good understanding of the material
covered so far, the next chapter should be fairly simple. If you are a bit shaky,
however, you could help yourself, by rereading this chapter before going on.
The steps we have learned can now be put to use in helping us examine such
things as average values and the variability of our sample.

EXERCISES

1. In your own words, define, describe, or discuss the following terms and give an
   hypothetical example of each:

   code system
   summation sign
   cross-classification
   percentage
   multivariate influence
   marginal frequency
   holding constant

  controlling for a test variable
  test variable
  general notation table

2. Express the following formulas for the sums in terms of the individual components.

  A. $\sum_{i=2}^{5} X_i^2$

  B. $\left(\sum_{i=1}^{4} X_i\right)^2$

  C. $\sum_{i=2}^{5} Y_i$

  D. $\sum_{9=1}^{5} X_i \cdot \sum_{i=3}^{6} Y_i$

3. Out of a sample of 1000 freshman students, half are male and half are female. 20% of the male students are education majors, while 20% of the female students are *not* education majors. Construct a 2 × 2 table which shows the numeric breakdown of students according to their sex and major.

4. Assume that you are interested in one's age performance in statistics, and overall academic performance. You have nine students with the following characteristics:

Name	Age	Grade Point Average	Statistics Examination Grade
Adrian	21	2.5	85.0
Carrie	19	2.5	81.0
Hal	20	3.5	87.5
Mike	21	3.0	92.0
Peter	19	3.5	85.0
Rose	20	2.5	78.0
Sarah	19	3.0	82.5
Wilbur	21	3.5	89.0
Zelda	20	3.0	85.0

  Construct a table arranging the students according to the general notation table; let age $= T$, grade point average $= X$, and statistics examination grade $= Y$.

5. Using the information on name of student, year in college, and number of extracurricular activities presented in Table 3.4.2, construct the following tables:
  A. Construct a frequency table. Let $X =$ year of college, $Y =$ number of extracurricular activities, and $T =$ student's name.
  B. Using the table constructed in Part A, give the general notation for John, Philip, Betty, Joan, and Nancy.
  C. Construct a numerical frequency table from the table constructed in Part A. Convert this table into percentage tables, percentaging the rows and columns.

6. The following table presents a distribution of marital adjustment scores by happiness rating categories. If you were going to percentage this table, in which direction would you do so? Would it be sensible to percentage in the other direction? Why or why not?

Frequency distribution of marriage adjustment scores by happiness rating categories

Adjustment Score	Very Unhappy	Unhappy	Average	Happy	Very Happy	No Rating	Total	Percentage Distribution
180–199				7	65		72	13.7%
160–179			5	31	118	1	155	29.4
140–159			12	34	32	1	79	15.0
120–139	2	9	27	24	6	2	70	13.3
100–119	3	13	23	8			47	8.9
80– 99	13	19	5	3		1	41	7.8
60– 79	15	17	3		3		38	7.2
40– 59	7	8		1			16	3.2
20– 39	2	5	1				8	1.5
Total	42	71	76	108	224	5	526	
Percentage Distribution	8.0%	13.5	14.4	20.5	42.6	1.0		100.0%

Mean score = 140.8; σ = 38.8

Source: Paul F. Lazarsfeld and Morris Rosenberg, *The Language of Social Research.* New York: The Free Press, 1955, p. 273, Table IV.

7. In the following three hypotheses, point out the independent variable, the dependent variable, and introduce a test or control variable.
   A. White college graduates who attended college in the South will be more prejudiced against Blacks than those who attended college in the North.
   B. Having a baby during graduate school reduces a student's chances of obtaining the Ph.D.
   C. Among adolescents, the greater the self-doubt, the greater the conformity to current fads.

part II

Univariate
Descriptive
Statistics

4

Measuring
Central
Tendency

In social statistics, as in everyday life, we are often interested in determining how much alike or different individuals or groups are compared to each other. Thus far we have discussed how we can study people according to various levels of measurement. We have also studied individuals and groups in terms of frequency distributions, which portray such things as people's responses to sociological questionnaires or their places in investigations of the Bureau of the Census. We also have seen how fairly complex information can be summarized in terms of one-way, two-way, or three-way frequency distributions.

It is common sense to seek single values that can be substituted for all the numbers in a frequency distribution. Fortunately, there are standard ways in which to classify people according to such single values, and we use some rather formal, but at the same time common-sense, techniques in order to do so. To determine how typical or how variable individuals or groups are, we utilize (1) measures of *central tendency*, that is, the tendency of the data to be clustered around a midpoint, and (2) *measures of dispersion*, that is, the tendency of the data to be spread out from the midpoint. This and the next chapter explain several of these measures of typicality and variability.

In an effort to summarize the information in a frequency distribution, we may want to describe that distribution in terms of its "central," or "middle," or "average," or "typical" value. This means that we want to find certain central values around which the frequency distribution is clustered. The measures that we use to describe such a value are called *averages*. Three kinds of averages are important for us: the mode, or the frequency average; the median, or the positional average; and the mean, or the arithmetic average. The aim of each type of average is to provide a single value for a distribution; this value summarizes or describes the distribution, or it enables us to estimate (guess) any individual case's score. Once we have such a value, it is always possible to compare a given case with the estimate and to measure how much they differ.

4.1 THE MODE

The mode is the most elementary form of average, and it is nothing more than the value that occurs most frequently. In order to determine the mode, we merely count the cases for all values of a variable. The one with the most cases, that is, the greatest frequency, is the modal value. For example, if we have 100 students, 30 of whom are 20 years old, 25 of whom are 19 years old, 20 of whom are 21 years old, 15 of whom are 18 years old, and 10 of whom are 22 years old, we would say that 20 years is the modal age, as can be seen in Table 4.1.1.

Table 4.1.1   Age of students

Age In Years	Frequency
18	15
19	25
[20]	[30]
21	20
22	10
Total (*N*)	100
Modal Age = 20 Years	

One of the advantages of using the mode is that it requires no calculation or formula when used in its crude form. There are refined methods of calculating the mode, which may be found in more detailed books on statistics. However, for our purposes all we need to do is order and group the data and then select the mode, as we just have done. A second advantage is that the mode can be used for summarizing nominal, ordinal, or interval variables. Thus it can provide a numerical value for both qualitative and quantitative variables. A third advantage of the mode is that if the cases are highly homogeneous, it is an easy-to-derive statistic that may facilitate describing the distribution in question. We may say when speaking of years of education, for example, that since the greatest percentage in a category or class is the same as the greatest frequency, the modal education is four years of high school, as can be seen in Table 4.1.2.

Table 4.1.2   Persons 25 years and over

Years of School Completed		Percentage
Elementary School	0–4 years	5.3%
	5–7 years	9.1
	8 years	13.4
High School	1–3 years	17.1
	[4 years]	[34.0]
College	1–3 years	10.2
	4 years	11.0
		100.0%
		(109,310)

Modal education = 4 years of high school

Source: Adapted from U.S. Bureau of the Census, *Statistical Abstract of the United States: 1971.* (92nd edition.) Washington, D.C., 1971, page 110, No. 165.

Two major disadvantages of the mode are (1) there may be no mode, or (2) there may be more than one mode. Either situation often shows that the cases are heterogenous in terms of the distribution. Another disadvantage is that the mode does not take into account information about the entire distribution. A third disadvantage is that the mode cannot be manipulated algebraically, thereby reducing effective use of it for certain more advanced statistical techinques.

In some ways the mode is useful, however. For example, teachers are concerned by the number of students they have in each class. The way in which a course is taught often depends more upon the modal size of a given class than it does on, say, the median or the mean sizes of all a teacher's classes.

4.2 THE MEDIAN

The median is the positional average. It is the value of the case that has an equal number of cases above it and below it. It is called "positional" because we use the value of the case that occupies the middle position. In order to determine the median, we first arrange the cases in rank order from low to high or vice versa. Then we select the case that is in the middle. The value that this middle case possesses is designated the median. The formula for the median tells us the *location* of the median case, but it does not tell us the median value itself. The general formula for the median is

$$Md = \frac{N + 1}{2}\text{th largest case.}$$

For example, in a family of seven people, including a 98-year-old great-grandmother, a 67-year-old grandmother, a 36-year-old father, a 35-year-old mother, and three children ages 9, 11, and 13, the median age is the age of the $(N + 1)/2 = (7 + 1)/2 = 8/2 = $ 4th person. Table 4.2.1 shows that if we start with the youngest (or oldest) and count to the 4th person, we find that the median age of this family is 35.

Table 4.2.1   Median age, family of seven

Rank		Age	
Old to Young	*Young to Old*		
7	1	9	
6	2	11	} 3 cases below
5	3	13	
4	Median Rank 4	35	Score (value) of the median case
3	5	36	
2	6	67	} 3 cases above
1	7	95	

The median is clearest when an odd number of cases is involved, as can be seen in Table 4.2.2. The middle case is the one occupying the position between the others; that is, it occupies the middle position.

When there is an even number of cases, the median is the value that falls halfway between the values of the two middle cases. To determine the median with an even number of cases, we add the value below the median to the value above the median and divide by 2; that is,

$$Md = \frac{\text{value below} + \text{value above}}{2}$$

Table 4.2.2   The median case given an odd number of cases

	Case No.	Case No.	Case No.	Case No.	
Case with lowest score	1	1	1	1	Case with lowest score
	[2]	2	2	2	
Case with highest score	3	[3]	3	3	
	$\frac{N+1}{2}=\frac{4}{2}=2$	4	[4]	4	
		5	5	[5]	
		$\frac{N+1}{2}=\frac{6}{2}=3$	6	6	
			7	7	
			$\frac{N+1}{2}=\frac{8}{2}=4$	8	
				9	Case with highest score
				$\frac{N+1}{2}=\frac{10}{2}=5$	

For example, if $N = 49$, then

$$Md = \frac{49+1}{2} = \frac{50}{2} = \text{25th largest case. Thus,}$$

1 2 3 4   5 6 7 8   9 10 11 12 13 14 15 16 17 18 19 20 21 22 23 24

[Md = Score (value) of the 25th case]

26 27 28 29 30 31 32 33 34 35 36 37 38 39 40 41 42 43 44 45 46 47 48 49

Thus, there are 24 cases below the median and 24 cases above it.

This determines the position of the "middle" value, even though there is not an actual case possessing such a value. Suppose that we add to the family described above a 67-year-old grandfather. The median age is the

$$\frac{\text{value below} + \text{value above}}{2} = \frac{35 + 36}{2} = \frac{71}{2} = 35.5.$$

Now the median age of this family is 35.5. If instead of a grandfather, the family had another child of, say, 7, the median age would be

$$\frac{\text{value below} + \text{value above}}{2} = \frac{13 + 35}{2} = \frac{48}{2} = 24.$$

Clearly, the median may be subject to considerable fluctuation, as can be seen in Table 4.2.3.

Table 4.2.3   Median age, family of eight

Rank		Age		
Old to Young	*Young to Old*	*An Additional Child Age 7*		*An Additional Grandfather Age 67*
8	1	7 ⎫		9 ⎧
7	2	9 ⎬ 4 cases below		11 ⎨
6	3	11		13
5	4	13 ⎭		35 ⎩
4.5	4.5	[24]	Score (value) halfway between the 2 middle cases	[35.5]
4	5	35 ⎫		36 ⎧
3	6	36 ⎬ 4 cases above		67 ⎨
2	7	67		67
1	8	98 ⎭		98 ⎩

Table 4.2.4 demonstrates the position of the median for an even number of cases.

So far, we have discussed the median as it is determined for a given number of cases for unique scores. However, very often sociologists work with grouped data. When data are grouped, each individual case loses its uniqueness, and the only thing we know about it is that it belongs in a certain category or class. Thus, when we calculate the median for grouped data we compromise knowledge of individual scores by treating the cases as if they are distributed evenly throughout the interval. For example, the information that appears as unique values in Table 4.1.1 can be restated in intervals, as can be seen in Table 4.2.5.

Two things have been added. In the first place, age in years has been stated in ranges rather than as specific values. Thus, we see that the first interval for age 18 is from 18.0 to 18.9; the second is from 19.0 to 19.9; and so forth. These are called *stated* limits and should be distinguished from what are called *true* limits. *Stated limits* are considered to be the midpoint of specific intervals; for instance, 18.0 is the midpoint for the interval 17.95 and 18.05. Thus the *true limits* of the class intervals are 17.95—18.95, 18.95—19.95, and so forth, as shown in Table 4.2.6.

For our purposes, and for most statistical work, stated limits are sufficient for tallying, computing, and reading as can be seen in Table 4.2.5. The right-hand column of Table 4.2.5. shows the cumulative frequency distribution, which is the value that appears in all of the intervals below as well as the one in question. Thus the cumulative frequency for the first interval is 15 because

Table 4.2.4   The median case given an even number of cases

	Case No.	Case No.	Case No.	Case No.	
Case with lowest score	1	1	1	1	Case with lowest score
	[1.5]	2	2	2	
Case with highest score	2	[2.5]	3	3	
		3	[3.5]	4	
		4	4·	[4.5]	
			5	5	
			6	6	
				7	
				8	Case with highest score

For example, if $n = 30$, then

$$Md = \frac{15 + 16}{2} = 15.5\text{th largest case. Thus,}$$

1   2   3   4   5   6   7   8   9   10   11   12   13   14   15

[Md = score (value) halfway between the values of 15th and 16th cases
= score (value) of the 15.5th case]

16   17   18   19   20   21   22   23   24   25   26   27   28   29   30

Thus, there are 15 cases below the median and 15 cases above it.

there are 15 cases in interval 1; for the second interval, the 25 cases in interval 2 are added, and the resulting cumulative frequency distribution shows that there are 40 cases in the first two intervals. The same is done for each interval.

We are seeking the median, which is that value with one-half of the distribution below it and one-half above it. Since we have an N of 100, we use that value in the median formula; thus,

$$Md = \frac{N + 1}{2} = \frac{100 + 1}{2} = \frac{101}{2} = 50.5\text{th case.}$$

Now we must calculate the value of the 50.5th case; that is, we are seeking the unique score with an equal number of cases above and below it.

Table 4.2.5 Age of students

Age in Years	Frequency	Cumulative Frequency
18.0–18.9	15	15
19.0–19.9	25	40
20.0–20.9	30	70
21.0–21.9	20	90
22.0–22.9	10	100
Total (N)	100	

Table 4.2.6   Stated and true limits

Stated Limits	True Limits
18.0–18.9	17.95–18.95
19.0–19.9	18.95–19.95
20.0–20.9	19.95–20.95
21.0–21.9	20.95–21.95
22.0–22.9	21.95–22.95

One of our assumptions is that the cases are spread out evenly over the intervals in question. Since the table indicates that there are 40 cases in the two intervals between the ages of 18.0 and 19.9, we know that the median will be in the next interval, that is, between ages 20.0 and 20.9. More specifically, it will be one of the 30 students in the interval 20.0 and 20.9, as can be seen in the frequency distribution. Remember, we are seeking the 50.5th case and we have 40 cases up to age 20.0, so we are interested in the score that falls between the 10th and the 11th cases of the 30 in the interval.

Without much difficulty, we can count the appropriate number of cases to arrive at the median value, as can be seen in Table 4.2.7A. There are 40 cases below the interval and 30 cases above it. We assume that the 30 cases in the interval 20.0–20.9 are evenly distributed. Now, by adding the appropriate number of cases to the cumulative frequency, we find that there are 49 cases whose ages are 20.2 or lower and that there are 48 cases whose ages are 20.4 and older. Having singled out the age 20.3, we now must divide it into equal intervals and locate the three cases in it. Once again we assume them to be evenly spread out. Table 4.2.7B shows this more detailed breakdown of the interval of 20.3 as being equal to 20.30–20.39. It is visually clear that there are 50 cases below the value 20.33 and 50 cases above it.

Deriving the median in this manner is both time-consuming and elaborate. Fortunately, we can substitute a formula that does the same thing with much

**Table 4.2.7A   Evenly distributing the 30 students
in the interval 20.0–20.9**

Age in Years	Frequency	20.0–20.99 Interval Frequency	Cumulative Frequency
18.0–18.9	15		15
19.0–19.9	25		40
20.0–20.09		3	43
20.1–20.19		3	46
20.2–20.29		3	49
[20.3–20.39]		3	52
20.4–20.49	30	3	55
20.5–20.59		3	58
20.6–20.69		3	61
20.7–20.79		3	64
20.8–20.89		3	67
20.9–20.99		3	70
21.0–21.9	20		90
22.0–22.9	10		100
Total (N)	100		

less effort. The formula uses the following values:

l = the lower limit of the interval
f = the frequency in that interval
F = the cumulative frequency
$\dfrac{N}{2}$ = one-half of the total frequency
i = the interval size

The median of grouped data may be found by the formula:

$$Md = l + \left(\frac{\frac{N}{2} - F}{f}\right) \cdot i$$

Thus,

$$Md = 20.0 + \left(\frac{50 - 40}{30}\right) \cdot 1 = 20.0 + \frac{10}{30}$$
$$= 20.0 + 0.33 = 20.33$$

At times data are displayed as whole numbers that do not allow us to make distinctions as fine as we might like. However, it is possible to treat the data as if they were continuous and to use the median as a summary measure.

Table 4.2.7B   Evenly distributing the 30 students
in the interval 20.30–20.39

Age in Years	Frequency	20.0–20.99 Interval Frequency	20.30–20.39 Interval Frequency	Cumulative Frequency
18.0–18.9	15			15
19.0–19.9	25			40
20.0–20.09		3		43
20.1–20.19		3		46
20.2–20.29		3		49
20.30–20.309				
20.31–20.319			1	50
20.32–20.329				
[20.33–20.339]			median interval 50.5	
20.34–20.349				
20.35–20.359		3	1	51
20.36–20.369	30			
20.37–20.379				
20.38–20.389			1	52
20.39–20.399				
20.4–20.49		3		55
20.5–20.59		3		58
20.6–20.69		3		61
20.7–20.79		3		64
20.8–20.89		3		67
20.9–20.99		3		70
21.0–21.9	20			90
22.0–22.9	10			100
Total (N)	100			

For example, when the number of years of school completed is considered, there is no "interval" for those completing four years of high school. However, by using interval limits as if they were continuous, we can list four years of high school as being the interval 12.0–12.9, as can be seen in Table 4.2.8.

The initial step in the formula tells us that we are seeking the 54,655.5th case. Using the information in the table about frequencies and limits, we calculate the median years of school completed for the 109,310 persons 25 years of age or more and find that is it 12.2. Without going through the necessary calculations, Table 4.2.9 shows a bivariate use of the median. We compare the years of school completed by age, listing the median year of school for each of the age categories for persons 25 years of age or older.

Naturally, it is possible to combine information for three variables (or more) and to use the median as a summary statistic. Table 4.2.10 shows the median earnings of people 14 years of age and over in 1969 by sex and occupation.

You should recognize that by increasing the number of grouped data

Table 4.2.8   Years of school completed, persons 25 years and over (in thousands)

Years of School Completed		Limits As If Continuous	Frequency	Cumulative Frequency
Elementary School	0–4 years	0.0–4.9	5,747	5,747
	5–7 years	5.0–7.9	9,924	15,671
	8 years	8.0–8.9	14,595	30,266
High School	1–3 years	9.0–11.9	18,682	48,948
	4 years	12.0–12.9	37,134	86,082
College	1–3 years	13.0–15.9	11,164	99,246
	4 years or more	16.0 or more	12,063	109,310
			109,310	

The general formula for the median is:

$$Md = \frac{N+1}{2} = \frac{109,310+1}{2} = 54,655.5\text{th case.}$$

The specific formula for grouped data is:

$$Md = l + \left(\frac{\frac{N+1}{2} - F}{f}\right) i = 12.0 + \left(\frac{54,655.5 - 48,948}{37,134}\right) \cdot 1$$

$$Md = 12.0 + \left(\frac{5,707.5}{37,134}\right) = 12.0 + 0.15$$

$$Md = 12.2$$

Table 4.2.9   Years of school completed by age

Age	Median Years of School	Number of Cases
25–34 years	12.6	24,865
35–44 years	12.4	23,021
45–54 years	12.2	23,298
55 years and over	9.3	38,126
Total	12.2	109,310

intervals we simultaneously decrease the interval width and increase the accuracy of using the grouped-data formulas as an approximation of the true median. With large samples this often is adequate, and it avoids the awesome task of sorting all the values.

One of the major weaknesses of the median is that it may be subject to considerable variability, as was seen in the example about the extended family. Another weakness is that, like the mode, it cannot be used in algebraic manipulation.

Table 4.2.10 Current population report of 1969 on earnings of people 14 years old and over, median earnings by sex and occupation (in dollars)

	Male	Female
White-Collar Workers		
Professional, technical workers	$10,516	$ 5,244
Managers, officers, proprietors	10,300	4,766
Clerical	6,804	3,603
Sales workers	6,829	1,241
Blue-Collar Workers		
Craftsmen and foremen	7,944	4,004
Operators	6,248	3,054
Nonfarm laborers	2,483	1,755
Service workers	3,684	1,497
Farm workers	821	341
Median for All Categories	$ 6,899	$ 2,564
Total (N)	(48,818)	(29,084)

4.3 THE ARITHMETIC MEAN

The arithmetic mean is another measure of central tendency. It is intended to indicate the typical numerical value of a group of observations. Most people are familiar with the arithmetic mean because it is commonly the "average;" thus, it is generally a breeze to work with. The "mean," as it is also called, is designated by the capital letter used for a variable covered by a bar, e.g., $\bar{Y}$ or $\bar{X}$, and is called Y-bar or X-bar. Using the notation we learned in Chapter 3, the mean of Y_1, Y_2, Y_3, ... Y_N may be defined as:

$$\bar{Y} = \frac{Y_1 + Y_2 + Y_3 \cdots + Y_N}{N}$$

or the mean of X_1, X_2, X_3 ... X_N may be defined as:

$$\bar{X} = \frac{X_1 + X_2 + X_3 \cdots + X_N}{N}$$

We can abbreviate the formula by using the summation sign (sigma, $\sum$) $\sum_{i=1}^{N} Y_i$, to show the summing of variable Y. This results in the following formula:

$$\bar{Y} = \frac{\sum_{i=1}^{N} Y_i}{N}$$

Suppose we have five students ($N = 5$) with the following grades on a midterm exam (variable Y): 50, 60, 40, 90, and 20, and that we are interested in finding the mean grade on this exam. Since Y_i refers to the generalized individual, each student is given a code number from 1 to 5. The summation sign tells us to sum the scores on variable Y from individual 1 through individual 5 and divide by 5. The formula tells us to count each student according to his score on the test, sum the grades, and divide by the number of students.

$$\bar{Y}_i = \frac{\sum_{i=1}^{N} Y_i}{N} = \frac{50 + 60 + 40 + 90 + 20}{5} = \frac{260}{5} = 52$$

Thus, the arithmetic mean or the "average" grade is 52. We can summarize the steps in a very simple fashion, as shown in Table 4.3.1.

Table 4.3.1   The arithmetic mean

Student	Symbol	Midterm Grade
1	Y_1	50
2	Y_2	60
3	Y_3	40
4	Y_4	90
5	Y_5	20
	$\sum_{i=1}^{N} Y_i = 260$	
	$\bar{Y} = 52$	

The calculation of the mean is straightforward and easy to understand when we deal with single cases individually enumerated according to their unique scores. When the number of cases to examine becomes large, however, the computation of the mean is much more complex and time-consuming. It may be simplified considerably by grouping the data into categories and calculating the mean from the frequency distribution of the grouped data. Instead of adding the scores separately, we assume that all the cases grouped within an interval have the same value, the midpoint, or that they are symmetrically distributed within each category. This assumption enables us to multiply the number of cases in the interval by their common value, the midpoint of the interval. For example, if there are 10 cases in the interval 70.0–79.9, the midpoint of which is 75, the product $10 \cdot 75 = 750$ is equal to the sum of 10 individual scores of 75. We carry out a similar operation for all intervals, sum the products, and then divide by the total number of cases to obtain the arithmetic mean. Thus, the formula for the mean from grouped data becomes:

$$\bar{Y} = \frac{\sum_{i=1}^{k} f_i m_i}{N}$$

where

i = the general category of the variable Y,
k = the number of categories,
f_i = the number of cases in the ith category,
m_i = the midpoint of the ith category,
N = the total number of cases in all categories.

Using the information we have on the age of our sample of students helps demonstrate the simplicity of this formula.

Table 4.3.2 Age of students

Age in Years	Midpoint m	Frequency f_i	$f_i m_i$
18.0–18.9	18.5	15	277.5
19.0–19.9	19.5	25	487.5
20.0–20.9	20.5	30	615.0
21.0–21.9	21.5	20	430.0
22.0–22.9	22.5	10	225.0
N		100	2035.0

$$\bar{X} = \frac{\sum_{i=1}^{N} f_i m_i}{N} = \frac{2035}{100} = 20.35$$

Even though the number of computations is simplified by grouping the data, we lose information by not dealing with each individual's unique score. This is largely because the simplification for the mean from grouped data locates all the individuals within each interval at the midpoint of their respective interval. Naturally the narrower the intervals, the less information we lose. An example makes this shortcoming clearer. Suppose that our 10 individuals in the interval 70.0–79.9 with the midpoint of 75 had scores of 70.0 or 79.9 or any combination of the two. The products, $10 \cdot 70.0 = 700$ or $10 \cdot 79.9 = 799$, show that the range of possible total values for the interval is from 700 to 799. Clearly, 750 is a compromise between them. Even though this is the case, in the long run the mean of grouped data still gives a good indication of the central tendency of the data.

Computer programs that calculate the mean usually indicate whether they are for grouped or ungrouped data. It is important for students using such work-saving machines to enter their data according to the instructions for the specific program in use.

The mean can be used to examine bivariate and multivariate situations as well as the univariate ones we have already discussed. For example, the comparative working conditions in terms of hours and earnings for various sorts of production workers can be analyzed by use of the mean for grouped data. Table 4.3.3 shows three mean values for each type of nonagricultural industry, one each for the number of workers annually, the number of hours worked per week, and the hourly earnings.

Table 4.3.3   Nonagricultural industries—number of production workers, hours, and earnings
Full- and part-time employees in 1970

Industry	Annual Number (in thousands)	Average Weekly Hours	Average Hourly Earnings
Mining	473	42.7	$3.84
Contract Construction	2,793	37.4	5.22
Manufacturing, Durable Goods	8,045	40.3	3.56
Manufacturing, Nondurable Goods	6,005	39.1	3.08
Transportation and Public Utilities	3,898	40.5	3.85
Wholesale and Retail Trade	13,293	35.3	2.71
Finance, Insurance, Real Estate	2,907	36.8	3.07
Services	10,521	34.5	2.84
Total	47,935	37.2	$3.23

Source: Adapted from U.S. Bureau of the Census, *Statistical Abstract of the United States: 1971*. (92nd edition.) Washington, D.C., 1971, page 219, No. 345.

To determine the total average number of workers (the first column in Table 4.3.3), we merely sum the workers for the eight industries. However, since the overall mean values for the "average weekly hours" and the "average hourly earnings," depend upon the number of workers, we treat these summarized data similarly to the way we did the raw (unsummarized) grouped data. Now, however, we substitute the "average" value for the industry for the midpoint of the interval, that is,

$$\text{overall} \quad \bar{X} = \frac{\sum_{i=1}^{N} f_i m_i}{N} = \frac{\sum_{i=1}^{N} f_i \bar{X}_i}{N}$$

Table 4.3.4 on page 78 shows the steps we use to calculate the mean for the overall (total) average weekly hours.

The arithmetic mean possesses two important characteristics that make it extremely useful to us in measuring associations between variables. First,

Table 4.3.4   Average weekly hours

Industry	Average Weekly Hours (Midpoint) $\bar{X}_i$	Annual Number (in thousands) Frequency f_i	$f_i\bar{X}_i$
Mining	42.7	473	20,197.1
Contract Construction	37.4	2,793	104,458.2
Manufacturing, Durable Goods	40.3	8,045	324,213.5
Manufacturing, Nondurable Goods	39.1	6,005	234,795.5
Transportation and Public Utilities	40.5	3,898	157,869.0
Wholesale and Retail Trade	35.3	13,293	469,242.9
Finance, Insurance, Real Estate	36.8	2,907	106,977.6
Services	34.5	10,521	362,974.5
Total	37.2	47,935	1,780,728.3

$$\bar{X} = \frac{\sum\limits_{i=1}^{N} f_i\bar{X}_i}{N} = \frac{1,780,728.3}{47,935} = 37.15 = 37.2$$

Source: U.S. Bureau of the Census, *Statistical Abstract of the United States: 1971*. (92nd edition.) Washington, D.C., 1971, page 219, No. 345.

the sum of the differences (*deviations*) of each individual score from the mean is zero. This can be expressed symbolically:

$$\sum_{i=1}^{N} (Y_i - \bar{Y}) = 0$$

At first glance, this may seem like a strange sort of claim to make for anything, let alone the mean. However, it makes more sense on closer examination. Let us return to our example of the midterm exam scores. This example helps in understanding this strange-appearing formula whose solution is *always* zero. We subtract the arithmetic mean from each score and sum the differences.

Table 4.3.5   The sum of the differences

$(Y_i - Y)$
$50 - 52 = -2$
$60 - 52 = 8$
$40 - 52 = -12$
$90 - 52 = 38$
$20 - 52 = -32$
$\sum\limits_{i=1}^{N} (Y_i - \bar{Y}) = 0$

Our results are:

$$(-2) + (-12) + (-32) = -46$$
$$(+\ 8) + (+38) = +46$$
$$(-46) + (+46) = \quad 0$$

Another way to think of this formula is to note that the sum of the individual scores is

$$\sum_{i=1}^{N} Y_i = 50 + 60 + 40 + 90 + 20 = 260$$

This is the same as the sum of giving each person the mean score

$$\sum_{i=1}^{N} \bar{Y}_i = 52 + 52 + 52 + 52 + 52 = 260$$

Clearly the individual scores result in a total that is merely "divided up equally" among all the cases in the form of the mean. Put differently, the mean is the value that would result if all of the cases in the sample possessed an equal value (score). This also means that if we multiply the mean by the number in our sample, we produce the same total value as we do when we sum the individual scores (Table 4.3.6), that is,

$$\sum_{i=1}^{N} Y_i = N\bar{Y}$$

Table 4.3.6   The sum of the scores
and the sum of means

Y_i	$\bar{Y}$
50	52
60	52
40	52
90	52
20	52
$\sum_{i=1}^{N} Y_i = 260$	$N\bar{Y} = 260$

All of this leads to the second and more important characteristic of the arithmetic mean: The sum of the *squared* differences (deviations) of each individual score from the mean is a minimum. This can be expressed symbolically:

$$\sum_{i=1}^{N} (Y - \bar{Y})^2 = \text{a minimum}$$

Taking this formula apart step by step helps clarify its meaning. Among other things, by squaring the deviations we convert the negative numbers to positive numbers. Thus when we add them together we no longer get zero. Furthermore, the sum of these squared deviations of each individual score from the mean is the lowest sum possible using the *same* individual scores with any *different* mean value or any other measure of central tendency unless it is identical to the mean. Thus, no matter what value is used *other than* the arithmetic mean, the sum of the squared deviations will be larger than the sum we get when using the mean. This is called the *least-squares* characteristic of the mean.

Table 4.3.7 Three mean values and the sums of squares

The Correct Mean

Y_i	$\bar{Y}$	$(Y_i - \bar{Y})$	$(Y_i - \bar{Y})^2$
50	52	−2	4
60	52	8	64
40	52	−12	144
90	52	38	1444
20	52	−32	1024

$$\sum_{i=1}^{N} (Y_i - \bar{Y})^2 = 2680 = \text{a minimum}$$

One Value above the Correct Mean

Y_i	$(\bar{Y} + 1)$	$(Y_i - (\bar{Y} + 1))$	$(Y_i - (\bar{Y} + 1))^2$
50	53	−3	9
60	53	7	49
40	53	−13	169
90	53	37	1369
20	53	−33	1089

$$\sum_{i=1}^{N} (Y_i - (\bar{Y} + 1))^2 = 2685$$

One Value below the Correct Mean

Y_i	$(\bar{Y} - 1)$	$(Y_i - (\bar{Y} - 1))$	$(Y_i - (\bar{Y} - 1))^2$
50	51	−1	1
60	51	9	81
40	51	−11	121
90	51	39	1521
20	51	−31	961

$$\sum_{i=1}^{N} (Y_i - (\bar{Y} - 1))^2 = 2685$$

Let us examine three different mean values, the correct mean $\bar{Y} = 52$ and two incorrect means. If our least-squares claim is accurate, we ought to find that values further away from the mean produce an even larger value than those near the mean. Since we do not want to stack the deck too much in our own favor, let us add and subtract only 1 to the mean, $\bar{Y} = 52$, and use $(\bar{Y} + 1) = 53$ and $(\bar{Y} - 1) = 51$. We find that carrying out the formulas results in three sums of squares. The mean deviation sum is 2680, the minimum sum. The mean-plus-one and the mean-minus-one give sums of 2685, both of which are five points higher than the "minimum sum," as can be seen in Table 4.3.7. This is the "least-squares" property. Since the median of these scores is 50, it is clear that using the median instead of the mean would also produce a larger sum.

4.4 COMPARISON OF AVERAGES

Each measure of central tendency possesses properties that make it more or less usable depending upon our interest in describing a distribution in terms of a central value. All of the averages may be represented by a unique number whose purpose is to describe a set of data. Each average has certain advantages and disadvantages depending upon our research interests and the data at hand.

The mean takes into account the value or score of each individual case. At times this advantage may be outweighed by the fact that the mean is sensitive to extreme values, and if a distribution has an extreme value or values, the mean may be affected considerably. Extreme values do not affect the median unless the value of the middle case is also changed. It turns out that the mean varies less from sample to sample, and sample medians vary more from sample to sample.

The mean may be calculated for any kind of quantitative data, and each set of data has only one unique mean. This is different from the mode, which may have more than one value for a given distribution, or in a rectangular distribution where there may be no mode at all. The median is determined by its position in the set of data, and it does not reflect individual values as such.

The mean may be manipulated algebraically, and it is possible to combine sets of data calculating the overall mean. This allows us to use weighted parts to calculate the mean for several sets of data. Unfortunately, even if we know the medians of each part of a set, we still cannot calculate the median for the entire set. To calculate the mean, we must have an interval variable. The median only requires an ordinal one, and this gives a certain advantage to the median for examining data that do not have actual numerical scores. The mode has no numerical requirements.

The mean is relatively stable since, although we lose some precision, grouping the data does not seriously distort the mean. Incidentally, the median is also relatively unaffected by grouping the data. However, the mode, whose very basis depends upon frequency, is seriously influenced by grouping the data.

The median is a useful and appropriate measure in a number of sociological situations. For example, because incomes may be skewed toward the high side, median family income often is used rather mean family income. This leads to a general rule of thumb that says, if the distribution is relatively symmetrical, that is, when the values are distributed essentially the same above as below the mean, use the mean instead of the median. However, if the distribution is skewed, that is, when there are more extreme cases on one side of the mean than the other, use the median instead of the mean.

If we are concerned about the average size of an introductory sociology class in order to plan classroom space, it might be most useful to know the modal number of students based on the past five years of enrollments. This is because seats in a classroom are not determined by an abstract arithmetic mean but are for an actual number of people that will occupy them. If we are interested in social security benefits in current payments, our main concern might be with the average monthly amount paid to retired workers.

With a perfectly symmetrical distribution, the mode, median, and mean are identical. Importantly, this identical value is not as important as the theoretical use to which it is put. Thus, it makes the most sense to select the average which most adequately deals with the distribution and the substantive theory under investigation. In other words, not only is it necessary to compute a value of central tendency, but also it is necessary to consider the pattern of the distribution in deciding upon which measure to report.

Table 4.4.1 summarizes the properties and characteristics of averages.

Table 4.4.1   Properties and characteristics of averages

Property or Characteristic	Mode	Median	Mean
Type of Average	frequency	positional	arithmetic
Formula	none	$Md = \dfrac{N+1}{2}$	$\bar{X} = \dfrac{\sum\limits_{i=1}^{N} X_i}{N}$
The number possible in a distribution	any	one	one
Ordering Necessary	yes	yes	no
Grouping Necessary	yes	no	no
Influenced by Grouping	yes	no	no
Reflect individual scores	no	no	yes
Possible to Manipulate Algebraically	no	no	yes
Appropriate for Qualitative Data	yes	no	no
Possible to Calculate if Table Ends are Open	yes	yes	no
Influenced by Extreme Values	no	no	yes

To evaluate which average is appropriate for a given set of data, it is essential to determine one's research interests. For theoretical and substantive reasons one must make such a decision prior to analyzing one's data for it is possible to manipulate the data by selecting that average that tells one's story as one wishes rather than according to its appropriateness. Such practices are not acceptable for analytical statistics.

EXERCISES

1. In your own words, define, describe, or discuss the following terms and give an hypothetical example of each:
   measure of central tendency
   measure of dispersion
   frequency average
   positional average
   arithmetic average
   grouped data
   mean
   median
   mode
   minimum
   least-squares

2. When 20 grade school children were asked, "What would you like to be when you grow up?" they answered: Doctor, cook, nurse, fireman, nurse, policeman, teacher, doctor, soldier, doctor, President, movie star, animal doctor, fireman, doctor, dentist, singer, rock n' roll star, President, and doctor. Which is the modal choice?

3. This table presents the hypothetical incomes of 20 fathers as reported by students in a sociology class.

14	12	9	15	
16	8	19	8	
15	20	11	12	Income in Thousands
6	5	13	10	
8	10	13	4	

   A. What is the mean income of this group?
   B. What is the median?
   C. What does the difference between the mean and median suggest in terms of income distribution for this group?

4. Using the results of the statistics exams in Problem 4, Chapter 3, divide the exams into three classes using each column as a separate class.
   A. Calculate the mean, mode, and median for each class.
   B. Calculate the total mean for all classes.
   C. Draw a line graph showing the means of the three classes and the mean of the three combined.

5. Assume that you have gathered the following information about the wages of a sample of 52 work-study students in your college:

$25	$40	$23	$26
32	27	30	29
28	31	17	39
21	24	28	27
23	38	20	30
37	35	19	22
16	18	22	34
25	21	24	29
33	28	36	26
27	30	28	28
27	26	29	25
31	29	32	33
34	31	35	28

A. Construct a frequency distribution using intervals of $1.00.

B. Calculate the percentage distribution and report it with the frequency distribution of Part A.

C. Draw a histogram on graph paper.

D. Calculate the mean, median, and mode.

6. Using the information presented in Question 4 in Chapter 3, calculate the mean, median, and modal scores for the variables age, grade point average, and examination grade.

7. The number of members in ten local chapters of the Republican party in 1976 were: 46, 79, 60, 43, 104, 135, 56, 73, 89, 36. What is the mean, median, and modal size of these groups? Which measure of central tendency best describes these groups. Why?

8. Complete the following sentences.

A. The arithmetic mean is obtained by _____ the scores and dividing by the _____.

B. The number which occurs most frequently is the _____.

C. The mean is that number in a distribution of numbers about which the sum total of the _____ is equal to _____.

9. Which measure of central tendency best represents the following data? Why?

A. The height and weight of high school basketball teams.

B. The incomes of families in a specified geographic area.

C. The church attendance in the United States.

D. Membership in voluntary associations.

E. The number of people in a meal line.

5

Measuring
Variability

On several occasions we have mentioned that in order to carry out appropriate statistical analyses the variables we select must actually vary. Much of the measurement we have discussed has focused on averages and central tendency, but we have touched on the way in which cases are spread out over a distribution. In one sense, this interest in "spread-outiveness" is the opposite side of the central tendency coin: When we measure variation we want to assess the extent to which the cases are spread out from some central value.

Part of our interest is in measuring just how different individuals are compared to each other. As with averages, we seek single values that can be used to summarize a frequency distribution's spread. These statistics are called *measures of dispersion*, that is, the tendency of the data to be spread out from the midpoint.

5.1 MEASURES OF DISPERSION

The idea that variables actually vary depends on the fact that if scores are spread out along the distribution of a variable, then the cases are not homogeneous. Put differently, we assume that the scores are spread out from some low point to some high point along the classes of an operationally defined variable.

One rather straightforward measure that can be used to evaluate dispersion is the *range*. The range is that single value that encompasses all those values between the lowest and the highest values. People use the range as a general descriptive statistic quite commonly in everyday situations. For example, in a simple distribution of ten people whose ages are 17, 21, 18, 32, 19, 27, 23, 21, 23, and 22, the upper and lower limits of the range are 32 and 17. In this case, the range includes the values 17, 18, 19, 20, 21, 22, 23, 24, 25, 26, 27, 28, 29, 30, 31, and 32. In other words, the range of ages for these ten people is 16 years with a low of 17 and a high of 32.

An intuitively appealing measure of dispersion, the range is equal to the interval containing all the values in a set. Generally it is customary to state the range in terms of the lower and upper values; however, the range itself is the one value that spans all the cases. In order to calculate the total or inclusive range as we did above, we subtract the lowest value from the highest value and add 1. This may be expressed symbolically as

$$\mathrm{Rg} = (H - L) + 1$$

It is also possible to use the range in describing average values and percentages. For example, in Table 5.1.1 we could say that in 1970 the range of average hourly earnings for production workers in private industry was $2.52 with a low of $2.71 per hour for wholesale and retail trade workers and a high of $5.22 per hour for contract construction workers.

Table 5.1.1.   Nonagricultural industries—number of production workers, hours, and earnings
Full- and Part-Time employees in 1970

Industry	Annual Number (in thousands)	Average Weekly Hours	Average Hourly Earnings
Mining	473	42.7	$3.84
Contract Construction	2,793	37.4	5.22
Manufacturing, Durable Goods	8,045	40.3	3.56
Manufacturing, Nondurable Goods	6,005	39.1	3.08
Transportation and Public Utilities	3,898	40.5	3.85
Wholesale and Retail Trade	13,293	35.3	2.71
Finance, Insurance, Real Estate	2,907	36.8	3.07
Services	10,521	34.5	2.84
Total	47,935	37.2	$3.23

Source: Adapted from U.S. Bureau of the Census, *Statistical Abstract of the United States: 1971*. (92nd edition.) Washington, D.C., 1971, pages 219–221, No. 345.

Similarly, Table 5.1.2 shows the number and percentage of persons over 16 who were unemployed for each of the years 1967–1972. The range of the number of people unemployed is 2,178 with a low of 2,817 in 1968, and a high of 4,994 in 1971. The range of the percentage of people unemployed is 2.6 with a low of 3.9 in 1969 and a high of 6.4 in 1971.

Table 5.1.2.   Unemployed persons

	N (in thousands)	Percentage
1967	2,975	4.2
1968	2,817	4.0
1969	2,831	3.9
1970	4,088	5.4
1971	4,994	6.4
1972	4,840	6.0

$$Rg = (H - L) + 1$$

For the number of persons unemployed.

$$Rg = (4,994 - 2,817) + 1 = 2,177 + 1 = 2,178$$

For the percentage of persons unemployed.

$$Rg = (6.4 - 3.9) + .1 = 2.5 + .1 = 2.6$$

Source: Adapted from U.S. Bureau of the Census, *Statistical Abstract of the United States: 1971*. (92nd edition.) Washington, D.C., 1971, page 214, No. 335.

The range is not an adequate measure of dispersion for analytical purposes for two reasons. In the first place, it fails to deal with the distribution that falls between the lower and upper limits. Clearly, when we state the limits of the range, we have no idea whether the values between those limits are clustered near one end or the other, grouped near the middle, or spread out evenly between the ends.

Second, it is sensitive to extreme scores. Clearly, a single extreme score can make the range appear to be large when, in reality, the distribution is fairly tightly clustered. This inadequacy of the range may be dealt with by utilizing the *interquartile range* as a measure of dispersion. This statistic simply truncates and excludes from consideration the scores below the first quartile and the scores above the third quartile. To compute the interquartile range, we first divide the data into three quartiles, called Q_1, Q_2, and Q_3, between each of which falls 25 percent of the scores, that is,

$$25\% \quad Q_1 \quad 25\% \quad Q_2 \quad 25\% \quad Q_3 \quad 25\%$$

Then, we subtract the score at the first quartile from the score at the third quartile. It can be expressed symbolically as:

$$\text{Interquartile Range} = \text{I Rg} = Q_3 - Q_1$$

By definition the interquartile range includes the extreme values of the center fifty percent of the cases. In our example of average hourly earnings the interquartile range may be computed by arranging the cases into quartiles and computing the formula.

$$2.71 + 2.84 \quad Q_1 \quad 3.07 + 3.08 \quad Q_2 \quad 3.56 + 3.84 \quad Q_3 \quad 3.85 + 5.22$$

$$\text{I Rg} = Q_3 - Q_1$$
$$\text{I Rg} = 3.84 - 3.07 = 0.77$$

Another way we can measure dispersion is to calculate a measure of central tendency and then compare each case's value with it. The basic premise behind this approach is that if the values in the aggregate have an average or typical value, then we can measure how much each individual score varies from that overall summary value or score. In this sense, the whole concept of variation grows from the idea that for one thing to vary there must be a value *from which* that variation occurs.

On first thought it may appear that we can utilize the mode as a summary statistic from which the actual scores may vary. This is especially so when the distribution has a sizable, unique mode. However, when a distribution has no mode or when there is more than one mode in a distribution, using it as a comparable value is impossible or meaningless. For example, if we are analyzing five midterm examination grades that were 50, 60, 40, 90, and 20, there is no mode. Thus it is impossible to compare the actual individual observations with any value. Similarly, if the five midterm examination grades were 30, 50, 90, 90, and 30, there are two modes, making it impossible to subtract one value from all the scores.

In symmetrical distributions that are "even" in both directions and in which the mode, the median, and the mean are identical, any one of the averages may be used as a comparative figure to assess each case's score. No matter what the nature of the distribution, however, both the median and the mean are unique values and may be utilized in developing measures of dispersion that utilize comparative figures.

As we just saw, it is possible to compare each individual case's actual score with the summary statistic we use in describing the distribution. Both the median and the mean may be used in this capacity, each having certain strengths and weaknesses. In our discussion of the mean, we saw that if the arithmetic mean is subtracted from each individual's score and the deviations (differences) are summed, the resulting sum is 0. If we use the median and it is not identical with the mean, however, the resulting sum will be higher or lower than 0, depending upon the skewness of the distribution of actual scores. For example, when we consider the five midterm examination grades of 50, 60, 40, 90, and 20, we find that the sum of the algebraic deviations from the mean is 0, but the sum of the algebraic deviations from the median is 10. This can be seen in Table 5.1.3.

Table 5.1.3.   The algebraic deviations

			Algebraic Deviations	
X_i	M_d	$\bar{X}$	$X_i - M_d$	$X_i - \bar{X}$
50	50	52	0	−2
60	50	52	10	8
40	50	52	−10	−12
90	50	52	40	38
20	50	52	−30	−32
$\sum_{i=1}^{N}$ 260	250	260	10	0

This is only half the story, however, for it focuses purely on the sum of the *algebraic* deviations. It is also possible to use the absolute value of the deviations. That is, we can ignore the sign ($+$ or $-$) and treat all of the deviations as if they were positive numbers. This step may be indicated by the sign | deviation |, which tells us to ignore the sign of the deviation and treat the deviation by its absolute value. When this is done, it may be that the median arithmetic deviation is smaller then the mean arithmetic deviation. In our example, this is the case, as is shown in Table 5.1.4. The sums divided by the number of cases are the *arithmetic* deviations, and they have certain utility as measures of dispersion.

Although it is not commonly used in social statistics, the median deviation is a useful measure of dispersion if the distribution is skewed. For example, the distribution of income in the United States is highly skewed, with a relatively small proportion of the people having very high earnings. Thus it is common to report median annual earnings rather than mean annual earnings. In such an instance, the median deviation indicates more accurately the fact that most workers are clustered around the lower median than around the higher mean.

Table 5.1.4.   The arithmetic deviations

X_i	M_d	$\bar{X}$	Arithmetic Deviations	
			$|X_i - M_d|$	$|X_i - \bar{X}|$
50	50	52	0	2
60	50	52	10	8
40	50	52	10	12
90	50	52	40	38
20	50	52	30	32
260	250	260	90	92

$$\text{Median Deviation} = M_dD = \frac{\sum_{i=1}^{N} |X_i - M_d|}{N} = \frac{90}{5} = 18.0$$

$$\text{Mean Deviation} = \bar{X}D = \frac{\sum_{i=1}^{N} |X_i - \bar{X}|}{N} = \frac{92}{5} = 18.4$$

5.2 THE VARIANCE

So far we have seen that although the range has some utility as a measure of dispersion, it fails to take into account much of the information of the distribution. Similarly, the median deviation only measures variation about a location or position in general but not about variation in terms of its own original value. The mean deviation also falls short not only because it distorts skewed distributions, but also and more importantly because it is not amenable to algebraic manipulation, since it uses absolute values. Furthermore, even though the arithmetic mean deviation has a certain descriptive value, there is no useful interpretation of what it measures, and therefore its effectiveness is limited. Many of these problems are dealt with and resolved by the variance, which involves *squared* deviations from the mean.

Before we look at the computations for the variance, let us take a closer look at the least-squares property, which was mentioned in Chapter 4. You may remember that by summing the individual scores and dividing by N, we get the mean score. The mean tells us the typical numerical value of the scores, but it does not give us a *comparative* idea of their typicality, that is, how spread out they are from each other. However, when we compare the arithmetic mean with each individual score by subtracting the mean from that score, we do have a comparative idea of how much each of these scores is "like the mean." For example, of the five students we discussed earlier, the one whose score was 50 might be called "more typical" (closer to the mean) than the student whose score was 90.

When we subtract the mean from each person's score, square the difference, and sum the squares of the differences, we find the *total sum of squares*. This total sum of squares equals the total variation around the mean, ignoring the number of cases involved. It can be expressed symbolically as:

$$\text{Total SS} = \sum_{i=1}^{N} (Y_i - \bar{Y}) \cdot (Y_i - \bar{Y}) = \sum_{i=1}^{N} (Y_i - \bar{Y})^2 = \text{total variation}$$

Now let us take into account the number of cases involved. If we divide the total variation $\sum_{i=1}^{N} (Y_i - \bar{Y})^2$ by the total number of cases N we measure the variation per case or the mean squared difference. Thus the *variance* is a measure that describes how spread out a distribution is in terms of squared differences from the mean. Knowing this, we may define the variance as the arithmetic mean of the squared deviations from the mean. It can be expressed symbolically as:

$$\text{variance of } Y = \frac{\sum_{i=1}^{N} (Y_i - \bar{Y})^2}{N} = \frac{\text{total SS}}{N} = s_y^2$$

The steps are simple ones and can be summarized as follows:

1. Find the deviation of the mean from each score: $(Y_i - \bar{Y})$.
2. Square each deviation: $(Y_i - \bar{Y})^2$.
3. Sum the resulting products: $\sum_{i=1}^{N} (Y_i - \bar{Y})^2$.
4. Divide by the total number of cases: $\dfrac{\sum_{i=1}^{N} (Y_i - \bar{Y})^2}{N}$.

Again using the example of the five students, we can work out the steps as shown in Table 5.2.1.

The more spread out the scores are, the larger the variance; and the more concentrated the scores are, the smaller the variance. As was the case with

Table 5.2.1. Computation of the variance

$(Y_i - \bar{Y})$	$(Y_i - \bar{Y})^2$	$\sum_{i=1}^{N} (Y_i - \bar{Y})^2$	$\dfrac{\sum_{i=1}^{N} (Y_i - \bar{Y})^2}{N}$	s_y^2
-2	4			
8	64			
-12	144	2680	$\dfrac{2680}{5}$	536
38	1444			
-32	1024			

the mean, extreme scores are the most influential in determining the value of the variance. For example, a deviation of two squared equals four ($2^2 = 4$), while a deviation of ten squared equals one hundred ($10^2 = 100$). Thus if we have a number of extreme cases in our distribution, the variance may be misleading because it will be a fairly large number.

We need not be too dismayed that the extreme scores have greater influence on the variation. The math is complex, but the logic goes something like this. Quantitative sociological research generally examines data gathered from a probability sample of some larger population. Since any sample deals only with *some* of the population, it makes sense that the *range of possible scores* in the population is likely to be greater than the *range of scores actually found* in the sample.[1] This leads to our assumption that extreme cases in our sample are likely to be magnified in the population. By squaring the differences from the mean of our sample values, we attempt to take into account the *possible* range of scores in the population, even if by so doing we get a large variance. Whatever its problems, the variance is one of the most useful measures that we will encounter, and it is the basis for many other statistics.

5.3 PREDICTION WITH ONE VARIABLE

If we could assign numerical values to each one of a set of variables and put them into the appropriate formula, we ought to be able to measure the variables' influence and make predictions based on them. In the sense that we use the word "prediction," we make no implications about controlling the future. Rather, *statistical prediction* means that we utilize information about one or more variables to "predict" information about a dependent variable.

Let us consider what we can predict about *one* variable with summary knowledge about that variable itself. First let us connect the mean and variance to prediction with a guessing game. Imagine that we enter a contest to guess two *single* scores that best typify (1) the grade point averages (Y) and (2) the number of hours spent studying (X) of eight students, Alan, Arthur, Betty, Candy, Helen, Maxine, Roland, and Sam. The guesser who has the least amount of error will be rewarded with champagne cocktails for two and a three-course home-cooked dinner with white wine. We learn that "error" is calculated by subtracting the guessed score from the student's actual score, squaring the difference, and summing the squared differences. These steps give a total sum of squared differences that represents the overall error for a set of guesses.

The least-squares principle states that if you subtract the *arithmetic mean*

[1] A similar approach may be found in Paul A. Games and George R. Klare, *Elementary Statistics* (New York: McGraw-Hill Book Company, 1967), esp. pp. 125–26.

from each actual score, square the differences, and sum the squares, you will end up with the lowest (least-squares) number possible. Recalling this, you decide to try to learn these values and submit them as your guesses. Fortunately, the registrar's secretary accidentally tells you that the mean grade point average for these eight students is 2.25 and the mean number of hours spent studying is 4.5. Seizing at the golden opportunity this information gives you, you guess each student's grade point average to be 2.25 and the number of hours spent studying to be 4.5. The big day arrives and all the guesses are in. You are announced as the guesser with the least error on both items! You win the drinks and the dinner, and probably will pass statistics.

Tables 5.3.1 and 5.3.2 show that the guesses (predicted scores) result in total errors of 10.48 and 42.5. Since any values other than 2.25 and 4.5 would produce higher errors, our best bet if we wish to predict a student's grade point average or the number of hours spent studying *without any additional information* is to use the mean values for all the students. The least-squares principle tells us that this value would produce the least total sum of squares; therefore it results in the least amount of error. Carrying out the steps for grade point average, we find the total sum of squares:

$$\text{TSS} = \sum_{i=1}^{N} (Y_i - \bar{Y})^2 = 10.48$$

Dividing the total sum of squares by the number of students gives us the variance of Y:

$$s_Y^2 = \frac{\sum_{i=1}^{N} (Y_i - \bar{Y})^2}{N} = \frac{\text{TSS}}{N} = \frac{10.48}{8} = 1.31$$

Table 5.3.1. Computation of the variance of Y

Student	i	Grade Point Average (Y)	$\bar{Y}$	$(Y_i - \bar{Y})$	$(Y_i - \bar{Y})^2$
Alan	1	4.0	2.25	1.75	3.06
Arthur	2	1.5	2.25	−0.75	0.56
Betty	3	3.5	2.25	1.25	1.56
Candy	4	2.5	2.25	0.25	0.06
Helen	5	2.0	2.25	−0.25	0.06
Maxine	6	0.5	2.25	−1.75	3.06
Roland	7	1.0	2.25	−1.25	1.56
Sam	8	3.0	2.25	0.75	0.56

$$\sum_{i=1}^{N} Y_i = 18.00 \qquad\qquad \sum_{i=1}^{N} (Y_i - \bar{Y})^2 = 10.48$$

$$\bar{Y} = \frac{18.00}{8} = 2.25 \qquad\qquad \text{var } Y = s_Y^2 = 1.31$$

We also can calculate the total sum of squares for the number of hours spent studying:

$$\text{TSS} = \sum_{i=1}^{N} (X_i - \bar{X})^2 = 42$$

and the variance of X:

$$s_X^2 = \frac{\sum_{i=1}^{N} (Y_i - \bar{Y})^2}{N} = \frac{\text{TSS}}{N} = \frac{42.5}{8} = 5.25$$

Table 5.3.2.   Computation of the variance of X

Student	i	Weekly Number of hours Spent Studying (X)	$\bar{X}$	$(X_i - \bar{X})$	$(X_i - \bar{X})^2$
Alan	1	4	4.5	−0.5	0.25
Arthur	2	2	4.5	−2.5	6.25
Betty	3	7	4.5	2.5	6.25
Candy	4	5	4.5	0.5	0.25
Helen	5	6	4.5	1.5	2.25
Maxine	6	1	4.5	−3.5	12.25
Roland	7	3	4.5	−1.5	2.25
Sam	8	8	4.5	3.5	12.25

$$\sum_{i=1}^{N} X_i = 36 \qquad \sum_{i=1}^{N} (X_i - \bar{X})^2 = 42.00$$

$$\bar{X} = \frac{36}{8} = 4.5 \qquad \text{var of } X = s_X^2 = 5.25$$

At about this point, you might begin to suspect that since we have information about each student on *two* variables, it would be a good idea to try guessing a student's grade point average (variable Y) based on the knowledge we have of the number of hours each spends studying each week (variable X). Later, we will do just that. First, however, let us see how the variance can be further utilized as a measure of dispersion.

5.4 THE STANDARD DEVIATION AND THE NORMAL CURVE

Since the variance was derived by *squaring* the deviations of the individual scores from the mean, it is not easily comparable to the original scores. To make it comparable we take its square root, thereby standardizing it in terms of the original scores. Thus, the *standard deviation* is the square root of the variance.[2] In other words, the square root of the mean squared deviations standardizes the variance making it comparable to the original scores. Incidentally, even though we carry out the square root operations, the initial effect of squaring the deviations still gives more weight to the extreme scores.

Put formally, the standard deviation is the square root of the arithmetic mean of the squared differences from the mean. It can be expressed symbolically as:

$$\text{s.d.} = \sqrt{\frac{\sum_{i=1}^{N}(Y_i - \bar{Y})^2}{N}} = \sqrt{s_Y^2} = s_Y$$

To insure your familiarity with this statistic, we can follow the same steps as in the calculation of the variance and then take the square root.

1. Find the deviation of the mean from each score:

$$(Y_i - \bar{Y})$$

2. Square each deviation:

$$(Y_i - \bar{Y})^2$$

3. Sum the resulting products:

$$\sum_{i=1}^{N}(Y_i - \bar{Y})^2$$

4. Divide by the total number of cases:

$$\frac{\sum_{i=1}^{N}(Y_i - \bar{Y})^2}{N}$$

5. Take the square root:

$$\sqrt{\frac{\sum_{i=1}^{N}(Y_i - \bar{Y})^2}{N}}$$

We can illustrate how this works by expanding our example, as can be seen in Table 5.4.1.

In general, the standard deviation may be used to measure the dispersion along the distribution of a given variable because it gives information as to how much *per unit* the scores are spread out from the mean. If the distribution is spread out, the standard deviation will be large, and vice versa.

---

[2]There is no need to learn how to calculate square roots, for there are tables that summarize various square root values for different numbers. One such table may be found in the Appendix as "Table of Square Roots."

Table 5.4.1. Computations of the standard deviation

$(Y_i - \bar{Y})$	$(Y_i - \bar{Y})^2$	$\sum\limits_{i=1}^{N} (Y_i - \bar{Y})^2$	$\dfrac{\sum\limits_{i=1}^{N} (Y_i - \bar{Y})^2}{N}$	s_Y^2	$\sqrt{s_Y^2}$	s_Y
-2	4					
8	64					
-12	144	2680	$\dfrac{2680}{5}$	536	$\sqrt{536}$	23.15
38	1444					
-32	1024					

It is possible to examine a variable or a set of variables according to its mean score and standard deviation simultaneously. For example, Table 5.4.2 shows the comparative values for selected variables for a sample of males ages 25 to 64 in 1962. Among other things, respondents report a higher average amount of education and less variation than their fathers; they are older at the time of their first marriage than in their first job; and there is greater variation in age at their first marriage than in age at their first job.

Table 5.4.2. Mean scores and standard deviations for selected variables for native non-negro civilian males ages 25–64, with nonfarm background, reared by both parents, and living with spouse

Variable	Mean Score	Standard Deviation
Father's Education	8.61	3.66
Number of Siblings	3.89	2.92
Respondent's Education	11.87	3.22
Age at First Job	18.59	2.97
Age at First Marriage	24.20	4.73

Source: Adapted from Otis Dudley Duncan, David L. Featherman, and Beverly Duncan, *Socioeconomic Background and Achievement.* (New York: Seminar Press, 1972), page 234, Table 8.14.

The standard deviation has many possible applications, one of the most important of which is its use in describing and interpreting what we call the *normal curve.* The normal curve is a special kind of distribution that is valuable for (at least) three reasons.[3] In the first place, many empirical distribu-

---

[3]There are numerous helpful treatments of the normal curve. Among the clearest are those by John E. Freund, *Statistics, A First Course* (Englewood Cliffs, New Jersey: Prentice-Hall, Inc., 1970), pp. 133–58; Hubert M. Blalock, Jr., *Social Statistics*, 2nd ed. (New York: McGraw-Hill Book Company, 1972), pp. 93–105; and Helen M. Walker and Joseph Lev, *Elementary Statistical Methods*, 3rd ed. (New York: Holt, Rinehart and Winston, Inc., 1969), pp. 107–32.

tions are found to be approximately normal; thus they may be analyzed according to the properties of the normal curve. Second, the normal curve is of great importance when we are interested in making inferences from a sample to a population, that is, as in inductive statistics. Third, the normal curve can be used to help us interpret the variability of a variable.

The normal curve is determined by the values of two familiar summarizing statistics, the mean and the standard deviation. It is a smooth, bell-shaped, symmetrical curve about the mean of the variable being examined. Since the total area under the curve represents *all* of the *probable* values a variable may possess, that area is equal to 1.00 (100 percent). Since there are many possible values for the mean and the standard deviation, there are many different normal curves. However, there is one and only one normal curve for each mean and standard deviation for a given variable. Thus it is possible to have two normal curves with identical standard deviations but with different means, as is shown in Figure 5.4.1.

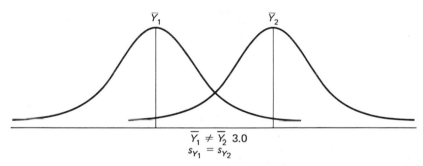

$$\overline{Y}_1 \neq \overline{Y}_2 \quad 3.0$$
$$s_{Y_1} = s_{Y_2}$$

Figure 5.4.1. Two normal curves with identical standard deviations but different means.

It is also possible to have two normal curves with different standard deviations but identical means, as is shown in Figure 5.4.2.

The extent to which a distribution is other than strictly normal is an important matter of concern and can be measured or demonstrated in many ways. For example, the difference between a histogram, which represents the observed distribution, and the normal curve, which represents the *theoretical* distribution, can be seen graphically in a suspended (or hanging) histogram.[4] Figure 5.4.3 displays the smooth normal curve superimposed over a histogram. In this case the histogram has been rearranged to show a symmetrical, relatively smooth upper surface of frequencies above the horizontal axis and

[4]See John W. Tukey, "Some Graphic and Semi-Graphic Displays," in T. A. Bancroft, ed., *Statistical Papers in Honor of George W. Snedecor*. (Ames, Iowa: The Iowa State University Press), pp. 293–316.

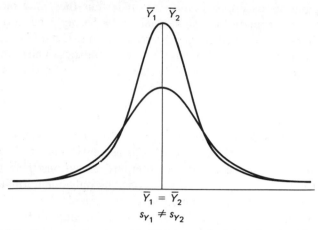

Figure 5.4.2. Two normal curves with different standard deviations but identical means.

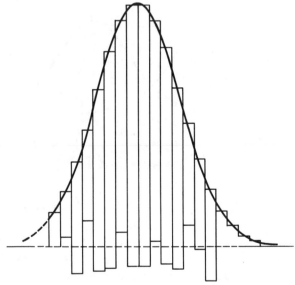

Figure 5.4.3. (*From John W. Tukey, "Some Graphic and Semi-Graphic Displays," in T.A. Bancroft, ed.,* Statistical Papers in Honor of George W. Snedecor. *Ames, Iowa: The State University Press.*)

a nonsymmetrical, relatively rough surface of frequencies below the horizontal axis.

The normal curve is intuitively appealing, and most of us have seen it displayed or heard it mentioned in various contexts throughout our lives.

98

One of its major uses takes it beyond the abstract description just given—it involves using the normal curve in a comparative framework. We often want to compare cases on several variables, some or all of which have different units of measure. When we measure, say, education, number of siblings, age, income, and occupational prestige for a group of respondents, we must contend with the fact that each is measured by a different unit, in these cases, amount completed, numerical count, years, dollars, and points on a scale. To deal with this problem, we use the standardized normal curve.

The *standardized normal curve* is a theoretical distribution of a hypothetically constructed variable that is conventionally called Z. Thus the standardized normal curve is determined by the values of variable Z, which is normally distributed with a mean of zero, $\bar{Z} = 0$, and a standard deviation of one, $s_z = 1$. This provides us with a set of standard scores that we can use with the standardized normal curve. Together they help us analyze normal distributions that are unstandardized because of differently measured variables and their means and standard deviations. It enables us to examine the area under any normal but unstandardized curve by converting the scores of our research variables to the standard scores of the Z variable.

The standardized normal curve is different from the empirical ones we previously described by drawing polygons from frequency histograms. The standardized normal curve represents the density of a *theoretical* frequency distribution polygon based on an *infinite* number of cases. This means, of course, that the normal curve may be approximated by an *actual* frequency distribution, which is based on a *finite* number of cases. However, since the standardized normal curve does not emerge from empirical "reality," we use it only as an abstraction and as an analytical device for examining actually observed distributions.[5]

As the number of observations increases, many empirical distributions bear close resemblance to the normal distribution. As a matter of fact, there is an important mathematical theorem, which we will discuss in more detail later on, that shows that the more cases observed in one's sample, the more likely it is that the resulting curve will be nearly normal.

Let us turn from this general discussion of the standardized normal curve to a more specific one. Imagine that we start with a normal curve such as the one shown in Figure 5.4.4. Since the total area under the normal curve is equal to 100 percent or 1.0000, and since the normal curve is symmetrical,

---

[5]The normal curve represents the mathematical function

$$f(x) = \frac{1}{\sqrt{2\Pi}} \cdot e^{-x^2/2}$$

For a fuller discussion see John G. Kemeny, J. Laurie Snell, and Gerald L. Tompson, *Introduction to Finite Mathematics*, 2nd ed. (Englewood Cliffs, New Jersey: Prentice-Hall, Inc., 1966), pp. 201–209.

The application of this function to the normal distribution we use in social statistics may

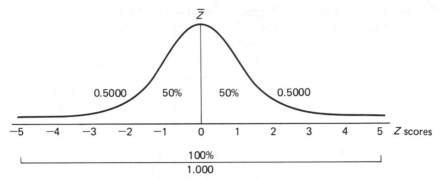

Figure 5.4.4. The area under the normal curve.

we can see that 50 percent of the area falls above the mean and 50 percent below the mean. We also can see that the standardized variable Z can take on any value from 0 to ± 5; thus Z scores range from 0.00 to ± 5. This means that we are interested in proportional areas under one-half of the normal curve from 00.00 percent of the area to 50.00 percent of the area; thus the proportion of cases beneath the curve ranges from 0.0000 to 0.5000, as can be seen in Figure 5.4.5. These proportions represent areas under a standardized normal curve calculated from the mathematical function mentioned in footnote number 5 above.

These proportional areas are summarized in tables such as the one in the Appendix called "Areas Under the Normal Curve." This table contains Z scores and corresponding proportions, which can be used to help interpret curves that are approximately normal. The entries in the normal curve table

---

be expressed by a formula that allows us to plot the line of a normal curve by using information of the mean and standard deviation. Thus,

$$Y = \frac{1}{s\sqrt{2\Pi}} \cdot e^{-(X-\bar{X})^2/2s^2}$$

Since a value with a negative exponent can be expressed as the reciprocal of that value raised to a positive power, the formula may be expressed symbolically as:

$$Y = \frac{1}{s_X\sqrt{2\Pi}} \cdot \left(\frac{1}{e^{(X-\bar{X})^2/2s_X^2}}\right)$$

in which

$\quad Y$ = the height of the curve for a given value of X.
$\quad \Pi$ = a constant approximately equal to 3.14.
$\quad e$ = a constant approximately equal to 2.72.
$\quad s_X$ = the standard deviation of the distribution of X.
$\quad s_{\bar{X}}^2$ = the variance of the distribution of X.
$\quad \bar{X}$ = the mean of the distribution of X.
$\quad X$ = the given value in the distribution of X.

For a fuller discussion see Blalock, *Social Statistics*, pp. 96–98.

100

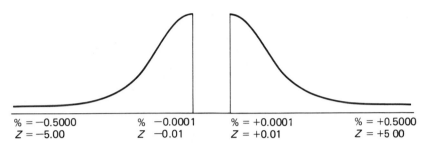

| % = −0.5000 | % −0.0001 | % = +0.0001 | % = +0.5000 |
| Z = −5.00 | Z −0.01 | Z = +0.01 | Z = +5 00 |

Figure 5.4.5. The areas under the two halves of the normal curve and the Z equivalent.

represent areas under the standardized normal curve between the mean $\bar{Z} = 0$ and a specific Z score. The proportions in the body of the normal curve table show the proportion of the area between the mean $\bar{Z} = 0$ and the score that corresponds with the Z score in the margins of the table.

For example, if $Z = \pm 1.00$, we find the Z value by going down the left-hand column to 1.0 and moving across the top row to 0.00, that is,

$$1.00 + 0.00 = 1.00$$

The proportion 0.3413 indicates that 34.13 percent of the total area under the normal curve falls between the mean and one standard deviation from it. This can be seen in Fig. 5.4.6. The table does not show negative values for Z:

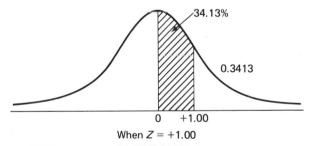

When $Z = +1.00$

Figure 5.4.6. The proportion of the normal curve between the mean and one standard deviation above it.

Since the normal curve is symmetrical, a negative Z score merely indicates the proportion between the mean and the appropriate Z score below the mean. That is, if $Z = -1.00$, we find the Z value exactly as we did above, except that now the proportion 0.3413 indicates that 34.13 percent of the total area under the normal curve falls between the mean and 1.00 standard deviation below it. This can be seen in Figure 5.4.7.

101

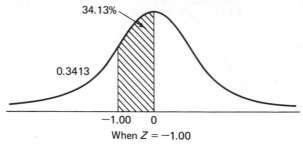

Figure 5.4.7. The proportion of the normal curve between the mean and one standard deviation below it.

Similarly, if $Z = \pm 2.00$, we go down the left-hand column to 2.0 and across the top row to 0.00. In the body of the table we find the proportion 0.4773, which indicates that 47.73 percent of the total area under the normal curve fall between the mean and ± 2 standard deviations from it. This can be seen in Figures 5.4.8 and 5.4.9. Finally, if $Z = \pm 3.00$, we find the propor-

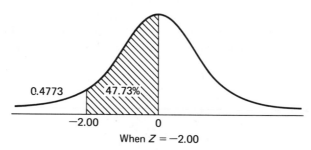

Figure 5.4.8. The proportion of the normal curve between the mean and two standard deviations below it.

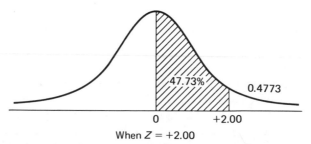

Figure 5.4.9. The proportion of the normal curve between the mean and two standard deviations above it.

102

tion 0.4987, which indicates that 49.87 percent of the cases falls between the
mean and ± 3 standard deviations from it. See Figures 5.4.10 and 5.4.11.

Naturally, we can add together two proportions. Thus Figures 5.4.12,
5.4.13, and 5.4.14 show that 68.26 percent of the cases are included between
the two Z scores that fall one standard deviation on either side of the mean;

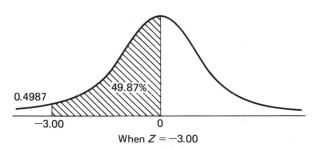

Figure 5.4.10. The proportion of the normal curve between the mean and three standard
deviations below it.

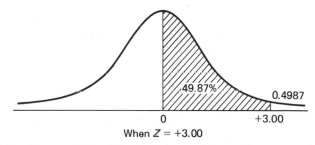

Figure 5.4.11. The proportion of the normal curve between the mean and three standard
deviations above it.

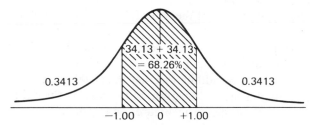

When the proportions between the mean and $\pm Z = 1.00$ are added together

Figure 5.4.12. The proportion of the normal curve between the two Z scores one standard
deviation below and above the mean.

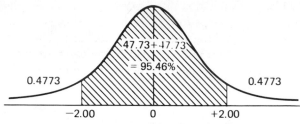

When the proportions between the mean and $\pm Z = 2.00$ are added together

Figure 5.4.13. The proportion of the normal curve between the two Z scores two standard deviations below and above the mean.

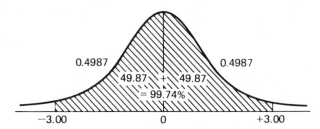

When the proportions between the mean and $\pm Z = 3.00$ are added together

Figure 5.4.14. The proportion of the normal curve between the two Z scores three standard deviations below and above the mean.

95.46 percent of the cases are included between the two Z scores that fall two standard deviations on either side of the mean; and 99.74 percent of the cases are included between the two Z scores that fall three standard deviations on either side of the mean.

We can calculate the proportion of cases having a Z score less than -1.00 by subtracting the proportion between the mean and that Z score from 50 percent, as can be seen in Figure 5.4.14; thus $0.5000 - 0.3413 = 0.1587$. In other words, 15.87 percent of the cases fall in the shaded portion of curve Figure 5.4.15.

A straightforward extension of the addition principle enables us to calculate the proportion of cases having a Z score less than $+1.00$. Getting a Z less than $+1.00$ is found by adding together the 50.00 percent of the cases to the left of the mean plus the 34.13 percent of the cases to the right of the mean up to the Z of $+1.00$, as shown in Figure 5.4.16; thus

$$0.5000 + 0.3413 = 0.8413$$

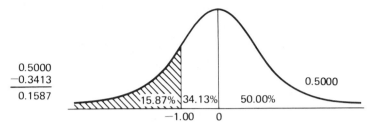

0.5000
−0.3413
_____
0.1587

15.87% 34.13% 50.00%

0.5000

−1.00 0

When the proportion below $Z = -1.00$ is noted

Figure 5.4.15. The proportion of the normal curve below one standard deviation below the mean.

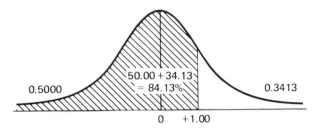

0.5000

50.00 + 34.13 = 84.13%

0.3413

0 +1.00

When the proportion below $Z = +1.00$ is noted

Figure 5.4.16. The proportion of the normal curve below one standard deviation above the mean.

This means that the shaded portion of Figure 5.4.16 indicates that 84.13 percent of the cases fall within it.

It is not always possible to find an area under the curve directly from the table. We may be interested in the area between $Z = +1.00$ and $Z = +2.00$. We merely find the area under the curve between the two values and take the difference between the two; thus $47.73 - 34.13 = 13.60$, as can be seen in Figure 5.4.17.

To calculate the proportion of cases that falls above a Z of $+2.00$, we subtract the known proportion of 47.73 percent between the mean and $+2.00$ from the 50.00 percent that falls above the mean; thus

$$0.5000 - 0.4773 = 0.0227$$

In other words, 2.27 percent of the area under the normal curve fall in the shaded portion of Figure 5.4.18.

We can do the same thing for all the proportions and Z scores in the normal curve table. Thus the standardized normal curve allows us to interpret

105

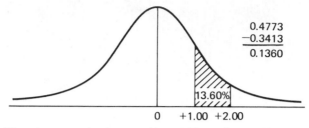

When the proportion between $Z = +1.00$ and $Z = +2.00$ is noted

Figure 5.4.17. The proportion of the normal curve between one and two standard deviations above the mean.

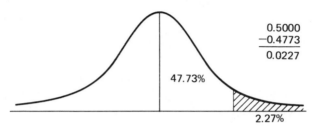

When the proportion above $Z = +2.00$ is noted

Figure 5.4.18. The proportion of the normal curve above two standard deviations above the mean.

the standard deviation as a comparative measure of dispersion, thereby helping us to examine the variability of a variable.

5.5 THE STANDARD SCORE CONVERSION

As we mentioned earlier, there are many empirically obtained frequency distributions that approximate the normal distribution closely enough to allow us to use the relationship between the area under the curve and the standard score for analytical purposes. For example, we may be interested in determining what proportion of the cases falls within specified intervals along the variable, so that we can compare sets of scores according to the normal standard distribution. With a nearly normal curve this is fairly simple, for no matter what mean or standard deviation it possesses, there will be a constant proportion of cases between the mean and a given score. A given score may be specified in standard deviation units from the mean because of the mathematical distributive characteristics discussed earlier.

106

In order to transform the numerical values for any normal curve into ones that can be used in determining the proportion of cases for a given interval in the normal curve, we make use of the *standard score conversion* to the variable Z. Sometimes it is called the Z *statistic* or Z *score conversion*. It can be expressed symbolically as:

$$Z = \frac{(Y_i - \bar{Y})}{s_Y}$$

where

Y_i = the individual score on the original variable Y,
$\bar{Y}$ = the mean of variable Y,
s_Y = the standard deviation of variable Y.

What this transformation from the original variable Y to the now standardized one Z amounts to is this: Assume that the distribution of the variable Y is normal with a mean $\bar{Y} \neq 0$ and a standard deviation $s_Y \neq 1$, and that we want to convert this variable to the Z scale with a mean of $\bar{Y} = \bar{Z} = 0$ and a standard deviation of $s_Y = s_Z = 1$. This standard form of conversion provides us with a standardized score, which can be analyzed according to our table "Areas Under the Normal Curve."

The Z score conversion is clearest when it is actually carried out. Assume that the variable high school average is approximately normally distributed, and that we want to convert the original scores of scale A in Figure 5.5.1 to

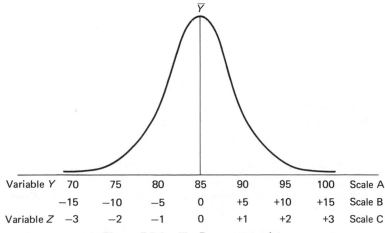

Variable Y	70	75	80	85	90	95	100	Scale A
		−15	−10	−5	0	+5	+10	+15	Scale B
Variable Z	−3	−2	−1	0	+1	+2	+3	Scale C

Figure 5.5.1. The Z score conversion.

the standard Z scores of scale C. To do this we use the Z score conversion formula

$$Z = \frac{(Y_i - \bar{Y})}{s_Y}$$

The process goes like this: We first determine the mean of all the students' high school averages, which turns out to be 85. Then we subtract from each individual's average the mean average, square the difference, divide by N, and take the square root. It turns out that the standard deviation is 5.

In order to standardize the scale of averages to the standard score, we use the Z conversion formula. First we subtract the mean $\bar{Y}$ from each Y_i. This operation can be seen in scale B, which shows scores

$$-15, -10, -5, 0, 5, 10, 15$$

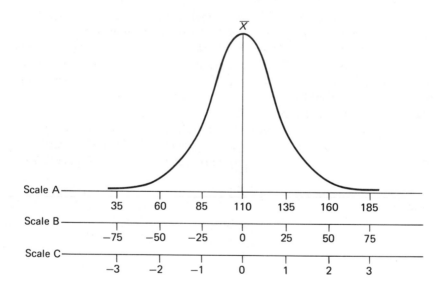

1. Suppose we have a normal distribution such that $\bar{X} = 110$ and $s = 25$; see scale A.

2. If we subtract 110 (the mean) from each raw score, we would have a normal distribution such that $\bar{X} = 0$ and $s = 25$; see scale B.

3. If we then divide each raw score by 25 (the standard deviation), we would obtain a normal distribution such that $\bar{X} = 0$ and $s = 1$; see scale C. A normal distribution with "zero mean and unit variance" is in *standard form*. The values on scale C are known as *standard scores*.

4. Therefore, the transformation of raw scores into standard form is accomplished by the following:

$$Z = \text{standard score} = \frac{\text{raw score} - \text{mean}}{\text{standard deviation}} = \frac{X - \bar{X}}{s}$$

Figure 5.5.2.

We have merely translated the original scores of scale A to those of scale B, which indicates that we now have a mean of $\bar{Y} = 0$ and a standard deviation of $s_Y = 5$. Next we divide these scores by the standard deviation as in scale C,

$$-3, -2, -1, 0, 1, 2, 3$$

We have now converted the original scores into standard scores, obtaining a nearly normal distribution with a mean of $\bar{Y} = 0$ and a standard deviation of $s_Y = 1$.

Depending on how closely our data approximate a normal distribution, we can similarly transform any scale to standard scores. By so doing, we can compare sets of scores according to the normal standard distribution. Or we can use the standard score to calculate the proportion of the students whose average is more or less than a specific score. For example, we know that 50 percent of the students had averages less than 85, and that 34.13 percent had averages between 85 and 90, because 90 is one standard deviation from the mean, $\bar{Y} = 85$, $s_Y = 5$. This means that 84.13 percent of the students had averages of 90 or lower in high school.

Suppose that we have another normal distribution with a mean $\bar{X}$ of 110 and a standard deviation of s_X of 25. Figure 5.5.2 summarizes the steps in the Z score conversion.

Try to imagine how *you* might be able to use the Z statistic. For example, if you know your percentile rank on the SATs, figure out how it was derived by looking at your scores and the Z table. Or try to figure out what percentile a given score is for midterm grades that have a mean $\bar{Y} = 91$ and a standard deviation $s_Y = 2.5$. By the way, this is how teachers grade "on the curve."

EXERCISES

1. In your own words, define, describe, or discuss the following terms and give an hypothetical example of each:

    range
    mean deviation
    median deviation
    total sum of squares
    variance
    standard deviation
    normal curve
    the Z statistic

2. The grades on a statistics examination ranged from 50 to 100. How would the range probably have been changed if:
    A. ten extra points had been added to each score?
    B. the lower scoring students had successfully completed a tutorial program?
    C. 200 questions had been included in the exam instead of 100?

    D. an easier exam had been given?
    E. a more difficult exam had been given?

3. Assume that you have given seven students a 150 item social awareness test and that they received the following scores: 140, 110, 96, 120, 113, 100, and 105. Using these scores, calculate the following descriptive statistics:
    A. the mean
    B. the variance
    C. the standard deviation

4. Using the sums and the frequencies of the rows and columns of the table you constructed in Problem 5, Chapter 3, carry out the following:
    A. Compute the means and the variances for each row and each column.
    B. Compute the overall mean, the total variance, and the standard deviation of variable Y (exam grade).

5. Suppose that a statistics examination was taken by 120 students. Assume that the scores are normally distributed with a mean of 85 and a standard deviation of 5.
    A. What are the Z scores for the grades 85, 92, 71, 81, and 100?
    B. What proportion of the cases falls between 85 and 95?
    C. What proportion of the cases falls between 75 and 80?
    D. What proportion of the cases falls between 77 and 91?

6. The following are two normal curves which approximate the distribution of grades for students in two different sociology classes. Study the diagrams and then answer the questions below.

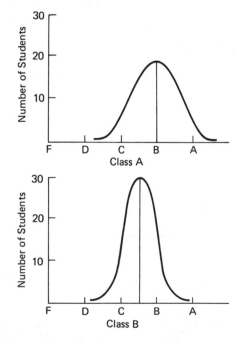

A. Which class had the highest mean?
B. Which class had the highest standard deviation?
C. Some teachers assume that grades should be distributed on a standardized normal curve, with the mean set at C. Comment on this concept in terms of the above graphs.

7. In a study, you gather the following information for twelve students.

Student Number	GPA (Y)	Father's Education (years of school completed) (X)	Father's Income (annual dollars earned) (T)
1	2.5	14	15,000
2	4.0	15	10,000
3	3.6	20	30,000
4	2.5	10	25,000
5	2.0	12	10,000
6	2.4	16	30,000
7	3.2	12	5,000
8	2.2	13	20,000
9	3.5	17	15,000
10	3.8	18	20,000
11	3.5	20	25,000
12	2.8	9	5,000

A. Prepare a table which locates each student according to GPA and father's education and income.
B. Calculate the following summary statistics for each of three variables.
   1. the mean
   2. the variance
   3. the standard deviation.

8. Ten students received the following test scores on an exam: 65, 68, 70, 73, 84, 85, 88, 90, 93, 94
   A. Calculate s and s^2.
   B. Add a constant of 5 to every value and recalculate s and s^2. What differences are there? What differences appear when subtracting 10 points from each exam score?
   C. Multiply each test score by 3. Explain the differences.
   D. Divide each test score by 2. Explain the differences.

9. A committee of a "closed" organization wishes to determine the potential growth and activity level of its group. The present membership has an average age of 36 with a standard deviation of 4. If no new members are invited to join and if no present members remove their membership, what will be the average age and standard deviation of the organization in ten years?

part III

Bivariate

Descriptive

Statistics

6

Association Between Two Nominal or Ordinal Variables

Thus far, we have learned how to measure various aspects of single variables. Most of the time, sociologists are confronted with data in which they are interested in the relationship between two or more variables. These are called *bivariate* (two variable) *statistics*. In this chapter, we learn how to compute several measures of association for two nominal or ordinal variables.

Often, we want to measure the association between dichotomous variables; so, we first learn how the association between two dichotomous variables can be measured. Then, we learn how the association between two ordinal variables can be measured.

6.1 STATISTICAL INDEPENDENCE AND DEGREE OF ASSOCIATION

The extent to which one variable has a predictable effect on another is a major aspect of the concept of association. If one variable has a predictable effect on another, we say that the variables are contingent upon each other, that is, they are *statistically dependent*. If one variable does not tell us anything that helps our predictions about another, we say that the variables are *statistically*

115

independent of each other, that is, they are not contingent upon each other. For example, if it is purely a random chance that someone of a certain age gets a certain score, then scoring is "independent" of the age.

Technically, *statistical independence* focuses on the difference between *expected* and *observed* frequencies. This difference can be determined by the use of *marginal frequencies*. Let us examine how this is accomplished with two dichotomous variables.[1] Dichotomies may be treated as nominal, ordinal, or interval variables; however, our present interest is in dichotomies representing categorical or qualitative ordinal characteristics and in seeking to measure the relationship between two such variables. When we cross-classify two dichotomous variables, the table which results has four inner cells in which frequencies are entered, and it is called a *fourfold* table. There are nine parts in a fourfold table—four cell frequencies, four marginal frequencies, and the total number of cases, or *N*, as shown in Table 6.1.1.

Table 6.1.1. Notation and nomenclature for a 2 × 2 (dichotomous variables) fourfold table

Variable X	Variable Y		Total
	$-$ (*Low*)	$+$ (*High*)	Total
$+$ (High)	Frequency a	Frequency b	Marginal Frequency $X+$ (High on X) $a + b$
$-$ (Low)	Frequency c	Frequency d	Marginal Frequency $X-$ (Low on X) $c + d$
Total	Marginal Frequency $Y-$ (Low on Y) $a + c$	Marginal Frequency $Y+$ (High on Y) $b + d$	Total Cases in the Table $a + b + c + d = N$

For convenience, the four inner cells are designated by the letters a, b, c, and d. By summing across a row or down a column, we can obtain the total frequency for the given category, and since it appears in the margins of the table, it is called a "marginal frequency." For dichotomies which are collapsed from ordinal, interval, or ratio variables, "high" is associated with + (plus) and "low" is associated with — (minus). Put differently, the value of the variables increases as we move up rows and from left to right across columns.

---

[1]The following discussion follows closely that of Davis, *Elementary Survey Analysis*, Englewood Cliffs, New Jersey: Prentice-Hall, Inc., 1971, pp. 33–39.

Statistical analysis backs into association rather than confronting it face to face. Even though this is logical, it often is confusing at first glance. The reasoning goes something like this:[2]

1. A fourfold table either shows some association or relationship (however weak) or it shows no association or relationship at all.
2. By manipulating the marginal frequencies we can estimate what the table would look like if X and Y show no association at all.
3. We can compare the actual fourfold table with the estimated "no association" table.
4. If the actual fourfold table differs from the "no association" table, we conclude that X and Y are associated.

Let us now link the notion of "no association" to the notion of statistical independence and the nine parts of a fourfold table.

Stated formally, we can say that X and Y are statistically independent if the observed cell frequencies equal the products of the relevant marginal frequencies divided by the total number of cases. An example helps demonstrate this statement. Suppose that in a sample of 100 people we have an independent variable X that has two categories, high and low, and that 60 score low and 40 score high. Also suppose that we have a dependent variable that also has two categories, high and low, and that 50 score high and 50 score low. We could construct a fourfold table that shows the two-way marginal frequencies of these two variables, as shown in Table 6.1.2.

Table 6.1.2. Table of two-way marginal frequencies for two variables

Variable X	Variable Y		Total
	Low	*High*	
High	a	b	40
Low	c	d	60
Total	50	50	100

If only chance factors were operating and there were no relationship, we would expect the cases to be distributed *within* the table in the same proportions as they are distributed in the margins. Since the entire sample is split 40/60 on variable X, those 50 who are low on variable Y also ought to be split 40/60 and vice versa. If the 60 people who are low on variable X are

[2]*Ibid.*, p. 36.

split 50/50, there would be 30 each in cells c and d, and if the 40 people who are high on variable Y are split 50/50, there would be 20 each in cells a and b. When this situation occurs, the two variables are said to be statistically independent; that is, they are not associated. Since our interest is in variables that *are* associated, we may test for independence as a test for "no association."

In order to test for no association we must calculate *expected frequencies* assuming statistical independence. To do this, we multiply the relevant marginal frequency of one category of variable X times the relevant marginal frequency of one category on variable Y, and divide the product by the total N. Thus, for cell c, we multiply the two marginals $(50) \cdot (60) = 3,000$, then divide $3,000/100 = 30 =$ the expected frequency for cell c. We state the *expected frequency f_e* for cell c formula symbolically as:

$$f_e \text{ (for cell } c) = \frac{(c + d) \cdot (a + c)}{(a + b + c + d)} =$$

$$\frac{(60) \cdot (50)}{100} = \frac{3,000}{100} = 30$$

We can do the same thing for each of the other cells, and a table can be constructed that shows all four expected frequencies. The values in Table 6.1.3 are the "expected" ones we would get if variables X and Y were statistically independent. When this is the case, variables X and Y are statistically *independent* and are considered to be "not associated."

Table 6.1.3. Table of two-way marginal and *expected* frequencies for two variables

Variable X	Variable Y		Total
	Low	*High*	
High	20	20	40
Low	30	30	60
Total	50	50	100

Now, suppose that instead of finding the expected values, we find that $a = 0$, $b = 40$, $c = 50$, and $d = 10$, as can be seen in Table 6.1.4.

By subtracting the expected frequency from the observed frequency, we can find the difference between them for any given cell. When the difference is greater than zero it can be either a positive or a negative number, and if we enter such differences into a summary table, there are two possible patterns in which the plus $(+)$ or minus $(-)$ signs can appear, as shown in Table 6.1.5.

Table 6.1.4.   **Table of two-way marginal and** *observed*
frequencies for two variables

	Variable Y		
Variable X	*Low*	*High*	Total
High	0	40	40
Low	50	10	60
Total	50	50	100

Table 6.1.5.   **Patterns of differences between observed and expected**
frequencies and signs of association

$$f_o - f_e = \pm \text{ difference}$$

a. Positive association

	$-$ (Low)	$+$ (High)
$+$ (High)	$-$	$+$
$-$ (Low)	$+$	$-$

b. Negative association

	$-$ (Low)	$+$ (High)
$+$ (High)	$+$	$-$
$-$ (Low)	$-$	$+$

Source: Adapted from James A. Davis, *Elementary Survey Analysis.*
(Englewood Cliffs, New Jersey: Prentice-Hall, Inc., 1971), p. 38.

In the case where high X and high Y tend to go together, we call it a positive association. If high X tends to go together with low Y and low Y tends to go together with high X, we call it a negative association.

In Table 6.1.4 we would say that these variables are statistically *dependent*. Since departure from statistical independence indicates some degree of association between the two variables, since we have a difference of ± 20 for each cell, and since being low on variable X tends to "go with" being low on variable Y and being high on variable X tends to "go with" being high on variable Y, we then called this a positive association between variables X and Y.

6.2 CHI-SQUARE AND PHI-SQUARE

In social statistics, we deal with sets of variables whose interrelationships are interesting because we assume that the values of one variable *depend upon* the values of the other ones. No matter what values we *expect*, it is only after we actually cross-classify the variables that we can know what the observed values are, and this value for each cell is called the *observed frequency* f_o.

What we now need is a statistic to assess whether or not the observed frequencies are significantly different from the expected frequencies. First, we subtract the f_e from the f_o. Second, we square this value. Third, the squared difference is compared to the expected frequency assuming statistical independence, f_e. Fourth, our concern is to measure the total differences between observed and expected frequencies, so we sum the values just computed for each cell. The statistic is called *chi-square* and it can be stated symbolically as

$$\chi^2 = \Sigma \frac{(f_o - f_e)^2}{f_e}$$

Instead of using this general formula, the principles that underlie the formula for expected frequencies (f_e) enable us to express *chi-square* for a fourfold table for two dichotomous variables symbolically as:

$$\chi^2 = \frac{N(bc - ad)^2}{(a + b) \cdot (a + c) \cdot (c + d) \cdot (b + d)}$$

Computing chi-square for Table 6.1.4, we find that

$$\chi^2 = \frac{100[(40 \cdot 50) - (0 \cdot 10)]^2}{(0 + 40) \cdot (0 + 50) \cdot (50 + 10) \cdot (40 + 10)}$$

$$= \frac{100[2000 - 0]^2}{(40) \cdot (50) \cdot (60) \cdot (50)} = \frac{100 \cdot [2000]^2}{2000 \cdot 3000}$$

$$\chi^2 = \frac{100 \cdot 4,000,000}{4,800,000} = \frac{400,000,000}{4,800,000} = 83.33$$

A look at the numerator tells us that chi-square is influenced by N, namely, the larger the sample, the larger the value of chi-square. Furthermore, in light of the numerator and given the steps of first, addition and *then* multiplication in the denominator, chi-square may easily exceed 1.00. This characteristic means that chi-square is not an appropriate measure for assessing the *degree* of association.

Even though chi-square cannot be used directly to measure the degree of association, there are coefficients that are based on it. One commonly used coefficient is called phi-square (ϕ^2), and it is computed by dividing the chi-square value by N. It can be stated symbolically as:

$$\phi^2 = \frac{\chi^2}{N}$$

In our example ϕ^2 can be computed as follows:

$$\phi^2 = \frac{\chi^2}{N} = \frac{66.66}{100} = \underline{\underline{0.667}}$$

Phi-square has a lower limit of 0.00 when there is no association and in the case of any two dichotomous variables phi-square has an upper limit of $+1.00$. Unfortunately, when there are more than two categories for both variables, phi-square has no meaningful upper limit. In a fourfold table phi-square reaches unity ($+1.00$) only with perfect association, that is, when there are two diagonally opposite cells with zero cases. Our example shows that with only one zero cell, phi-square is not equal to unity.

It is possible to take the square root of phi-square and derive a value called phi (ϕ) that can be used to measure the association between two dichotomous variables. However, since chi-square by definition cannot be a negative value, we can state phi for a two-by-two table symbolically as:

$$\phi = \frac{bc - ad}{\sqrt{(a + b) \cdot (a \mid c) \cdot (c + d) \cdot (b + d)}}$$

The numerator contains the difference between the cross-products and the denominator contains the square root of the product of the marginals. Phi tells us the extent to which the cases are diagonally concentrated. It has limits ranging from -1.00 to $+1.00$, for perfect association, and a lower limit of 0.00 for no association. For two dichotomous interval variables, phi provides us with a 2×2 version of a measure called the product moment correlation coefficient, which will be discussed when we deal with interval variables.

In the case of Table 6.1.4, ϕ may be computed using the formula:

$$\phi = \frac{bc - ad}{\sqrt{(a + b) \cdot (a + c) \cdot (c + d) \cdot (b + d)}}$$

$$= \frac{(40 \cdot 50) - (0 \cdot 10)}{\sqrt{(0 + 40) \cdot (0 + 50) \cdot (50 + 10) \cdot (40 + 10)}}$$

$$\phi = \frac{2000 - 0}{\sqrt{40 \cdot 50 \cdot 60 \cdot 50}} = \frac{2000}{\sqrt{6,000,000}} = \frac{2000}{2449} = \underline{\underline{0.817}}$$

6.3 THE PERCENTAGE DIFFERENCE, d_{YX}

Another way to measure the size of an association between two variables is by computing the difference between two percentages. The percentage difference is a straightforward statistic which is understandable because "it makes sense." Moreover, people need not have advanced statistical training to translate the meaning of a percentage difference fairly easily. Consider the percentage difference in Table 6.3.1 which compares *education* and *voting*.

Table 6.3.1.   Participation in national elections,
by population characteristics: 1972

Education	Percent Voting in the National Election of 1972 $N = 136{,}203$
12 years or more	71% (85,861)
11 years or less	50% (50,342)

Source: Adapted from U.S. Bureau of the Census, *Statistical Abstract of the United States: 1974*. (95th edition.) Washington, D.C., 1974, page 437, No. 703.

We need not make a statistical interpretation of the difference of 21 percentage points to recognize that the percentage of people voting "depends upon" the level of education. That is, the percentage difference enables us to say that there is an association between the variables *education* and *voting*.

A major feature of a correctly presented percentage difference table is that we can utilize the percentage values, the case bases, and the total cases reported to reconstruct the original figures. Thus, interested readers may examine the original data for further evaluation.

By convention, the categories of the dependent variable appear as percentages and the categories of the independent variable appear as case bases.[3] To calculate the percentage difference for the dependent variable, called d_{YX}, we merely need to subtract one percentage value from another. Consider, for example, the hypothetical data shown in Table 6.3.2, which is the cross tabulation of respondents' occupational status and parents' educational attainment.

There are two ways we can percentage a fourfold table, and the direction of percentaging is determined by our causal assumptions. First, we can percentage the table for use in computing d_{YX}. Table 6.3.3 presents respondents'

[3]*Ibid.*, p. 70.

Table 6.3.2.   Parents' education and respondents' occupation (hypothetical data)

		Respondents' Occupational Status Y		
		Low	*High*	*Total*
Parents' Educational Attainment	High	84	70	154
	Low	180	106	286
	Total	264	176	440

Table 6.3.3.   Parents' education and respondents' occupation (Percent)

		Respondents' Occupational Status Y		
		Low	*High*	
Parents' Educational Attainment	High	54.5%	45.5%	100% (154)
	Low	62.9%	37.1%	100% (286)
		60.0%	40.0%	100% (440)

$$d_{YX} = \text{larger } \% - \text{smaller } \%$$
$$d_{YX} = \quad 45.5 \quad - \quad 37.1 \quad = 8.4\%$$

or

$$d_{YX} = \quad 62.9 \quad - \quad 54.5 \quad = 8.4\%$$

occupational status as the dependent variable Y, indicating that it is "caused by" the independent variable X, parents' educational attainment. We have percentaged in the direction of the causal variable, showing that 45.5 percent of those whose parents had a high education have high occupational status, and 37.1 percent of those whose parents had a low education have high occupational status. Thus, occupational status can be said to be *associated with* parents' educational attainment.

The percentage difference may be computed merely by subtracting the percent of respondents with high occupational status among parents with low educational attainment (37.1) from the percent of respondents with high occupational status among parents with a high educational attainment (45.5). Thus,

$$d_{YX} = 45.5 - 37.1 = 8.4\%$$

This may be interpreted as indicating that there is an association between parents' educational attainment and respondent's occupational status.

Second, we can percentage the table for use in computing d_{XY}, by running the percentage table the other way. Table 6.3.4 shows that the percentage difference is 8.0 percent rather than 8.4 percent. In this case the difference is small, but it demonstrates an important aspect of the percentage difference statistic. Namely, the percentage difference is assymetric, and it is important to distinguish between d_{YX} and d_{XY}.

Table 6.3.4.   Parents' education and respondents' occupation (Percent)

| | | Respondents' Occupational Status Y | | |
		Low	High	Total
Parents' Educational Attainment	High	31.8%	39.8%	35.0%
	Low	68.2%	60.2%	65.0%
	Total	100.0% (264)	100.0% (176)	100.0% (440)

d_{XY} = larger % − smaller %

d_{XY} =   39.8   −   31.8   = 8.0%

or

d_{XY} =   68.2   −   60.2   = 8.0%

Because of the ease of computation, we hardly need a formula. However, the technical properties of the percentage difference are such that the formulas are useful to consider. The formulas for d may be expressed symbolically as

$$d_{YX} = \frac{bc - ad}{bc + ad + ac + bd}$$

$$d_{XY} = \frac{bc - ad}{bc + ad + ab + cd}$$

The difference between the two formulas occurs only in the last two terms in the denominators. The denominator of d_{YX} uses the terms ac and bd, that is, we multiply vertically the cells that are not cross-products. The denominator of d_{XY} uses the terms ab and cd, that is, we multiply horizontally the cells that are not cross-products. The step of multiplying ac and bd for d_{YX} enables us to include the pairs that are tied on Y but not on X, while the multiplication of ab and cd in the formula for d_{XY} enables us to include the pairs that are tied on X but not on Y.

We can compute the percentage differences using the formulas and Table 6.3.2 of original values, rather than using the summarized values in Tables 6.3.3 and 6.3.4. To compute the percentage difference for respondent's occupational status (Y) given parents' educational attainment (X), we use d_{YX}.

$$d_{YX} = \frac{bc - ad}{bc + ad + ac + bd} = \frac{(70 \cdot 180) - (84 \cdot 106)}{(70 \cdot 180) + (84 \cdot 106) + (84 \cdot 180) + (70 \cdot 106)}$$

$$d_{YX} = \frac{12{,}600 - 8{,}904}{12{,}600 + 8{,}904 + 15{,}120 + 7{,}420} = \frac{3{,}696}{44{,}044} = \underline{\underline{8.4\%}}$$

To compute the percentage difference for parents' educational attainment (X) given respondents' occupational status (Y), we use d_{XY}.

$$d_{XY} = \frac{bc - ad}{bc + ad + ab + cd} = \frac{(70 \cdot 180) - (84 \cdot 106)}{(70 \cdot 180) + (84 \cdot 106) + (84 \cdot 70) + (180 \cdot 106)}$$

$$d_{XY} = \frac{12{,}600 - 8{,}904}{12{,}600 + 8{,}904 + 5{,}880 + 19{,}080} = \frac{3{,}696}{46{,}464} = \underline{\underline{8.0\%}}$$

One difficulty of the percentage difference for fourfold tables is that the further the marginals of the independent variable are from a 50/50 cut, the larger will be the percentage difference. Skewed marginals on the independent variable increase the percentage difference. Another difficulty is that the percentage difference may be considerably influenced by changing the cutting points of the variables, thereby affecting the marginals.

6.4 YULE'S Q

Another commonly used measure of association for dichotomous nominal or ordinal variables is called Yule's Q. Q is a measure of association that focuses on the order in which pairs of cases are distributed given the rankings of two dichotomous ordinal variables. Yule's Q has an upper limit of ± 1.00 when X and Y are conditionally associated, that is, when there is one out of four cells with zero cases, and a lower limit of 0.00 when X and Y are not associated, that is, when they are statistically independent. Most important, we can give the values a meaningful interpretation based on their intrinsic meaning. Before we tackle its meaning, let us consider its formula.

Fourfold tables can be treated so that the *order* on X and Y is measured for pairs of cases. In Table 6.1.1, the cases and any pair of cases in cells b and c are *consistent* in their order on X and Y, that is, they are $X+$ and $Y+$ or $X-$ and $Y-$; whereas the cases and any pair of cases in cells a and d are *inconsistent* in their order on X and Y, that is, they are $X+$ and $Y-$ or $X-$ and $Y+$. In order to compute Q, we use the difference between consistent cross products less inconsistent cross products, divided by the sum of the cross products. Q can be expressed symbolically:

$$Q_{XY} = \frac{\text{consistent cross products} - \text{inconsistent cross products}}{\text{consistent cross products} + \text{inconsistent cross products}}$$

Using the notation in Table 6.1.1, Q can be expressed symbolically as:

$$Q_{XY} = \frac{bc - ad}{bc + ad}$$

Our example from Table 6.1.4 shows a perfect positive association for Q. Thus,

$$Q_{XY} = \frac{bc - ad}{bc + ad} = \frac{(40 \cdot 50) - (0 \cdot 10)}{(40 \cdot 50) + (0 \cdot 10)}$$

$$Q_{XY} = \frac{2000 - 0}{2000 + 0} = \frac{2000}{2000} = \underline{\underline{+1.00}}$$

Our example from Table 6.3.2 shows a low positive association for Q. Thus,

$$Q_{XY} = \frac{bc - ad}{bc + ad} = \frac{(70 \cdot 180) - (84 \cdot 106)}{(70 \cdot 180) + (84 \cdot 106)}$$

$$= \frac{12{,}600 - 8{,}904}{12{,}600 + 8{,}904} = \frac{3{,}696}{21{,}504} = \underline{\underline{+0.172}}$$

Like chi-square, Q uses the difference between cross products in the numerator; however, the formula for chi-square calls for us to square this difference. This will, of course, eliminate negative values. To measure the association between variables we want to know the direction of influence, and negative differences should be considered, and that is just what Q does.

Aside from being an easy measure to compute, Q has a distinct advantage over ϕ for ordinal variables in that it has an easily stated intrinsic meaning. Specifically, Q tells us the proportionate improvement in prediction on Y if we know the order of the cases on X.

For example, a Q_{XY} value of 0.27 indicates that we would do 27 percent better than random chance prediction by predicting the order on the dependent variable from the order on the independent variable by always predicting that those high on X (+) also are high on Y (+), and those low on X (−) also are low on Y (−) or vice versa.[4]

The computation of Yule's Q for actual data when N is large is just as straightforward but the values become difficult to handle. Consider, for example, Table 6.4.1 which shows a fourfold table in which education and participation in the national election of 1972 have been cross-tabulated for U.S. census data. Our concern is with computing the value for Q.

Table 6.4.1.   Participation in national elections and education: 1972 persons of voting age in ten-thousands

	Variable Y Vote in National Election in 1972		
Variable X Education	*No*	*Yes*	*Total*
High 12 Years or more	2,499	6,087	8,586
Low 11 Years or less	2,544	2,490	5,034
Total	5,043	8,577	13,620

Source: Adapted from U.S. Bureau of the Census, *Statistical Abstract of the United States: 1974*. (95th edition.) Washington, D.C., 1974, page 437, No. 703.

The computation for Q involves only the frequencies from the inner cells of the cross tabulated fourfold table. Thus,

$$Q_{XY} = \frac{bc - ad}{bc + ad}$$

$$Q_{XY} = \frac{(6087 \cdot 2544) - (2499 \cdot 2490)}{(6087 \cdot 2544) + (2499 \cdot 2490)} = \frac{15,485,328 - 6,222,510}{15,485,328 + 6,222,510}$$

$$= \frac{9,262,818}{21,707,838} = \underline{0.427}$$

The Q of +0.427 for Table 8.2.1 means that we would do 42.7 percent better than random chance prediction if we always predict that the voter is a

[4] *Ibid.*, pp. 48, 49.

person with a high education and the non-voter is a person with a low education (or alternatively, if we always predict that the person with a high education is the voter and the person with a low education is the non-voter).

Yule's Q is related to the percentage difference in that Q_{XY}, d_{YX}, and d_{XY} always will have the same sign. Furthermore, when X and Y are not associated, that is, when they are statistically independent, Q_{XY}, d_{XY}, and d_{YX} all will be equal to zero.

6.5 GAMMA

Gamma (γ) is an ordinal measure of association in which we measure improvement in prediction based on the order of the independent variable to predict order on the dependent variable. Gamma is the general statistic of which Yule's Q is the special 2×2 case. γ is computed in a manner analogous to Q, but the steps are more complex, since we must calculate several sets of cross-products to enable us to compute the total number of consistent and inconsistent pairs. In order to compute these, however, we need a notation system.

Using the layout for two trichotomous (three level) ordinal variables shown in Table 6.5.1, in which we have assigned the low-low cases to the lower left hand cell and the high-high cases to the upper right hand cell, we can designate the number of cases in each cell by the letter n and the location of the cell by the subscript letters showing the level of the ordinal variables. Thus, the number of low-low cases is given the abbreviation n_{LL} and the number of high-high cases is given the abbreviation n_{HH}. The other cell frequencies can be similarly designated with the first subscript representing the row in which

Table 6.5.1. Notation and nomenclature for a 3 × 3 (ordinal variables) ninefold table

Variable X	Variable Y			Total
	Low	*Medium*	*High*	*Total*
High	n_{HL}	n_{HM}	n_{HH}	Marginal High on X
Medium	n_{ML}	n_{MM}	n_{MH}	Marginal Medium on X
Low	n_{LL}	n_{LM}	n_{LH}	Marginal Low on X
	Marginal Low on Y	Marginal Medium on Y	Marginal High on Y	Total Cases N

the cell is located and the second subscript representing the column in which the cell is located.

Using the same logic regarding order that we did with Q, we can now compute consistent cross-products. First, we treat the cases in the upper right-hand cell n_{HH} as consistent on both X and Y as we did those in cell b of a fourfold table. Now, we need an equivalent to cell c. There are several such cells, and we must incorporate all of them in our computations. Consider first those in contrast with n_{HH}.

If any of the n_{HH} cases were paired (two cases drawn at random) with any of the other n_{HH} cases or any of the n_{HL} or n_{HM} cases (high row) or any of the n_{MH} or n_{LH} cases (high column), there would be a tie either on the row, the column, or with n_{HH} itself on the row and column. Such pairs of tied cases are *not* included in our computations for gamma. Our concern, instead, is with those cases that are below and to the left of the upper right-hand cell. That is, any of the n_{MM}, n_{ML}, n_{LM}, and n_{LL} cases are not tied with n_{HH} on X and Y. Thus, if any of the n_{HH} cases were paired with any of the ($n_{MM} + n_{ML} + n_{LM} + n_{LL}$) cases, we would have a pair of cases in which the case lower on variable X also will be lower on variable Y. Put differently, we would now have a set of pairs with the same ordering on the two ordinal variables. The step just described is shown in Table 6.5.2.

Table 6.5.2. **Computations for first step in determining consistent cross products**

Variable X	Variable Y		
	Low	*Medium*	*High*
High			n_{HH}
Medium	n_{ML}	n_{MM}	
Low	n_{LL}	n_{LM}	

High X-High Y cases $\times$ the sum of all those cases less than high X and high Y

$$n_{HH} (n_{MM} + n_{ML} + n_{LM} + n_{LL})$$

A close look at Table 6.5.1 will reveal that the step just described does not take into account the complete set of pairs with the same ordering on the two ordinal variables X and Y. By using the same logic just developed, we can move across, first, the top row (high X) and, second, the middle row (medium X), multiplying the frequencies in the cells which are higher on both variables, that is, cases n_{HH}, n_{MH}, and n_{MM}, times the sums of the cells which are lower on both variables, that is, all those cases below and to the left of cases n_{HH}, n_{MH}, and n_{MM}. Tables 6.5.3–6.5.5 show these steps and the appropriate computations.

Now we can put all this together to compute the total number of pairs called consistent cross products, abbreviated ccp. It can be expressed symbolically as:

$$\text{ccp} = n_{HH}(n_{MM} + n_{ML} + n_{LM} + n_{LL}) + n_{HM}(n_{ML} + n_{LL})$$
$$+ n_{MH}(n_{LM} + n_{LL}) + n_{MM}(n_{LL})$$

Since our concern is parallel to that when computing Yule's Q, we also must determine the value of the inconsistent cross products. To do this, we merely reverse the procedures just described and work our way across and

Table 6.5.3.   Computations for second step in determining consistent cross products

Variable X	Variable Y		
	Low	*Medium*	*High*
High		n_{HM}	
Medium	n_{ML}		
Low	n_{LL}		

High X-Medium Y cases × the sum of all those cases less than high X and medium Y

$$n_{HM}(n_{ML} + n_{LL})$$

Table 6.5.4.   Computations for third step in determining consistent cross products

Variable X	Variable Y		
	Low	*Medium*	*High*
High			
Medium			n_{MH}
Low	n_{LL}	n_{LM}	

Medium X-High Y cases × the sum of all those cases less than medium X and high Y

$$n_{MH}(n_{LM} + n_{LL})$$

Table 6.5.5.   Computations for fourth step in determining consistent cross products

Variable X	Variable Y		
	Low	*Medium*	*High*
High			
Medium		n_{MM}	
Low	n_{LL}		

Medium X-Medium Y × the sum of all those less than medium X and medium Y

$$n_{MM} (n_{LL})$$

down the table starting from the upper left-hand cell. These cross products are analogous to those we get by multiplying $a \times d$ in a fourfold table, namely, they are inconsistent pairs. Clearly, the cases in the upper left-hand cell n_{HL} are inconsistent on X and Y. If any of these cases were paired with any of the other n_{HL} cases or with any of the n_{HM} or n_{HH} cases (high row) or any of the n_{ML} or n_{LL} cases (low column), there would be a tie. However, any of the n_{MM}, n_{MH}, n_{LM}, or n_{LH} cases are not tied on X and Y with any of the n_{HL} cases. Thus, if any of the n_{HL} cases were paired with any of the $(n_{MM} + n_{MH} + n_{LM} + n_{LH})$ cases we would have a pair of cases in which the case lower on variable X will be higher on variable Y. Put differently, we would now have a set of pairs with the ordering reversed on the two ordinal variables. The step just described is shown in Table 6.5.6.

Table 6.5.6.   Computations for first step in determining inconsistent cross products

Variable X	Variable Y		
	Low	*Medium*	*High*
High	n_{HL}		
Medium		n_{MM}	n_{MH}
Low		n_{LM}	n_{LH}

High X-Low Y × the sum of all those cases less than high X and greater than low Y

$$n_{HL} (n_{MM} + n_{MH} + n_{LM} + n_{LH})$$

The other sets of pairs with reverse ordering may be computed by moving across the table from left to right and down from top to bottom. Tables 6.5.7–6.5.9 show these steps and the appropriate computations.

Using the steps we can compute the total number of pairs called inconsistent cross products, abbreviated icp. It can be expressed symbolically as:

$$\text{icp} = n_{HL}(n_{MM} + n_{MH} + n_{LM} + n_{LH}) + n_{HM}(n_{MH} + n_{LH})$$
$$+ n_{ML}(n_{LM} + n_{LH}) + n_{MM}(n_{LH})$$

Table 6.5.7.   Computations for second step in determining inconsistent cross products

Variable	Variable Y		
X	*Low*	*Medium*	*High*
High		n_{HM}	
Medium			n_{MH}
Low			n_{LH}

High X-Medium Y × the sum of all those cases less than high X and greater than medium Y

$$n_{HM}\,(n_{MH} + n_{LH})$$

Table 6.5.8.   Computations for third step in determining inconsistent cross products

Variable	Variable Y		
X	*Low*	*Medium*	*High*
High			
Medium	n_{ML}		
Low		n_{LM}	n_{LH}

Medium X-Low Y × the sum of all those cases less than medium X and greater than low Y

$$n_{ML}\,(n_{LM} + n_{LH})$$

Table 6.5.9.   Computations for fourth step in determining inconsistent cross products

Variable X	Variable Y		
	Low	*Medium*	*High*
High			.
Medium		n_{MM}	
Low			n_{LH}

Medium X-Medium Y × the sum of all those cases less than medium X and greater than medium Y

$$n_{MM}\,(n_{LH})$$

We finally are ready to state the formula for gamma using the same format as we did with Yule's Q. Gamma can be expressed symbolically as:

$$\gamma = \frac{\text{consistent cross products} - \text{inconsistent cross products}}{\text{consistent cross products} + \text{inconsistent cross products}}$$

$$\gamma = \frac{\text{ccp} - \text{icp}}{\text{ccp} + \text{icp}}$$

The steps in computing gamma demonstrate that it must be interpreted in terms of pairs and not in terms of individual cases. This is because we have a measure which deals with the predictability of order on two variables for pairs and not for the predictability of the rank for each case. Fortunately, gamma may be interpreted exactly as Yule's Q was. For example, a γ_{XY} of 0.27 indicates that we would do 27 percent better than random chance prediction by predicting the order on the dependent variable from the order on the independent variable.

Gamma is not restricted to trichotomous variables, but it is restricted to ordinal ones. Thus, when we wish to refine the data beyond dichotomies we must be careful to adhere to the level of measurement assumption. For example, consider the data in Table 6.4.1. We could treat education as a four-level ordinal variable, and we could operationally define "Yes" on voting as indicating "higher" participation. Gamma can be computed as we learned in the 3 × 3 notation table for the data shown in the 4 × 2 Table 6.5.10. The notation and computational steps are shown in Table 6.5.11.

Table 6.5.10.   Participation in national elections and education: 1972

Persons of Voting Age in Ten-Thousands

Variable X Education	Variable Y Vote in National Election in 1972		
	No	*Yes*	*Total*
More than 12 years	743	2,768	3,511
12 years	1,756	3,319	5,075
9–11 years	1,069	1,159	2,228
8 years or less	1,475	1,331	2,806
Total	5,043	8,577	13,620

Abbreviation Symbols for Notation Table 6.5.11

Education		*Vote*	
More than 12 years	$= C$	Yes $= H$	
12 years	$= H$	No $= L$	
9–11 years	$= S$		
8 years or less	$= E$		

Source: Adapted from *Statistical Abstract*, 1974, Table 703, p. 437

$$\gamma = \frac{\text{ccp} - \text{icp}}{\text{ccp} + \text{icp}} = \frac{22{,}055{,}461 - 10{,}111{,}366}{22{,}055{,}461 + 10{,}111{,}366} = \frac{11{,}944{,}095}{32{,}166{,}827} = \underline{\underline{0.371}}$$

The gamma of .371 may be interpreted as indicating that we would do 37 percent better than random chance prediction by predicting the order on the dependent variable from the order on the independent variable.

The question often arises as to whether we should use Q or gamma, but the issue actually has more to do with the way the data are organized. The following considerations are important when tackling the problem of whether or not to refine the data beyond dichotomies:[5]

1. Gamma requires a lot more work, particularly in multivariate analysis treating several variables at one time. Indeed, multivariate analysis using gamma is almost impossibly tedious without an electronic computer and, even with one, the programming is complex.

[5] *Ibid.*, p. 75.

Table 6.5.11.   Notation and nomenclature for a 4 × 2 (ordinal variable) table and computations for Table 6.5.10 participation and education

Variable X	Variable Y	
	Low	High
Highest $= C$	n_{CL}	n_{CH}
High $\;\;\;= H$	n_{HL}	n_{HH}
Low $\;\;\;\;= L$	n_{LL}	n_{LH}
Lowest $= E$	n_{EL}	n_{EH}

$\mathrm{ccp} = n_{CH}(n_{HL} + n_{LL} + n_{EL}) + n_{HH}(n_{LL} \times n_{EL}) + n_{LH}(n_{EL})$

$\mathrm{ccp} = 2,768(1,756 + 1,069 + 1,475) + 3,319(1,069 + 1,475) + 1,159(1,475)$

$\mathrm{ccp} = 2,768(4,300) + 3,319(2,544) + 1,159(1,475)$

$\mathrm{ccp} = 11,902,400 + 8,443,536 + 1,709,525 = \underline{22,055,461}$

$\mathrm{icp} = n_{CL}(n_{HH} + n_{LH} + n_{EH}) + n_{HL}(n_{LH} + n_{EH}) + n_{LL}(n_{EH})$

$\mathrm{icp} = 743(3,319 + 1,159 + 1,331) + 1,756(1,159 + 1,331) + 1,069(1,331)$

$\mathrm{icp} = 743(5,809) + 1,756(2,490) + 1,069(1,331)$

$\mathrm{icp} = 4,316,087 + 4,372,440 + 1,422,839 = \underline{10,111,366}$

2. Gamma uses more of the information in the data, allowing more of the possible pairs to contribute to the conclusions.
3. Gamma will be more sensitive to how the relationship operates at the extremes. If, for example . . . , college graduation does not raise income beyond the income associated with 1 to 3 years of college, this would "pull down" the value of gamma but would not affect Q.
4. Gamma may be used only if all your variables are ordinal, while Q may be used on any variable regardless of level of measurement.[6]
5. Gamma will generally run a little lower in magnitude on the same data. This is because it includes more of the small differences in the variables and they generally are less consistent (less positive if Q is positive, less negative if Q is negative). You are more likely to get a consistent income difference when you compare college graduates with those completing less than eighth grade than when comparing those completing 2 years of college with those completing 1 year of college.

Verbally describing statistical values often is a problem because of arbitrarily reporting various magnitudes. To prevent such practices, Table 6.5.12 presents conventions that can be used to describe ranges of values for Q and γ. These phrases enable us to verbally cite Q and/or γ values in a consistent fashion. Thus our Q of 0.427 and/or our γ of 0.371 represent a moderate positive association between education and voting.

[6]As Davis points out "You can, of course, calculate gamma when one or both items are interval or ratio scales because both have the ordinal property plus their various distance properties." *Ibid.*, p. 75.

Table 6.5.12.   Conventions for describing values for Yule's Q and γ

Coefficient Value	Appropriate Phrase
+1.00	A perfect positive association
+0.70 or higher	A very strong positive association
+0.50 to +0.69	A substantial positive association
+0.30 to +0.49	A moderate positive association
+0.10 to +0.29	A low positive association
+0.01 to +0.09	A negligible positive association
0.00	No association
−0.01 to −0.09	A negligible negative association
−0.10 to −0.29	A low negative association
−0.30 to −0.49	A moderate negative association
−0.50 to −0.69	A substantial negative association
−0.70 or lower	A very strong negative association
−1.00	A perfect negative association

Source: Adapted from James A. Davis, *Elementary Survey Analysis* (Englewood Cliffs, New Jersey: Prentice-Hall, Inc., 1971), p. 49, Table 2.8.

EXERCISES

1. In your own words, define, describe, and discuss the following terms and give an hypothetical example of each:
   marginal frequencies
   fourfold table
   no association
   expected frequency
   f_o and f_e
   phi-square
   d_{YX}
   percentage difference
   consistent cross products
   perfect negative association

2. Explain to a non-statistical friend the utility of determining association and of measuring the size of that association.

3. Tables can be percentaged in two directions. Explain the advantages and disadvantages in percenting in both directions.

4. Compare Yule's Q and Chi-square. What are their internal (computational) differences? What are their differences in statistical use?

5. In his study of *The Great Books Program*, Davis analyzes the discussion process, asking "whether new groups have trouble, or new individuals tend to drop out, or both." One aspect of this question involves the association between the age of the reading group (X) and the individual longevity in the reading program (Y). Table A shows the cross-tabulation of these two variables.

Table A.   Age and longevity

| Age of the Reading Groups (X) | Individual Longevity in Reading Program (Y) | | | |
	Short (less than 1 year)	Medium (1 or 2 years)	Long (3 years or more)	Total
Old	37	273	272	582
Fairly Old	53	120	77	250
Middle Age	31	51	21	103
Fairly Young	70	45	31	146
Young	455	51	20	526
Total	646	540	421	1607

Source: Adapted from James A. Davis, *Great Books and Small Groups* (New York: The Free Press of Glencoe, Inc., 1961), pp. 100–104 and Chart 4.1 p. 224.

A. Compute γ for this 5 × 3 table.
B. Discuss and interpret this statistical finding verbally.

6. Combine the rows for the three "medium" age groups (Fairly Young + Middle Age + Fairly Old). This creates a 3 × 3 table.
   A. Compute γ for this 3 × 3 table.
   B. Describe and discuss the difference between the γ values for the 5 × 3 table in problem 5 and the 3 × 3 table just created.

7. First, combine the rows for the three "medium" age groups (Fairly Young + Middle Age + Fairly Old): second, combine the columns for the "high" longevity groups (Medium + Long). This creates a 3 × 2 table.
   A. Compute γ for this 3 × 2 table.
   B. Describe and discuss the difference between the γ values for the 5 × 3 table in problem 5, the 3 × 3 table in problem 6, and the 3 × 2 table just created.

8. First, combine the rows for the three "low" age groups (Young + Fairly Young + Middle Age): second, combine the rows for the "high" age groups (Fairly Old + Old): third, combine the columns for the "high" longevity groups (Medium + Long). This creates a 2 × 2 table.
   A. Compute γ (Yule's Q) for this 2 × 2 table.
   B. Describe and discuss the differences between the γ values for the 5 × 3 table in problem 5, the 3 × 3 table in problem 6, the 3 × 2 table in problem 7, and the 2 × 2 table just created.

9. Compute χ^2 for the 2 × 2 table in problem 8 and discuss its meaning compared to the Q_{XY} for the same table.

10. Compute d_{YX} and d_{XY} for the 2 × 2 table in problem 8 and discuss their meaning compared to the Q_{XY} and χ^2 for the same table.

7

Prediction and
Regression with
Two Interval Variables

We are concerned in this chapter with the problem of measuring and describing associations or relationships between two interval variables. Now we are interested in the way in which variables vary together; that is, the way they covary. Importantly, it is through the concept of *covariation* that we examine the joint (together) distribution of two variables. Generally our purpose is to estimate or predict the individual contributions of several independent variables on one dependent variable.[1] Since estimation and prediction are among the basic goals of sociology, it is common for sociologists to utilize statistics when they are interested in such matters. From this perspective, social statisticians attempt to estimate or predict the value of one variable from knowledge they possess about other variables.

[1]There are several useful treatments of this topic. For particularly helpful ones see John E. Freund, *Statistics, A First Course* (Englewood Cliffs, New Jersey: Prentice-Hall, Inc., 1970), pp. 251–83; Hubert M. Blalock, Jr., *Social Statistics*, 2nd edition. (New York: McGraw-Hill Book Company, 1972), pp. 361–95; James Fennessey, "The General Linear Model: A New Perspective on Some Familiar Topics," *American Journal of Sociology*, 74 (July 1968), pp. 1–27; and Roland K. Hawkes, "The Multivariate Analysis of Ordinal Measures," in the *American Journal of Sociology*, 76 (March 1971), pp. 908–926.

7.1 PREDICTION ABOUT ONE VARIABLE FROM ANOTHER

Earlier we discussed an example involving eight students and discovered that knowledge of the number of hours they spent studying helped us predict their grade point averages (Section 5.3.). Our interest now is in *joint* observations, "joint" referring to the two variables for each student. In a general technical sense we are concerned with situations in which we want to predict values of the dependent variable, and in which we measure the independent variable because knowledge of its value may help us predict (or describe) the distribution of values for the dependent variable.

We often use a common-sense version of linear regression on the everyday level to predict such things as hourly, weekly, monthly, and annual wages; seconds, minutes, hours, and time in general; automobile and airplane speed and distance; and many other daily concerns. Assume, for instance, that a regular full-time job means working eight hours a day, five days a week. It is possible to state how many hours someone works each week if we know the number of hours of overtime worked, as can be seen in hypothetical Table 7.1.1. In this sort of prediction, we can use a formula to express the relationship. Thus if weekly hours is variable Y and hours of overtime is variable X, then to "predict" the value for weekly hours we could say

$$Y = 40 + X$$

Table 7.1.1.   Hours of overtime and weekly hours

Hours of Overtime X	Weekly Hours Y
1	41
2	42
3	43
4	44
5	45
6	46
7	47
8	48
9	49
10	50

Weekly hours = 40 + hours of overtime

$$Y = 40 + X$$

Similarly, if the minimum hourly wage for 1970 was \$1.80, and if we know the average hourly earnings *above* the minimum wage for workers in the seven general, private, nonagricultural industries listed by the Bureau of Labor Statistics, we could "predict" the average hourly earnings by adding the two amounts together, as can be seen in Table 7.1.2. The formula now would be

$$Y = \$1.80 + X$$

Table 7.1.2.   Average hourly earnings above minimum wage and average hourly earnings (1970)

Industry	Hourly Earnings Above Minimum Wage X	Hourly Earnings Y
Mining	$2.04	$3.84
Contract Construction	3.42	5.22
Manufacturing	1.56	3.36
Transportation and Public Utilities	2.05	3.85
Wholesale and Retail Trade	.91	2.71
Financial, Insurance, Real Estate	1.27	3.07
Services	1.04	2.84

Hourly earnings = $1.80 + hourly earnings above minimum wage

$$Y = \$1.80 + X$$

Source: Adapted from U.S. Bureau of the Census, *Statistical Abstract of the United States, 1971.* (92nd edition.) Washington, D.C., 1971, pages 219–221, No. 345.

Now, suppose that for these workers the Bureau of Labor Statistics uses fifty weeks as the base for calculating annual earnings. If we know the average weekly earnings for private, nonagricultural industries for 1970, then we can predict with absolute precision what the average annual earnings will be for each industry. This can be seen in Table 7.1.3. In this sort of prediction problem, what we are saying is that if annual earnings is variable Y and weekly earnings is variable X, then to predict Y we use the formula

$$Y = 50 \cdot X$$

One doesn't need advanced mathematical skills to understand these and similar problems: We use such formulas in much of our daily lives to determine the cost of such things as food, clothing, and gasoline. It is clear that there are instances in which we can have "perfect" prediction of one variable from another. In our example involving average annual earnings, we also can see in Figure 7.1.1 that the relationship (in the form of "predicted" values) is *linear* in that the predictions fall exactly on a *straight line*.

In the first two examples (involving hours worked and average hourly earnings), we predict using information about one variable in an *additive*

Table 7.1.3. **Weekly earnings and annual earnings (1970)**

Industry	Weekly Earnings	Annual Earnings
Mining	$164	$8,200
Contract Construction	195	9,750
Manufacturing	134	6,700
Transportation and Public Utilities	156	7,800
Wholesale and Retail Trade	96	4,800
Financial, Insurance, Real Estate	113	5,650
Services	98	4,900

$$\text{Annual earnings} = 50 \cdot \text{weekly earnings}$$

$$Y = 50 \cdot X$$

Source: Adapted from U.S. Bureau of the Census, *Statistical Abstract of the United States, 1971.* (92nd edition.) Washington, D.C., 1971, pages 219–221, No. 345.

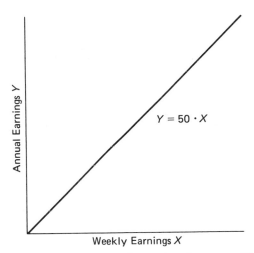

Figure 7.1.1. Linear regression of annual earnings on weekly earnings.

formula. In the third example involving average annual earnings, we predict using information about one variable in a *multiplicative* formula. Both types of formula contain *constants*, in these cases 40 hours, $1.80, and 50 weeks, which are combined with the *variable* levels of X to yield the variable values for Y.

There are many instances when we combine these two procedures to make predictions. For example, consider the common situation of someone who works a set amount of time each week for a set amount of pay. Let us use the average amounts for all industrial workers in 1970 of $3.23 per hour for 37.2

hours each week, giving average earnings of $120 each week. Imagine that we find a person who earns and works exactly the average. If this person is paid time and a half for overtime, the rate per overtime hour becomes $4.85. To predict such a person's salary is an easy task if we expand our formula so that total weekly earnings (variable Y) is computed with *two constants*, first, the base pay per week of $120, and second, the standard hourly rate for overtime of $4.85, and *one variable* (X), the amount of overtime hours, as can be seen in hypothetical Table 7.1.4 and Figure 7.1.2. The formula becomes

$$Y = 120 + (4.85 \cdot X)$$

Table 7.1.4.   Hours of overtime and total weekly earnings

Hours of Overtime X	Total Weekly Earnings Y
1	$124.85
2	129.70
3	134.55
4	139.40
5	144.25
6	149.10
7	153.95
8	158.80
9	163.65
10	168.50

$$Y = 120 + (\$4.85 \cdot X)$$

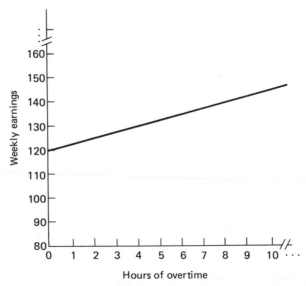

Figure 7.1.2.   Linear regression of weekly earnings on hours of overtime.

Now we need to develop a method of accurately (if not precisely) predicting the values of a dependent variable Y given the information we have about Y itself and about X and Y together. Since our interest is in the association between two variables, we must somehow blend our information about their individual variation with information about their joint variation. This leads us to an interest in measuring the way in which two variables vary together or covary.

7.2 THE COVARIANCE

Earlier we compared the scores of each student on variables X and Y to the respective means for each variable and then calculated the total variances for each variable. We can put all this together in a simple, straightforward extension of the variance. Namely, we can measure the extent to which X and Y covary by cross-classifying the variables and looking at the joint (simultaneous) location of each individual.

Our computations include the familiar comparison of each individual's score with the mean for the variable in question, that is

$$(X_i - \bar{X}) \quad \text{and} \quad (Y_i - \bar{Y}).$$

In order to get the total variances we square these differences for each individual,

$$(X_i - \bar{X})^2 \quad \text{and} \quad (Y_i - \bar{Y})^2$$

sum the squares,

$$\sum_{i=1}^{N} (X_i - \bar{X})^2 = \text{TSS of } X \quad \text{and} \quad \sum_{i=1}^{N} (Y_i - \bar{Y})^2 = \text{TSS of } Y$$

and divide by N,

$$\frac{\sum_{i=1}^{N} (X_i - \bar{X})^2}{N} = s_X^2 \quad \text{and} \quad \frac{\sum_{i=1}^{N} (Y_i - \bar{Y})^2}{N} = s_Y^2$$

Hopping back one step, we should realize that squaring the difference amounts to multiplying the difference by itself, that is,

$$(X_i - \bar{X})^2 = (X_i - \bar{X}) \cdot (X_i - \bar{X})$$

and

$$(Y_i - \bar{Y})^2 = (Y_i - \bar{Y}) \cdot (Y_i - \bar{Y})$$

Instead of multiplying the differences by themselves (squaring them), we can use the cross-classified differences and multiply the *two differences* by each other, producing a *cross product* as follows:

$$(X_i - \bar{X}) \cdot (Y_i - \bar{Y})$$

This operation enables us to consider how *each individual* is different from the mean on variable X and the mean on variable Y simultaneously. This can be seen in Figure 7.2.1.

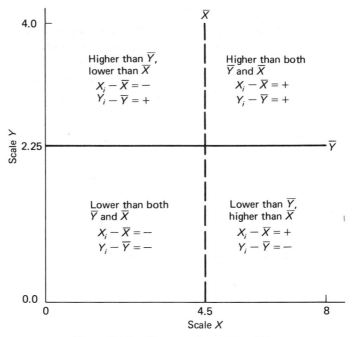

Figure 7.2.1. The mean lines of Y and X.

If an individual has scores higher than both means, the two differences will be positive numbers, giving a positive cross product. Similarly, if an individual has scores lower than both means, the two differences will be negative numbers, also giving a positive cross product. However, if an individual score is higher than one mean and lower than the other, one difference will be positive and one negative, thereby giving a negative cross product, as can be seen in Figure 7.2.1. Thus unlike the variance, the covariance may be either positive or negative: It is positive when most individuals are consistent, that is, higher and higher or lower and lower; and it is negative when most individuals are inconsistent, that is, higher and lower or lower and higher.

If we sum these cross products, we have a measure of the covariation of X and Y:

$$\text{covariation of } X \text{ and } Y = \sum_{i=1}^{N} (X_i - \bar{X}) \cdot (Y_i - \bar{Y}) = \frac{\text{total sum of cross}}{\text{products}}$$

When we divide this expression by N, we obtain the covariance. It can be expressed symbolically:

$$\text{cov} = s_{XY} = \frac{\sum_{i=1}^{N} (X_i - \bar{X}) \cdot (Y_i - \bar{Y})}{N}$$

The formula tells us to take individual i's score on X and subtract the mean $\bar{X}$, take i's score on Y and subtract the mean $\bar{Y}$, multiply the two differences, sum the products of the individual differences, and divide by the number of observations or cases. Thus the *covariance* may be defined as the mean of the sum of the cross products of the differences from the mean.

Let us put the steps for calculating the covariance together for the eight students discussed earlier on variables X and Y. We start with each individual's raw scores on variables X and Y. From the raw score we subtract the mean of X and the mean of Y. Each of these subtractions gives us a difference from the mean for that individual. We multiply the two differences by each other and get the cross product of the deviations.

Consider Alan and Betty. Alan spends four hours a week studying and has a grade point average of 4.0, while Betty spends seven hours a week studying and has a grade point average of 3.5. When we subtract the mean of each variable from the two scores, we find that Alan studies 0.5 less hours than the average student, but that he has a grade point average 1.75 points higher than the average, that is, $(X_i - \bar{X}) = -0.5$, and $(Y_i - \bar{Y}) = 1.75$. The product of these two differences indicates that Alan contributes a negative value to the total covariation, that is,

$$(X_i - \bar{X}) \cdot (Y_i - \bar{Y}) = -0.88$$

Betty, on the other hand, studies 2.5 hours more than the average and has a grade point average 1.25 units higher than the average, that is, $(X_i - \bar{X}) = 2.5$ and $(Y_i - \bar{Y}) = 1.25$. The cross product of these two values equals a positive contribution to the covariation, that is,

$$(X_i - X) \cdot (Y_i - Y) = 3.13$$

Table 7.2.1 shows these steps for all the students, indicating that the total covariation is equal to 14.52. This total is divided by the number of students to produce the covariance between variables X and Y. Thus,

$$\text{cov} = s_{XY} = \frac{\sum_{i=1}^{N}(X_i - \bar{X})\cdot(Y_i - \bar{Y})}{N} = \frac{14.52}{8} = 1.81.$$

Table 7.2.1

Student	i	X_i	$\bar{X}$	$(X_i - \bar{X})$	Y_i	$\bar{Y}$	$(Y_i - \bar{Y})$	$(X_i - \bar{X})(Y_i - \bar{Y})$
Alan	1	4	4.5	(−0.5)	4.0	2.25	(1.75)	−0.88
Arthur	2	2	4.5	(−2.5)	1.5	2.25	(−0.75)	1.88
Betty	3	7	4.5	(2.5)	3.5	2.25	(1.25)	3.13
Candy	4	5	4.5	(0.5)	2.5	2.25	(0.25)	0.13
Helen	5	6	4.5	(1.5)	2.0	2.25	(−0.25)	−0.38
Maxine	6	1	4.5	(−3.5)	0.5	2.25	(−1.75)	6.13
Roland	7	3	4.5	(−1.5)	1.0	2.25	(−1.25)	1.88
Sam	8	8	4.5	(3.5)	3.0	2.25	(0.75)	2.63

$$\sum_{i=1}^{N}(X_i - \bar{X})(Y_i - \bar{Y}) = 14.52$$

$$\text{cov } XY = s_{XY} = 1.81$$

Our concern is with estimating values for variable Y, from the values of another variable. To do this we must extend the least-squares principle by formalizing a method to fit the data with a *best fitting* equation; that is, an equation that fits the least-squares principle and is "best" because it minimizes the sum of the squares.

7.3 LINEAR REGRESSION COEFFICIENTS

In line with our earlier effort at prediction, we want an equation that uses information about both the dependent variable Y and the independent variable X. Since we wish to predict the values of Y *from* the values of X, it seems sensible for us to take into account how they covary. Put differently, it would be helpful if we could use specific values to predict a Y score for each X score. If we did, we would be able to predict a unique Y value for each X value.

Recall that the types of equations we developed in Section 7.2 allowed us to use two kinds of information to predict values for Y. In the first place, there are some constants that *add* the same information no matter what the level of X. For example, if we have as givens 40 hours per week, $1.80 minimum hourly wage, or $120 average weekly earnings, we can *add* these to varying amounts of time, varying total hourly wages, or varying amounts of overtime work.

Secondly, there are some constants that contribute information depending upon the level of X. For example, if the weekly earnings for all industries are *multiplied* by 50 weeks, then the annual earnings will vary depending on varying amounts of weekly earnings.

Let us state a general equation to take into account both types of constants as well as the variables X and Y:

$$Y = \begin{pmatrix} \text{constant value} \\ \text{regardless of} \\ \text{the value of } X \end{pmatrix} + \begin{pmatrix} \text{constant value} \\ \text{specific to varying} \cdot X_i \\ \text{levels of } X \end{pmatrix}$$

Now we want to develop formulas to calculate these constant values. Of course, the first thing we need to know is the given value of X. Then we must know the size of the change in Y values that each X score brings; in other words, we want to calculate the constant value specific to varying levels of X. Fortunately, there is a measure that tells us just what is this change in Y given X. It is called a *regression coefficient*, or the *total b* of Y on X, abbreviated b_{YX}.

To calculate the total b, we divide the covariation of X and Y by the variation in X. The constant quotient resulting from this is the b, and it indicates the size of the change in Y for a specified change in X. The total b can be expressed symbolically:

$$b_{YX} = \frac{\sum\limits_{i=1}^{N} (X_i - \bar{X})(Y_i - \bar{Y})}{\sum\limits_{i=1}^{N} (X_i - \bar{X})(X_i - \bar{X})}$$

$$= \frac{\text{total sum of the cross products of } X \text{ and } Y}{\text{total sum of squares of } X}$$

A close look at the formula for the total b is helpful in understanding it. We just said that the numerator for b_{YX} is the covariation of X and Y. When it is divided by N, it becomes the *covariance* of X and Y. Likewise, the denominator is the variation of X, which if divided by N is the *total variance* of X. Thus we can see that the total b_{YX} is nothing more than the *ratio of the covariance of variables X and Y to the total variance of variable X*. Therefore it also can be expressed symbolically:

$$b_{YX} = \frac{\dfrac{\sum\limits_{i=1}^{N} (X_i - \bar{X})(Y_i - \bar{Y})}{N}}{\dfrac{\sum\limits_{i=1}^{N} (X_i - \bar{X})^2}{N}} = \frac{s_{XY}}{s_X^2}$$

Because it depends on the covariance, b_{YX} can be either plus or minus ($\pm$). A plus ($+b_{YX}$) indicates that as X goes up, Y goes up. A minus ($-b_{YX}$) indicates that as X goes up, Y goes down.

Naturally, since we can calculate the value of b_{YX}, that is, the b of Y *given* X, we also can calculate the value of b_{XY}, the b of X *given* Y. Look closely at the subscripts for b, for there is a distinct difference depending on the order of the subscript letters. The first subscript always represents the dependent variable and the second subscript the independent variable. For example, with b_{XY}, we are predicting values of X from values of Y. This means that the denominator contains the total variance of Y. b_{XY} can be expressed symbolically as:

$$b_{XY} = \frac{\dfrac{\sum\limits_{i=1}^{N} (X_i - \bar{X})(Y_i - \bar{Y})}{N}}{\dfrac{\sum\limits_{i=1}^{N} (Y_i - \bar{Y})^2}{N}} = \frac{s_{XY}}{s_Y^2}$$

The essential elements for computing regression coefficients can be summarized in a covariance *matrix*. "Matrix" is derived from the Latin word for mother and means the uterus, that part within the mother that encloses the unborn child. In general, matrix means that which encloses something, and in statistics we use it to mean a rectangular enclosure of rows and columns of scores or statistics for any combination of variables. Thus a statistical matrix may resemble a checkerboard, because there are rows and columns. Since the covariance matrix shows all the variances and covariances for a set of variables, we can use it to calculate the total regression coefficients (b's) between each pair of variables. Since the total b's are unique for each combination, we calculate b_{YX} for variable Y given X, b_{XY} for variable X given Y, and so forth. Returning to our example of the eight students, we can carry out the calculations for each pair of variables. We end up with two regression coefficients, as is shown in Table 7.3.1.

It is customary to summarize the b's in a matrix of zero-order regression coefficients. The reason we say "zero-order" is that we are examining the relationship between two variables not controlling for any others; that is,

Table 7.3.1. Calculations for zero-order regression
coefficients for variables *Y, X,* and *T*

$$b_{YX} = \frac{s_{XY}}{s_X^2} = \frac{1.81}{5.25} = 0.34$$

$$b_{XY} = \frac{s_{XY}}{s_Y^2} = \frac{1.81}{1.31} = 1.38$$

we have *zero* (0) extra independent or control variables. Thus b_{YX} is the zero-order regression coefficient of Y given X, and b_{XY} is the zero-order regression coefficient of X given Y. These values can be seen in Table 7.3.2.

Table 7.3.2.   Matrix of zero-order regression
coefficients for variables *Y* and *X*
($N = 8$)

	Y	X
Y	1.00	0.34
X	1.38	1.00

In a matrix of *b*'s, read the row variable as the first subscript and the column variable as the second subscript, e.g., $b_{YX} = 0.34$ and $b_{XY} = 1.38$, etc.

Table 7.3.2 helps demonstrate a notable characteristic of the two total regression coefficients for a given pair of variables; namely, the two *b*'s are not necessarily symmetrical. That is, $b_{YX} \neq b_{XY}$. If we read the regression coefficients in the table by using the column abbreviation as the first subscript and the row abbreviation as the second subscript, we see that $b_{YX} = 0.34$ versus $b_{XY} = 1.38$. In order to calculate either coefficient, we use the covariance of X and Y in the numerator. The only distinguishing aspect of the two total coefficients is that we use the total variance of X to predict Y in the denominator of b_{YX}, and the total variance of Y to predict X in the denominator of b_{XY}.

In order to grasp the utility of regression coefficients, let us restrict our attention to estimating the value of Y *given* X by use of b_{YX} itself. Putting together all that has gone so far, and using the covariance of X and Y, $s_{XY} = 1.81$, and the variance of X, $s_X^2 = 5.25$, we can come up with b_{YX} for our eight students:

$$b_{YX} = \frac{s_{XY}}{s_X^2} = \frac{1.81}{5.25} = 0.34$$

This may be interpreted as meaning that a change of 1 unit in X produces a change of 0.34 of a unit in Y. Thus a change of 1 unit in X produces a change of $(1 \cdot 0.34)$ of a unit in Y.

$$Y = 1.00 + (1 \cdot 0.34) = 1.34.$$

A change of 2 units in X produces a change of $(2 \cdot 0.34)$ of a unit in Y,

$$Y = 1.00 + (2 \cdot 0.34) = 1.68$$

and so forth.

Since we know the constant value specific to varying levels of X from the term b_{YX}, it is a simple matter to multiply it times any specific value of X_i. The product of the one constant term b_{YX} and the one unique term X_i varies depending on the value of X_i. Thus we can predict a value for Y given a value of X. It can be expressed symbolically as:

$$\text{specific variable contribution} = (b_{YX} \cdot X_i)$$

Since we use the ratio of the covariance in X and Y to the variance in X, b_{YX} is a *constant* value that summarizes the amount of change in Y given X. We now must derive the constant value regardless of the value of X. This coefficient is called the a of Y given X abbreviated $a_{Y \cdot X}$.

We can use b to calculate a. a is the average contribution of X on the average value of Y, and to compute it we use the mean values of X and Y, $\bar{X}$ and $\bar{Y}$, and the b of Y given X, b_{YX}. To calculate the average contribution of X on the average value of Y, we must take into account whether b is plus or minus. We subtract the product of b times the mean of X from the mean of Y, that is, $\bar{Y} - (b_{YX} \cdot \bar{X})$. In this way we combine our knowledge of the two means for prediction purposes.

Let us trace the steps. If we predict Y with the mean of Y, we would say that each $Y_i = \bar{Y}$. However, if X and Y are *not* independent, that is, if Y depends upon X, the value of Y ought to be estimated more accurately by knowledge of X. Thus we can calculate the average contribution that X makes to the average value of Y. We start with the value of $\bar{Y}$ and subtract from it the value of $\bar{X}$ times the constant amount of change in Y given X, b_{YX}. This gives us $\bar{Y} - (b_{YX} \cdot \bar{X})$, which is a constant value indicating the average contribution of X to the average value of Y. This average *constant* is called the a of Y given X, $a_{Y \cdot X}$. It can be expressed symbolically:

$$a_{Y \cdot X} = \bar{Y} - (b_{YX} \cdot \bar{X})$$

We know all the terms from earlier in the chapter, so let us substitute the appropriate values. Thus,

$$a_{Y \cdot X} = \bar{Y} - (b_{YX} \cdot \bar{X})$$
$$a_{Y \cdot X} = 2.25 - (0.34 \cdot 4.5) = 2.25 - 1.53$$
$$a_{Y \cdot X} = 0.72$$

It may not be obvious at first glance, but this formula results in an average constant value that does *not* enable us to use knowledge of *specific* X scores but rather of the mean value $\bar{X}$. The above calculations are therefore helpful insofar as they provide us with the first type of constant value. However, what we also need is the *b* that enables us to calculate the *unique* value for each Y_i. The calculations for the *a*'s are shown in Table 7.3.3.

Table 7.3.3.   Calculations for zero-order a
coefficients for variables Y and X

$$a_{Y \cdot X} = \bar{Y} - b_{YX} \cdot \bar{X} = 2.25 - (0.34 \cdot 4.50) = 2.25 - 1.53 = 0.72$$
$$a_{X \cdot Y} = \bar{X} - b_{XY} \cdot \bar{Y} = 4.50 - (1.38 \cdot 2.25) = 4.50 - 3.10 = 1.40$$

7.4 TWO-VARIABLE LINEAR REGRESSION

We are finally ready to tie all this together, but first let us briefly trace our steps through this chapter. We began by observing a sample of respondents and measuring them according to a set of operationally defined variables. Once we had accomplished this, we tried to summarize and describe their singular and joint characteristics considering the variables in our set. When we had gathered sufficient information, it became possible to "predict" the influence of one characteristic or variable upon another. This form of prediction bears close correspondence to the way we predict on the everyday level of common sense. Our problem now is to develop a method of prediction that takes into account as much information about two variables as is possible.[2] To do this we must compromise between "perfect prediction" and "average prediction." Two-variable linear regression provides us with an equation that does just that.

In Chapter 4, we used the symbol $\bar{Y}$ to represent the mean, which, as we came to learn in Section 5.3, can also be thought of as a predicted value of Y given information of itself alone. We can now use the symbol $\hat{Y}$ (called Y predicted) to represent the predicted value of Y_i given information about X_i. Combining the regression coefficients we just learned, we can express the two-variable-linear-regression equation symbolically:

$$\hat{Y} = a_{Y \cdot X} + (b_{YX} \cdot X_i)$$

There are (at least) two reasons why linear equations are so useful from an analytical viewpoint. In the first place, linear equations often approximate relationships that we find in actual research. Secondly, linear equations

[2]There are many versions of the general linear model. The present one is based primarily on James Fennessey, "The General Linear Model: A New Perspective on Some Familiar Topics," *American Journal of Sociology*, 74 (July 1968), pp. 1–27; Blalock, *Social Statistics;* and Robert A. Bottenberg and Joe H. Ward, Jr., *Applied Multiple Linear Regression.* 6570th Personnel Research Laboratory, Technical Documentary Report PRL-TDR-63-6 (Lackland Air Force Base, Tex.: March 1963).

enable us to summarize and describe relationships that are difficult to get at with other analytical procedures.

In Chapters 4 and 5, we learned that the principle of least squares enables the mean and variance to be very useful statistics. The same principle can be restated in terms of the linear regression predicted values $\hat{Y}$. The least-squares criterion is that the sum of squares of the differences of the observed Y values (Y_i) from the predicted Y values ($\hat{Y}$) is a minimum. Thus the sum of squares is the lowest possible sum of any set of estimated values produced by a linear combination of X and Y, that is,

$$\sum_{i=1}^{N} (Y_i - \hat{Y})^2 = \text{a minimum}$$

One of the reasons we refer to this kind of analysis as "linear regression" is that when we plot the predicted values of our respondents on Y *regressed* on X on graph paper according to the regression equation, each predicted score will fall on a *line* that meets the least-squares criteria. Therefore, the estimated values $\hat{Y}$ comprise a unique linear combination, which minimizes the sum of squares of the vertical deviations when plotted on a graph. In the present case it will be a straight line.

We can demonstrate how this comes about by drawing a graph that represents any two of the variables in which we are interested. On this graph we can plot a series of points, each of which represents an individual's actual joint score on the two variables. For example, consider two of the students in the group we have been studying. Figure 7.4.1 shows that Helen spends six hours studying each week and has a grade point average of 2.0. Her joint score is the unique point where the value 6 on the X scale intersects with the value 2.0 on the Y scale. Similarly, Roland can be located at $X = 3$ and $Y = 1.0$.

If we locate all students according to their joint locations on the two variables, we get a *scattergram*, which displays all the data according to X and Y. As a practical hint when carrying out research, before you compute values for $\hat{Y}$, prepare a scattergram of your data to get a general picture of what is "going on" between the variables. Usually you can detect the general form of the relationship, possibly eliminating unnecessary work.

Figure 7.4.2 is a scattergram of the eight students. The wavy general direction arrows reveal a roughly linear relationship, with those who study more hours each week tending to have higher grade point averages. There is the obvious exception of Alan, who studies only four hours but has a 4.0 average, i.e., $X = 4$, $Y = 4.0$. Even so, the wavy arrows show that the general flow is upward and to the right, indicating a *positive* relationship.

We want to come up with predicted Y values that will enable us to plot a *unique set* of values and that become a straight line summarizing and de-

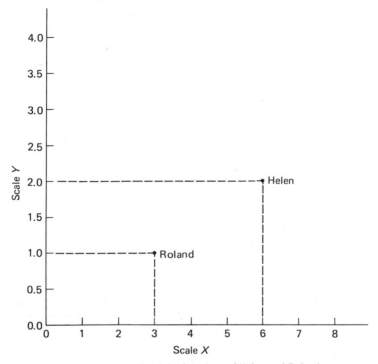

Figure 7.4.1. The joint locations of Helen and Roland.

scribing the students. This set of values is similar to our use of the mean $\bar{Y}$ as a unique (single) value to summarize and describe the same students. Think of it this way: We can represent the mean with a single point $\bar{Y}$. We could also represent it as a line, with each point on the line being equal to $\bar{Y}$, as was shown in Figure 7.2.1.

Now, however, we want to plot a line to represent the unique set of values defined by $\hat{Y}$. In section 7.3, we calculated the regression coefficients for X and Y, $a_{Y.X}$ and b_{YX}. Thus,

$$a_{Y.X} = \bar{Y} - (b_{YX} \cdot \bar{X})$$

$$a_{Y.X} = 0.72$$

and

$$b_{YX} = \frac{s_{XY}}{s_X^2}$$

$$b_{YX} = 0.34$$

Using a and b, we will use the equation to predict the values for Y given the values of X from 0 to 8. The predicted values can be seen in Table 7.4.1.

153

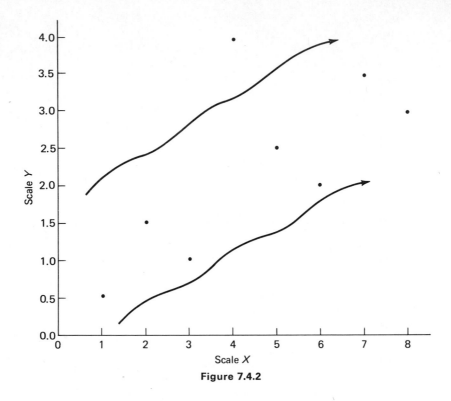

Figure 7.4.2

Table 7.4.1. Estimated values of Y, Y_i, for specific values of X, X_i, given that $a_{Y \cdot X} = 0.72$ and $b_{YX} \cdot X_i = 0.34$

$$\hat{Y} = a_{Y \cdot X} + b_{YX} \cdot X_i$$

if $X_i = 0$, then $\hat{Y}_i = 0.72 + (0.34 \cdot 0) = 0.72$
if $X_i = 1$, then $\hat{Y}_i = 0.72 + (0.34 \cdot 1) = 1.06$
if $X_i = 2$, then $\hat{Y}_i = 0.72 + (0.34 \cdot 2) = 1.40$
if $X_i = 3$, then $\hat{Y}_i = 0.72 + (0.34 \cdot 3) = 1.74$
if $X_i = 4$, then $\hat{Y}_i = 0.72 + (0.34 \cdot 4) = 2.08$
if $X_i = 5$, then $\hat{Y}_i = 0.72 + (0.34 \cdot 5) = 2.42$
if $X_i = 6$, then $\hat{Y}_i = 0.72 + (0.34 \cdot 6) = 2.76$
if $X_i = 7$, then $\hat{Y}_i = 0.72 + (0.34 \cdot 7) = 3.10$
if $X_i = 8$, then $\hat{Y}_i = 0.72 + (0.34 \cdot 8) = 3.44$

These values for X may be hypothetical (no student in our sample studies zero hours) or observed (the eight respondents' hours of study). These are *predicted* values and are not necessarily the same Y values as the eight respondents actually possess, just as none of them possesses the exact mean value.

Since $a_{Y \cdot X}$ is where $X = 0$, the coefficient $a_{Y \cdot X} = 0.72$ indicates that we start our linear slope on the Y axis at the grade point average 0.72. This value tells us that, based on these eight respondents, we would predict that

154

a student who studies not at all would have a grade point of 0.72. Of course, this grade point could be the result of a combination of native intelligence, family background, previous educational training, or any of many other unmeasured variables. It is important to remember that these variables exist in reality, but they are not part of our system of measured variables. *a* may be important in clarifying what minimal value to predict for variable Y given zero for variable X.

The coefficient $b_{yx} = 0.34$ tells us to plot the linear slope by multiplying the unique values of X_i by 0.34 and marking the point. These points are identical to the predicted values for variable Y as indicated by $\hat{Y}$. We can carry out the steps of the equation for our example and summarize them by plotting the linear regression line $\hat{Y} = a_{Y \cdot x} + b_{yx}X_i$. The results are shown in Figure 7.4.3.

If you were to guess each student's grade point average Y, based on the number of hours he or she spends studying weekly X, you would guess that a student who studies 1 hour per week ($X = 1$) would have a grade point average $\hat{Y} = 1.06$, one who studies 2 hours per week ($X = 2$) would have a grade point average $\hat{Y} = 1.40$, and so forth. These estimated values fall on

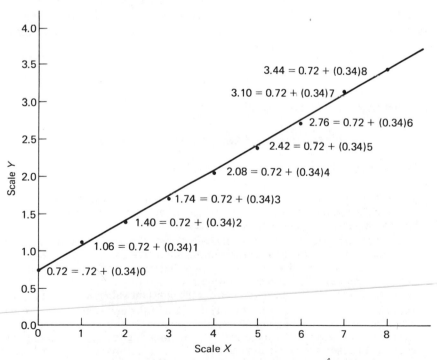

Figure 7.4.3.   Estimated values of Y given $X, \hat{Y}$.

the unique straight line which meets the least-squares criterion. That is, if we subtract from each observed value Y_i, the estimated value $\hat{Y}$, square the difference, and sum the squares, we would find the unique sum of squares which is less than any other sum of squares with any other straight line. If instead of guessing the mean grade point average $\bar{Y}$, you use knowledge of hours spent studying X, and guess the values along that straight line $\hat{Y} = a_{Y \cdot X} + b_{YX} \cdot X_i$, you would have improved your guesses over the mean. Throughout, our efforts are to minimize our errors in prediction.

The formula for $\hat{Y}$ enables us to use the constant comparative values by estimating Y_i using the means of Y and X and the unique value of X_i. However, since Y seldom will be perfectly estimated from X, we also need a correction factor that enables us to take into account our error in estimation. The correction factor is called an *error term*, and it is a function of how much off our predicted value $\hat{Y}$ is from the actual value Y_i. It can be expressed symbolically as:

$$e_{Y \cdot X_i} = Y_i - \hat{Y}$$

Earlier we predicted values for eight students using the two-variable linear-regression coefficients. The term for each individual is the difference from the predicted score, or the regression line to the actual score. We can put the three terms together in one equation, in which we can calculate Y_i values exactly by adding each person's predicted value with that person's error term. This can be expressed symbolically:

$$Y_i = \hat{Y} + e_{Y \cdot X_i}$$

or, in terms of our linear regression formula:

$$Y_i = \hat{Y} + e_{Y \cdot X_i} = a_{Y \cdot X} + (b_{YX} \cdot X_i) + (e_{Y \cdot X_i}).$$

Linear regression is based on four straightforward assumptions:

Assumption one: The regression coefficients are *constant* values; that is, $a_{Y \cdot X}$ and b_{YX} do not vary, and the values of variables X and Y are randomly distributed. This enables us to estimate various Y values for various levels of X.

Assumption two: The relationship between X and Y may be represented by a *linear combination* of the two. Linear combinations are combinations belonging to a line (as in a family line of X's) of values that can be multiplied by the constant coefficient b_{YX}. This means that we assume that the value of X adds estimative information that can be multiplied by the constant value of b, i.e., $b_{YX} \cdot X_i$.

Assumption three: The independent variable X and the error term $e_{Y \cdot X_i}$ are *not* dependent upon each other, and the sum of squares of the error terms is a minimum. Thus we assume that the error in estimating Y given X is randomly distributed. In other words, we assume that the error in predicting Y from X is independent of the X values and that it meets the least-squares criterion.

Assumption four: The relationship between variables X and Y may be summarized by a best-fitting set of predicted values for Y because X gives us predictive information in addition to what Y gives in general. Thus we assume that the net effects of the independent variable X are additional to the general effect of Y itself. By best-fitting, we mean that *set* of scores that meets the least-squares criterion; that is, the set that minimizes the sum of the squared deviations of the observed scores less the estimated ones.

Two-variable linear regression is not limited to samples such as the one we have been using. As a matter of fact, it is much more useful as the number of respondents increases, especially if one has access to a computer. For example, Duncan, Featherman, and Duncan examine the relationship between the students' scores on selected "scholastic aptitude" and the status of the parents' occupation on Duncan's socioeconomic index of occupational status.[3] They present the following information:

		Covariance Matrix	
Mean		Y	X
100.607	Y	400	90.47
32.41	X	90.47	500.99

We can use the standard calculations for a and b. Thus,

$$b_{Y \cdot X} = \frac{s_{XY}}{s_X^2} = \frac{90.47}{500.99} = 0.1806$$

$$a_{Y \cdot X} = \bar{Y} - b_{YX}\bar{X} = 100.607 - (0.1806)(32.41) = 94.75$$

This information may be summarized in a *scatter plot*, which also displays the two-variable linear regression line

$$\hat{Y} = a_{Y \cdot X} + b_{YX} \cdot X_i = 94.75 + 0.1806 \cdot X_i$$

In one sense, the linear regressions we have seen may be thought of as sets of predicted scores that "average" the mean Y scores for respondents at various levels of X. Figure 7.4.5 demonstrates this concept as it was illustrated in the Blau and Duncan study.[4]

The one straight regression line represents an "average" slope of predicted scores. The mean Y values depending upon level of education move irregularly but somewhat imperfectly around the regression slope. In Chapter 8

[3]Otis Dudley Duncan, David L. Featherman, and Beverly Duncan, *Socioeconomic Background and Achievement* (New York: Seminar Press, 1972), pp. 79–82.

[4]Peter M. Blau and Otis Dudley Duncan, *The American Occupational Structure.* (New York: John Wiley & Sons, Inc., 1967), p. 144.

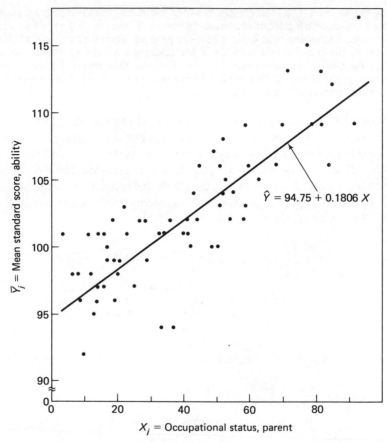

Figure 7.4.4. Scatter plot, mean standard scores of ability on occupational status of parent, Wisconsin high school seniors, 1929–1933 (after Byrns & Henmon 1936). (*From Otis Dudley Duncan, David L. Featherman, and Beverly Duncan,* Socioeconomic Background and Achievement. *New York: Seminar Press, 1972, p. 82.*)

we will examine ways to measure just how well a set of linear regression predicted scores fits the data it describes.

EXERCISES

1. In your own words, define, describe, or discuss the following terms and give an hypothetical example of each:
   linear regression
   statistical prediction

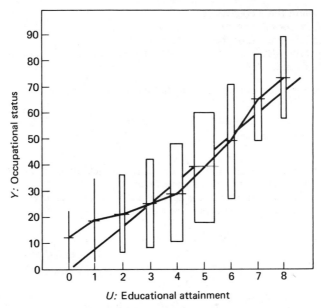

Figure 7.4.5. Mean of Y + one standard deviation for each category of educational attainment (U) with linear regression computed on the assumption of interval measurement of U, for males 20 to 64 years old with nonfarm background. (*From Peter M. Blau and Otis Dudley Duncan,* The American Occupational Structure. *New York: John Wiley & Sons, Inc., 1967, p. 144.*)

        independence
        joint observation
        constants
        covary
        matrix
        covariance
        b_{yx}
        scattergram
        best fitting equation

2. True and False Questions:
   A. The difference between the covariation and the covariance is that one uses the variance of X in the denominator and the other uses the variance of Y in the denominator.
   B. One of the assumptions of linear prediction is that the form of the relationship is curvilinear.
   C. The covariance may be expressed as a straight line.
   D. The covariance is the square root of the covariation.
   E. A positive relationship between X and Y is shown by two unique straight lines whose formulas are $\hat{Y} = a_{Y.x} + b_{YX}X$.
   F. A scattergram measures the variance of a joint distribution.
   G. Linear prediction is the same thing as linear regression.

159

3. A. Explain the a coefficient.
   B. What does the covariance indicate?
   C. Explain the use of the b coefficient.
   D. Explain what the difference between Y predicted ($\hat{Y}$) and the mean of Y ($\bar{Y}$) indicates.

4. Compute the following regression coefficients for the data presented in problem 4 in Chapter 5.
   A. The covariance of all pairs of variables.
   B. b_{YX}, b_{XY}, b_{YT}, b_{TY}
   C. $a_{Y \cdot X}$, $a_{Y \cdot T}$

5. Assume that you are conducting a research project investigating the relationship between student age (X) and academic aspiration (Y). Age is measured in years on last birthday and academic aspiration is measured by a test score on a scholastic aspiration test. The data produce the following findings:

$$
\begin{array}{ll}
N = 100 & s_X^2 = 25 \\
\bar{X} = 21 & s_Y^2 = 49 \\
\bar{Y} = 80 & s_{XY} = 28
\end{array}
$$

Calculate the following using the data above:
   A. b_{YX}
   B. b_{XY}
   C. $a_{Y \cdot X}$
   D. $a_{X \cdot Y}$

6. Assume that you have conducted a research study involving a comparison between the age of students enrolling for continuing education courses (L) and the amount of time spent studying (M). Your research produces the following findings:

$$
\begin{array}{ll}
N = 100 \\
\bar{L} = 36 \\
\bar{M} = 5 \\
s_L^2 = 9 \\
s_M^2 = 4 \\
s_{LM} = 3
\end{array}
$$

Calculate the following using the data presented above:
   A. b_{ML}
   B. b_{LM}
   C. $a_{L \cdot M}$
   D. $a_{M \cdot L}$

7. Assume that you are interested in studying the relationship between leisure time activities and academic effort. You carry out a research project which investigates this relationship by studying the amount of time spent watching TV weekly (X) and the amount of time spent studying weekly (Y).

Hours spent watching TV X	Hours spent studying Y
8	1
10	2
6	3
2	4
4	5

A. Using the data above calculate the following statistics:
1. The covariance
2. The b coefficient of Y
3. The b coefficient of X
B. What is the form of the relationship?
C. What do you conclude from the findings?

8. Your friend, an avid TV viewer, challenges the findings of your survey described in problem 7 above. To satisfy him you replicate your project with a different set of respondents and collect the following data:

Hours spent watching TV X	Hours spent studying Y
6	1
3	2
9	3
15	4
12	5

A. Calculate the following using the data above:
1. The covariance
2. The b coefficient of Y
3. The b coefficient of X
B. What is the form of the relationship?
C. What do you conclude from the findings?
D. Compare the results from the two samples.

9. Suppose you have been asked to conduct a research project to help uncover the reasons why some students do poorly in college. Based on a review of the literature, you decide to examine the influence of TV on the academic performance of students. TV influence is measured by an interval scale representing the number of hours each student watches television daily. Academic performance is measured by an interval scale representing the number of D's and F's received last semester by each student. A sample of ten respondents is

drawn from the student body. Using the following information, answer the statements below.

Name	Hours Spent Watching TV (X)	Number of D's and F's (Y)
Arthur	5	4
Betty	2	1
Clifton	3	3
Elsworth	4	5
Georgette	1	2
Harry	3	3
Ignatz	2	3
Monroe	1	1
Sally	3	3
Wilma	4	4

A. Draw a scattergram of the data on graph paper.
B. What are the values of a and b in the regression equation?
C. Plot the regression slope on the graph.

10. A recreation director was asked to submit an estimate of pool users for the following summer. Sampling six days for the two previous years, he got the following number of pool users:

X = year 1	Y = year 2
70	60
260	320
150	230
100	120
20	50
60	60

Calculate the least-squares line showing how year 1's number of pool users (X) is related to year 2's (Y).

8

Explanation and Correlation with Two Interval Variables

In this chapter we discuss the general concepts of explanation and correlation. To do so we deal with the familiar total variance, which is described as being comprised of two as-yet-unmentioned basic elements, the explained and the unexplained variance. Our focus is primarily on our predictive improvement, using linear combinations to estimate the values of the dependent variable Y. We also discuss the general idea of variables varying together in terms of "how well" our linear predictions fit the data. Technically this is called correlation. We will learn not just the general idea of correlation, but how to calculate it as well.

This chapter is a bridge between the statistics we have been discussing until now and their uses and applications, which we will be treating throughout the rest of this book. We need a bridge rather than a springboard between the two for several reasons; some of these reasons are related to the issues discussed in this chapter. A thorough grasp of the material in this chapter is essential in understanding those that follow. So read carefully, *thinking it through*.

8.1 CAUSATION, PREDICTION, AND THE CONCEPT OF EXPLANATION

Earlier we learned that the statistical version of "cause and effect," in the framework of social research, means that we treat one variable as being "dependent" upon one or more others, which are "independent" of it. In other words, the independent variables influence the dependent variable. Then we learned that it is common to list percentages in the "direction" of causal influence. This makes the most sense when we have variables that can be arranged in a coherent temporal order, so that some precede others. In such temporal sequencing, the preceding, or antecedent, variables are said to "cause" (meaning "influence") the succeeding, or consequent, variables. Later we learned that statistical independence may be interpreted as demonstrating no association between variables. In other words, if two variables are independent, then we can say that they are not "related" to each other.

Throughout these early discussions, we focused on the association or relationship between variables, leading up to the idea that if one variable "causes" another, then knowledge of the "cause" ought to help us predict the "effect." Most recently, we learned that if there is a linear relationship, then information about the values of an independent variable can help us predict the values of a dependent variable.

Another way of looking at this is to say that values on the independent variable help "account for" values on the dependent variable. For example, in the study *Voting*, an index of socioeconomic status is used as an independent variable, and the final two-party vote decision according to the percentage Republican is used as a dependent variable.[1] It was found that 65 percent of the 641 people analyzed voted Republican, so the chances that a person selected at random would have voted Republican would be 65 out of 100. When the two variables are cross-tabulated, we can see some sizable differences in terms of the percentage of Republican votes. Figure 8.1.1 clearly shows that upper status people vote Republican more than lower status people. In this instance we can say that social status to some extent "accounts for" the way people vote. Thus we have a link between association and "accounts for-ability." In other words, since social status to some extent accounts for voting decisions, if we know someone's social status level, then we are *better able* to explain that person's final vote decision than if we do not know his social status level. As the authors of *Voting* put it "The higher the socioeconomic status (SES), the more Republican the vote; put crudely, richer people vote Republican more than poorer people."[2]

[1] Bernard R. Berelson, Paul F. Lazarsfeld, and William N. McPhee, *Voting*. (Chicago: The University of Chicago Press, 1954).
[2] *Ibid.*, p. 56.

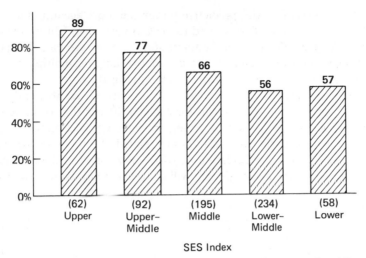

Figure 8.1.1. Socioeconomic status index differentiates vote: Percentage Republican of two-party vote. (*From Bernard R. Berelson, Paul F. Lazarsfeld, and William N. McPhee,* Voting. *Chicago: The University of Chicago Press, 1954, p. 55.*)

At the most general level, when we use the words "accounts for" we mean in terms of the variables and their values. The word *explains* is analogous with "accounts for," but its meaning extends to the clarification and elaboration of the covariation and relationship between more than two variables as well as to the more elementary aspect of clarifying the extent of variation in one variable given another. For present purposes we confine our attention to the specific issue of explanation in terms of variance. Later on we will learn about explanation in terms of multivariate analysis. The most important thing for now is to realize that we are using the word "explain" in the same common-sense way that we have been using such words as "cause," "direction," and "account for." Even though we use such words, it also is crucial to be aware that in no way do our statistics *prove* causation, prediction, or explanation. Instead, these terms are meant to help us describe mathematical results verbally.

The concepts of prediction and explanation are often used interchangeably; however, there is a major difference between the meanings of the two terms. Our interest in prediction is in determining which independent variables project the most accurate estimate of the value of the dependent variable. Our interest in explanation is in determining which independent variable(s) best clarifies variability in the dependent variable. For example, as regards the *Voting* study, if we wanted to predict vote, we might decide to use the earlier mentioned objective index of socioeconomic status *plus* a subjective measure of the respondent's own status identification *plus* ethnic status on the grounds that the more variables with predictive power the more accurate the prediction. On the other hand, if we wanted to *explain* vote, we might decide to use only the objective index of socioeconomic status on the grounds that (1) it is the best single indicator, and (2) the use of overlapping multiple variables creates a sharing of the explanatory power among them.

Until now our estimates have helped us summarize, describe, and predict the general form of the relationship for a given distribution of paired (two variables) observations. As yet, however, we have not measured the extent to which or *how well* the two variables are related. To be more specific, we do not know the combined relation or *correlation* between them. What we want to do now is find a statistic that will enable us to measure how strongly

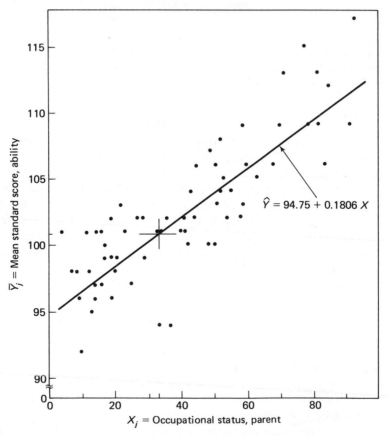

Figure 8.1.2. Scatter plot, mean standard scores of ability on occupational status of parent, Wisconsin high school seniors, 1929–1933 (after Byrns & Henmon 1936). (*From Otis Dudley Duncan, David L. Featherman, and Beverly Duncan,* Socioeconomic Background and Achievement. *New York: Seminar Press, 1972, p. 82.*)

the two variables are correlated to each other; that is, we want to know the strength or degree of the relationship.

For example, in Section 7.4, we saw Figure 7.4.4 presenting the linear regression of mean student ability $(\bar{Y})$ on parents' occupational status (X) for the study in *Socioeconomic Background and Achievement*. Rather than considering the slope, however, we now are concerned with *how closely clustered* the respondents are given their joint location on the two variables, as shown in Figure 8.1.2. Put differently, we want to know the interdependence between variables.

8.2 THE PRODUCT MOMENT CORRELATION COEFFICIENT

A *correlation coefficient* is a measure of the interdependence between two quantitative variables. In order to measure such interdependence, we can examine how the actual covariance between X and Y compares to the maximum possible covariance between the two.

Reviewing the terms, you will recall that the variance in X is equivalent to squaring the standard deviation of X, that is, $s_X^2 = s_X s_X$, and the variance in Y is composed of the square of the standard deviation in Y, that is $s_Y^2 = s_Y s_Y$. Since each variance equals the total variance in the particular variable, the product of the two standard deviations will equal the maximum possible total covariance between the two, that is, $s_X s_Y$. Also, recall that the covariance is the mean sum of the product of the differences and is composed of the error on Y, $(Y_i - \bar{Y})$, and the error on X, $(X_i - \bar{X})$, for each individual. We can represent the means of Y by a single horizontal line drawn across a scattergram at $\bar{Y}$ on the Y axis, and the mean of X by a single vertical line drawn up a scattergram at $\bar{X}$ on the X axis, as we have done in Figure 8.2.1.

The covariance is a measure that deals with the actual or observed location of each individual on the two variables simultaneously. Figure 8.2.2 portrays graphically how this happens for two of the eight students by showing the two differences for Betty $(X_3 - \bar{X}) = 2.5$ and $(Y_3 - \bar{Y}) = 1.25$, and the two differences for Arthur $(X_2 - \bar{X}) = 2.5$ and $(Y_2 - \bar{Y}) = -0.75$. The covariation of each is the product of these two differences from the means. Each respondent can be similarly located on a scattergram.

The connection between the variances and the covariances is that each one measures how much each individual deviates from the means of the variables. When we calculate the coefficient $b_{YX} = s_{XY}/s_X^2$, we compare the covariance of X and Y to the variance in X, the independent variable. Thus we would be predicting Y *from* X. Also, when we calculate the coefficient $b_{XY} = s_{XY}/s_Y^2$, we compare the same covariance of X and Y to the variance in Y, the dependent variable, but in this instance we would be interested in predicting X *from* Y.

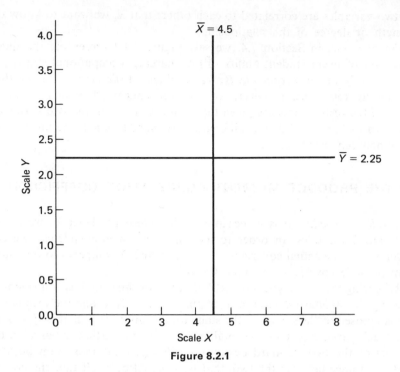

Figure 8.2.1

We want to know the ratio of the actual covariance of X and Y to the total possible covariance of X and Y. Thus we put the covariance of X and Y in the numerator and the product of the standard deviation of X times the standard deviation of Y in the denominator. This measure is called the *correlation coefficient* and is designated by the letter r.[3] It can be expressed symbolically as:

$$r_{XY} = \frac{s_{XY}}{s_X s_Y}$$

A simple comparison shows that $r_{YX} = s_{XY}/s_X s_Y = s_{YX}/s_Y s_X = r_{XY}$. Thus, unlike the regression coefficient b_{YX}, when we calculate the correlation coefficient $r_{YX'}$ we do not have to calculate r_{XY}, for the two are identical. Unlike

[3]It can be expressed in terms of differences:

$$r_{XY} = \frac{\sum\limits_{i=1}^{N} (X_i - \bar{X}) \cdot (Y_i - \bar{Y})/N}{\sqrt{\dfrac{\sum\limits_{i=1}^{N} (X_i - \bar{X})^2}{N}} \sqrt{\dfrac{\sum\limits_{i=1}^{N} (Y_i - \bar{Y})^2}{N}}}$$

168

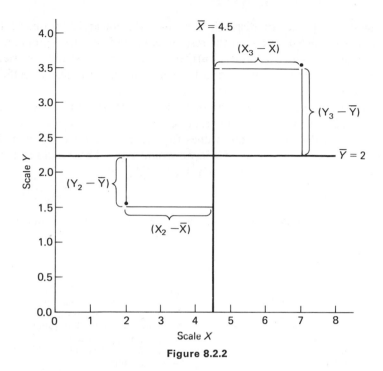

Figure 8.2.2

the subscripts of the regression coefficient, the order of subscripts is not important for the correlation coefficient.

This generalized version of the correlation coefficient is applicable to many situations. When we have interval data, the correlation coefficient is called Pearson's Product Moment, r_{YX}.[4] The correlation coefficient r_{XY} has characteristics that help make it analytically useful. In the first place, the limits of the coefficient are $+1.00$ and -1.00. The nearer the coefficient is to ± 1.00, the stronger is the relationship. Similarly, the nearer the coefficient is to 0.00, the weaker is the relationship.

Secondly, when the data are interval variables, and r_{YX} is a measure of the linear relationship between X and Y, it measures how well a single straight line fits the points of a scattergram. If all the points fall exactly on the line, r_{YX} will equal ± 1.00, and there is a perfect relationship. If all the points are randomly distributed, r_{YX} will equal 0.00, and there is no relationship.

It is important to understand that when we find the ratio of the covariance in X and Y to the total possible covariance between X and Y, we can measure

---

[4]The general correlation coefficient may also be applied to ordinal data, in which case it becomes Spearman's rank correlation coefficient r or Kendall's Tau correlation coefficient T. For a full discussion see Roland K. Hawkes, "The Multivariate Analysis of Ordinal Measures," in the *American Journal of Sociology*, 76 (March 1971), pp. 908–926.

the amount of spread, or dispersion, around the linear least-squares line. That is, we can measure how far the respondents are from the regression line formed by our estimated $\hat{Y}$ values. Furthermore, since we use the covariance, the ratio can be either $+$ or $-$, and the value of the correlation coefficient can be either positive or negative.

We find the conventions set forth in Table 8.2.1 to be helpful in describing correlation coefficient values. This table delineates which verbal combinations we use to describe various values for the correlation coefficient. For example, for a coefficient value of 0.56 we would say: "There is a large positive relationship ($r_{YX} = 0.56$) between X and Y."

Table 8.2.1   Conventions for describing correlation coefficient values

Value of the Correlation Coefficient	Appropriate Phrase
+1.00	A perfect positive relationship
+0.66 to +0.99	A very large positive relationship
+0.36 to +0.65	A large positive relationship
+0.16 to +0.35	A medium positive relationship
+0.06 to +0.15	A small positive relationship
+0.01 to +0.05	A negligible positive relationship
0.00	No relationship
−0.01 to −0.05	A negligible negative relationship
−0.06 to −0.15	A small negative relationship
−0.16 to −0.35	A medium negative relationship
−0.36 to −0.65	A large negative relationship
−0.66 to −0.99	A very large negative relationship
−1.00	A perfect negative relationship

Source: This table is adapted from James A. Davis, *Elementary Survey Analysis* (Englewood Cliffs, New Jersey: Prentice-Hall, Inc., 1971), p. 49; and Jacob Cohen, *Statistical Power Analysis for the Behavioral Sciences* (New York: Academic Press, 1969) pp. 72–104.

Since regression and correlation coefficients can be positive or negative, it is helpful to visualize the way they might look. Figure 8.2.3 shows five regression slopes and scattergrams for various combinations of relationships. It would be possible to calculate the exact coefficients if we had enough information; however, this diagram merely demonstrates the general ideas of regression and correlation.

A specific example of the calculation of the correlation coefficient clarifies this discussion. Duncan, Featherman, and Duncan show the regression of the mean score for a classification test (AGCT) on one's previous occupa-

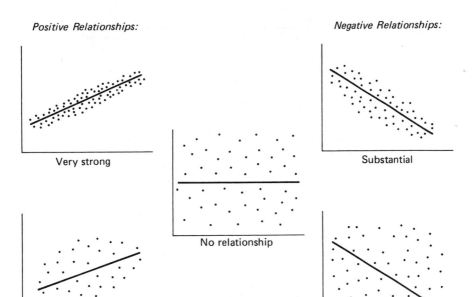

Positive Relationships:

Very strong

Moderate

No relationship

Negative Relationships:

Substantial

Low

Figure 8.2.3

tional status.[5] They present the following data and information in Figure 8.2.4.

$$\bar{Y} = 106.6 \qquad s_Y = 19.1 \qquad s_Y^2 = 364.8$$
$$\bar{X} = 31.8 \qquad s_X = 19.2 \qquad s_X^2 = 368.6$$
$$s_{XY} = 155.5$$

We can calculate the needed coefficients by substituting the appropriate values in our formula. Thus,

$$b_{YX} = \frac{s_{XY}}{s_X^2} = \frac{155.5}{368.6} = 0.4218$$

$$b_{XY} = \frac{s_{XY}}{s_Y^2} = \frac{155.5}{364.8} = 0.4262$$

$$r_{XY} = \frac{s_{XY}}{s_X s_Y} = \frac{155.5}{(19.2)(19.1)} - 0.4241$$

[5]Otis Dudley Duncan, David L. Featherman and Beverly Duncan, *Socioeconomic Background and Achievement*. (New York: Seminar Press, 1972), p. 86.

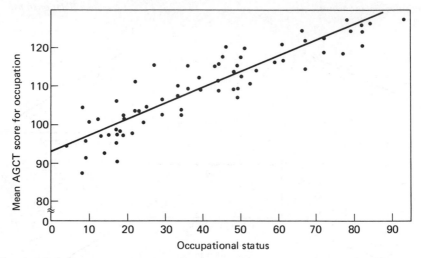

Figure 8.2.4. Regression of AGCT score on previous civilian occupational status, as measured by scores on Duncan (1961a) scale. (*From Otis Dudley Duncan, David L. Featherman, and Beverly Duncan,* Socioeconomic Background and Achievement. *New York: Seminar Press, 1972, p. 86.*)

We can say: "There is a large positive relationship between previous occupational status (variable X) and mean AGCT score (variable Y) ($r_{XY} = +0.42$)."

Let us return to the eight students and calculate the correlation coefficient for hours spent studying (X) and grade point average (Y). We know that the covariance of X and Y equals $s_{XY} = 1.81$. We also know that the standard deviations of X and Y are the square roots of the variances in X and Y. Thus,

$$s_X = \sqrt{s_X^2} = \sqrt{5.25} = 2.29$$

and

$$s_Y = \sqrt{s_Y^2} = \sqrt{1.31} = 1.14$$

Now we can substitute these values into the formula for r_{YX} and compute its value:

$$r_{XY} = \frac{s_{XY}}{s_X s_Y} = \frac{1.81}{(2.29) \cdot (1.14)} = \frac{1.81}{2.61} = +0.69$$

We can now use this r_{YX} value to put together a verbal statement about our findings: "There is a very large positive relationship between the number of hours spent studying (variable X) and a student's grade point average (variable Y) ($r_{XY} = +0.69$)."

172

8.3 EXPLAINED AND UNEXPLAINED VARIANCE

Correlation coefficients can be interpreted in terms of a special type of explanation called "explained variance." *Explained* and *unexplained variance* are two parts of the total variance. Thus,

total variance = explained variance + unexplained variance

Understanding what this general formula means is of the utmost importance in grasping much of what we are going to do throughout this book. So let us start out by examining what we mean by the total variance, and then move to its two parts.

Each variable has a mean and a total variance, and the total variance exists for one variable only, that is, s_Y^2, s_X^2, s_T^2, and so forth. The total variance of any variable is made up of the mean sum of the squared differences. The formula for this can be stated in several ways. Thus,

$$\text{var} = s_Y^2 = \frac{\sum_{i=1}^{N} (Y_i - \bar{Y})^2}{N} = \frac{\text{total variation}}{N} = \frac{\text{Total SS}}{N}$$

The sum of the squared differences is the *total variation*, and it becomes the numerator of the variance. The total variation also is called the *total sum of squares*, Total SS. When we divide up this total sum of squares evenly among all the respondents N, we get the *total variance*.

In this sense the total variance measures how close each score is to the arithmetic mean, and we learned earlier that the squared difference between Y_i and $\bar{Y}$, $(Y_i - \bar{Y})^2$ is a measurement of error. Clearly, as an error term, the total variance tells us something about how wrong we are in predicting or estimating using the arithmetic mean.

Instead of going directly to the matters of explained and unexplained variance, let us pause to consider the *total covariance*. The total covariance is calculated similarly to the total variance, and, technically, the covariance is the mean sum of the products of the differences. It can be expressed symbolically:

$$\text{cov} = s_{YX} = \frac{\sum_{i=1}^{N} (Y_i - \bar{Y})(X_i - \bar{X})}{N} = \frac{\text{total covariation}}{N}$$

$$= \frac{\text{total cross products}}{N}$$

Clearly, the covariance is composed of the error on Y, $(Y_i - \bar{Y})$, and the error on X, $(X_i - \bar{X})$, for each individual. When we multiply the two errors or differences together and sum the products, we get the total actual covariation between X and Y for all the respondents, and when we divide this sum by N, we get the total covariance.

Remember, in Chapter 7 we learned that we can use the covariances and the variances for a set of variables and can calculate estimates by use of linear regression. The formula for linear regression enables us to make a set of predictions with values *other than* the mean. Namely, instead of the mean, we use $\hat{Y}$. Now, if we compare $\hat{Y}$ to the mean $\bar{Y}$, we can see the amount of improvement we have made in predicting by using the linear regression score for each level of X instead of the mean. All of this meshes some familiar measures in a new and useful way.

To begin with, our interest is in the total variance and the mean of the dependent variable Y, that is, s_Y^2 and $\bar{Y}$. As an example let us consider the scores of the eight students we plotted in Chapter 7. We can represent the mean by a single horizontal line drawn across a scattergram at $\bar{Y}$ on the Y axis. Then we can measure the vertical distance between each individual Y_t and the mean line, and square it, as can be seen in Figure 8.3.1. By summing

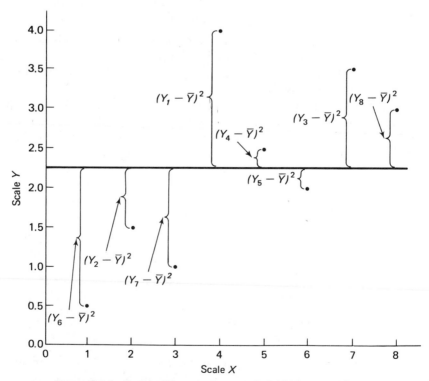

Figure 8.3.1 Square differences between individual score and mean.

all the squared differences, we obtain the total variation of Y, that is, $s_Y^2 = \sum_{i=1}^{N} (Y_i - \bar{Y})^2$.

Now we can represent the estimated $\hat{Y}$ values with the regression line of Y on X, which gives us a diagonal line across the scattergram, in the form of $\hat{Y} = a_{Y \cdot X} + (b_{YX} \cdot X_i)$. If there is a linear relationship between X and Y, then the regression values $\hat{Y}$ provide a more accurate description of the data than does the mean $\bar{Y}$, thereby improving our prediction. By use of subtraction we can calculate how much we have improved our predictions by using $\hat{Y}$ instead of $\bar{Y}$.

Instead of describing the steps theoretically, let us see how things work if we carry them out for three of the eight students. Suppose we look at the grade point average of, first, Betty (Y_3) and Maxine (Y_6) and second, Alan (Y_1) and Helen (Y_5). Betty's grade point average is $Y_3 = 3.5$, Maxine's is $Y_6 = 0.5$, Alan's is $Y_1 = 4.0$, and Helen's is $Y_5 = 2.0$. To calculate their contributions to the total variation, that is, the total sum of squares, Total SS $= \sum_{i=1}^{N} (Y_i - \bar{Y})^2$, we subtract the mean grade point average from each of their individual grade point averages, thus,

for Betty

$$(Y_3 - \bar{Y})^2 = (3.5 - 2.25)^2 = (1.25)^2 = 1.56$$

for Maxine

$$(Y_6 - \bar{Y})^2 = (0.5 - 2.25)^2 = (-1.75)^2 = 3.06$$

for Alan

$$(Y_1 - \bar{Y})^2 = (4.0 - 2.25)^2 = (1.75)^2 = 3.06$$

and for Helen

$$(Y_5 - \bar{Y})^2 = (2.0 - 2.25)^2 = (-0.25) = 0.06$$

Betty studies an average of 7 hours per week, thus $X_3 = 7$, and Maxine studies an average of 1 hour per week, thus $X_6 = 1$. Let us put this information together and predict the two grade point averages by using the least-squares equation

$$\hat{Y} = a_{Y \cdot X} + (b_{YX} \cdot X_i)$$

For Betty, $\hat{Y} = 0.72 + (0.34) \cdot 7 = 3.10$; therefore, given the knowledge that she studies 7 hours weekly, we would predict her grade point to be 3.10. For Maxine, $\hat{Y} = 0.72 + (0.34) \cdot 1 = 1.06$; therefore, we would predict her grade point to be 1.06.

Alan studies an average of 4 hour per week, thus $X_1 = 4$, and Helen studies an average of 6 hours per week, thus, $X_5 = 6$. We now predict Y

from this information using the least-squares equation. For Alan, $\hat{Y} = 0.72 + (0.34)\cdot 4 = 2.08$; therefore, given the knowledge that he studies 4 hours weekly, we would predict his grade point average to be 2.08. For Helen, $\hat{Y} = 0.72 + (0.34)\cdot 6 = 2.76$; therefore, given the knowledge that she studies 6 hours weekly, we would predict her grade point average to be 2.76. Without knowledge of X, we would predict all grade point averages to be 2.25, the mean of Y.

Figure 8.3.2 is a scattergram that portrays the mean line and the regression line for the eight students. Figure 8.3.2 shows graphically and Table 8.3.1 shows algebraically that the deviations of the actual scores of Alan Y_1, Betty Y_3, Helen Y_5, and Maxine Y_6 from the mean line $(Y_i - \bar{Y})$ may be thought of as the sum of two parts, the first being the deviation of their predicted regression score $\hat{Y}$ from the mean $\bar{Y}$, i.e. $(\hat{Y} - \bar{Y})$, and the second being the deviation of their actual score from their predicted regression score $(Y_i - \hat{Y})$. Thus,

$$(Y_i - \bar{Y}) = (\hat{Y} - \bar{Y}) + (Y_i - \hat{Y})$$

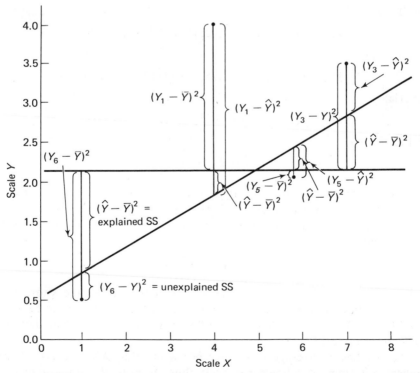

Figure 8.3.2. Differences between individual score and mean and individual score, predicted score, and mean.

Table 8.3.1 Algebraic deviations

$$(Y_i - \bar{Y}) = (\hat{Y} - \bar{Y}) + (Y_i - \hat{Y})$$

Betty	$(3.5 - 2.25) = (3.10 - 2.25) + (3.5 - 3.10)$ $(1.25) \quad = \quad (0.85) \quad + \quad (0.40)$ $1.25 \quad = \quad 1.25$
Maxine	$(0.5 - 2.25) = (1.06 - 2.25) + (0.5 - 1.06)$ $(-1.75) \quad = \quad (-1.19) \quad + \quad (-0.56)$ $-1.75 \quad = \quad -1.75$
Alan	$(4.0 - 2.25) = (2.08 - 2.25) + (4.0 - 2.08)$ $(1.75) \quad = \quad (-0.17) \quad + \quad (1.92)$ $1.75 \quad = \quad 1.75$
Helen	$(2.0 - 2.25) = (2.76 - 2.25) + (2.0 - 2.76)$ $(-0.25) \quad = \quad (0.51) \quad + \quad (-0.76)$ $-0.25 \quad = \quad -0.25$

Of course, we can multiply each of these deviations by themselves, producing squared deviations. You may recognize the first term as the total variation or Total SS and that which appears in the numerator of the variance of Y. The second term has not yet been described, but it is a logical extension of our previous knowledge. When the mean of Y, $\bar{Y}$, is subtracted from the predicted value of Y given X, $\hat{Y}$, as in the formula, it may be thought of as that part of the total variation in Y that is accounted for or explained by its linear relationship with X. It also ought to provide us with the improvement in predicting Y from X over predicting Y from its own mean. The second part may be thought of as the unexplained variation in Y, in other words, the *linear regression error*. Thus,

$(Y_i - \bar{Y})^2$ = the variation of individual i's score on Y from the mean $\bar{Y}$

$(\hat{Y} - \bar{Y})^2$ = (the squared difference between i's estimated $\hat{Y}$ score given the score on X and the mean $\bar{Y}$) = (the variation of i's score on Y which *is* accounted for by its linear relationship with X) = (the improvement in estimating Y given X over estimating with $\bar{Y}$). Called the explained variance.

$(Y_i - \hat{Y})^2$ = (the squared difference between i's actual score on Y and the estimated $\hat{Y}$ score given the score on X) = (the variation of i's score on Y which is *not* accounted for by its linear relationship with X) = (the error in estimating Y given X). Called the unexplained variance.

There are six main points and/or steps that concern us. First, the mean $\bar{Y}$ is the best unique value to describe the distribution on variable Y according to the least-squares principle. Second, we can represent the mean with a horizontal line. Third, if we know the value of X, and if the relationship between it and Y is linear, we can estimate the values of Y from the values of X by the linear regression equation $\hat{Y} = a_{Y.x} + (b_{YX} \cdot X_i)$. Fourth, we can represent the estimated values of this equation with a diagonal line. Fifth, if our assumptions are correct, this regression line will be closer to more of the points than is the mean line, thereby meeting the two-variable least-squares criterion. Sixth, we can do better estimating Y values by using the linear least-squares values of $\hat{Y}$ than we can by using the mean least-squares value of $\bar{Y}$.

Put somewhat differently, the difference between $\hat{Y}$ and $\bar{Y}$ $(\hat{Y} - \bar{Y})$ indicates the amount of improvement we make by using $\hat{Y}$ rather than $\bar{Y}$ to estimate the values of Y; that is, it is the amount "accounted for" or "explained by" X. Thus when we square the difference $(\hat{Y} - \bar{Y})^2$ and sum the squares, we get a sum that we call the "explained" sum of squares: $\sum_{i=1}^{N} (\hat{Y} - \bar{Y})^2$. Again, by "explained" we do not necessarily mean causally explained, but rather we mean *the amount of the variance in Y that is account-ed for by using knowledge of X to estimate Y.*

Similarly, the remaining difference between the $\hat{Y}$ values and the observed Y values $(Y_i - \hat{Y})$ indicates the amount of error we have made by using $\hat{Y}$ to predict the actual Y value. The sum of these squared differences is called the "unexplained" sum of squares. This means that we can measure the amount of total variation which is explained and unexplained by using the $\hat{Y}$ values. Thus we can measure the reduction in error by predicting Y given knowledge of X.

If we carry out the same steps for each of the eight students, we end up with three sums:

(1) the total variation $= \sum_{i=1}^{N} (Y_i - \bar{Y})^2 =$ the total sum of squares $=$ Total SS.

(2) the variation in Y that is *accounted for* by its linear relationship with $X = \sum_{i=1}^{N} (\hat{Y} - \bar{Y})^2 =$ the explained sum of squares $=$ Explained SS.

(3) the variation in Y that is *not accounted for* by its linear relationship with $X = \sum_{i=1}^{N} (Y_i - \hat{Y}) =$ the unexplained sum of squares $=$ Unexplained SS.

In Chapter 2 we learned that a proportion is nothing more than the ratio of one of the parts to the whole. Let us use this knowledge to set up the ratio of the explained variation in Y, Explained SS, to the total variation in Y, Total SS. This gives us the proportion of variation in Y that is explained by the linear relationship of X and Y. Thus,

$$\frac{\sum_{i=1}^{N} (\hat{Y} - \bar{Y})^2}{\sum_{i=1}^{N} (Y_i - \bar{Y})^2} = \frac{\text{Explained SS}}{\text{Total SS}} = \begin{matrix} \text{proportion of variation in } Y \text{ that is} \\ \text{explained by its linear relationship with } X \end{matrix}$$

In our example we start with the values in Table 5.3.1, and now we can carry out the sum-of-squares calculations for the proportion of variation explained, using the estimated values from the least-squares equation as shown in Table 8.3.2.

Table 8.3.2   Calculations for sums of squares (SS)

Student	Total SS $(Y_i - \bar{Y})^2$	Explained SS $(\hat{Y} - \bar{Y})^2$	Unexplained SS $(Y_i - \hat{Y})^2$
Alan	3.06	0.03	3.69
Arthur	0.56	0.73	0.01
Betty	1.56	0.73	0.16
Candy	0.06	0.04	0.01
Helen	0.06	0.26	0.59
Maxine	3.06	1.44	0.31
Roland	1.56	0.26	0.59
Sam	0.56	1.44	0.19
	$\sum_{i=1}^{N} (Y_i - \bar{Y})^2$	$\sum_{i=1}^{N} (\hat{Y} - \bar{Y})^2 +$	$\sum_{i=1}^{N} (Y_i - \hat{Y})^2$
	10.48 $=$	4.93 $+$	5.55
	Total SS $=$	Explained SS $+$	Unexplained SS

Let us now substitute the appropriate values into the formula for the proportion of variation explained:

$$\frac{\text{Explained SS}}{\text{Total SS}} = \frac{4.93}{10.48} = +0.47 = \text{proportion of variation in } Y \text{ that is explained by its linear relationship with } X$$

Since we divide the variation by N to get the variance, and since we could divide both the numerator and the denominator by N and not change the value of the proportion, this formula also tells us the proportion of the *variance* in Y that is explained by X. It is useful to reexamine the use of the variance in developing the linear regression values since calculating the sum of squares for many respondents is time-consuming and burdensome.

8.4 REGRESSION COEFFICIENTS, EXPLAINED VARIANCE, AND THE CONCEPT OF CORRELATION

In Chapter 7 we learned that in order to predict values for Y given X, we use the linear regression equation $\hat{Y} = a_{Y \cdot X} + (b_{YX} \cdot X_i)$. We also learned that even though $b_{YX} \neq b_{XY}$, they both provide us with useful information about the linear relationship between X and Y, b_{YX} telling the slope of Y given X, and b_{XY} telling the slope of X given Y. We can put the two together and examine the combined product by multiplying the two regression coefficients together, that is, $b_{YX} \cdot b_{XY}$. This can be shown symbolically:

$$b_{YX} \cdot b_{XY} = \frac{s_{XY}}{s_X^2} \cdot \frac{s_{XY}}{s_Y^2} = \frac{s_{XY}^2}{s_X^2 s_Y^2}$$

A look at the denominator $s_X^2 s_Y^2$ reveals that we multiply the variance of X, s_X^2, times the variance of Y, s_Y^2. Since the two variances for each pair of variables represent the mean total variation of each of the variables, their product represents the total *possible* covariation between the two. Thus the denominator $s_X^2 s_Y^2$ equals the maximum possible covariance between the two variables X and Y. The numerator, on the other hand, deals with the covariance squared s_{XY}^2, and the covariance measures the *actual* covariation between X and Y measured simultaneously.

It may not be completely clear at first, but a bit of thought will help you see that when we multiply the two total regression coefficients for a given set of variables, the numerator can never be larger than the denominator. Thus the maximum value that the product of the two total regression coefficients $b_{YX} b_{XY}$ can attain is 1.00. This would occur only if the numerator, that is, the covariance squared, were equal to the denominator, that is, the product of the two variances. In other words, if the actual covariance equals the total possible covariance the product of the two b's will equal 1.00.

These steps provide us with a value that may be interpreted as a proportion. Namely, it is the proportion of the *actual* covariance squared s_{XY}^2 to the maximum possible covariance as determined by the product of the two variances. For example, using the information in Table 7.2.1, we can compute the covariance of X and Y, the variance of X, and the variance of Y. Now, let us calculate the regression coefficient for variables X and Y. Thus,

$$b_{YX} \cdot b_{XY} = \frac{s_{XY}^2}{s_X^2 s_Y^2} = \frac{(1.81)^2}{5.25 \cdot 1.31} = \frac{3.28}{6.88} = 0.47$$

Now let us use the information in the matrix of regression coefficients, that is, $b_{YX} = 0.34$ and $b_{XY} = 1.38$. Thus,

$$b_{YX} \cdot b_{XY} = 0.34 \cdot 1.38 = 0.47$$

Since the product of the two zero-order coefficients $b_{YX} b_{XY}$ must be less than or equal to 1.00, it may be thought of as a proportion. Even though the logic is as yet unstated, a comparison of this product, 0.47, and the sum of squares quotient, 0.47, indicates that the proportion of variance explained is equal to the product of the two zero-order regression coefficients for a given pair of variables. We are now ready to mesh all that has gone on thus far to see why this is the case.

We have just learned that the product of the two regression coefficients for a given pair of variables equals the ratio of the covariance squared to the product of the two variances. In Chapter 7, we learned that the regression coefficient $b_{YX} = s_{XY}/s_X^2$ tells us the regression of Y on X, thereby measuring the influence of the variance of X on the covariance of X and Y. Similarly, the coefficient $b_{XY} = s_{XY}/s_Y^2$ tells us the regression of X on Y, thereby measuring the influence of the variance of Y on the covariance of X and Y. Now our interest is in the proportion of variance in one variable that is explained by its linear relationship with the second variable. This interest leads us back to the concept of correlation, which can be used to measure the explained variance.

Since

$$b_{YX} \cdot b_{XY} = \frac{s_{XY}}{s_X^2} \frac{s_{XY}}{s_Y^2} = \frac{s_{XY}^2}{s_X^2 \cdot s_Y^2}$$

and since

$$r_{YX} \cdot r_{XY} = \frac{s_{XY}}{s_Y s_X} \cdot \frac{s_{XY}}{s_X s_Y} = \frac{s_{XY}^2}{s_X^2 \cdot s_Y^2}$$

the product-moment correlation coefficient squared is equal to the product of the two regression coefficients, that is, $r_{XY}^2 = b_{YX} \cdot b_{XY}$, and they both equal the proportion of variance in Y that is explained by its linear relationship with X and vice versa.[6] Since $r_{YX}^2 = $ the proportion of variance explained, we need not go through the bother of carrying out all the sums-of-squares calculations. That is, we do not have to carry out the following formula and steps we learned earlier;

$$\sum_{i=1}^{N} (Y_i - \bar{Y})^2 = \sum_{i=1}^{N} (\hat{Y} - \bar{Y})^2 + \sum_{i=1}^{N} (Y_i - \hat{Y})^2$$

[6]The formulas can be stated in terms of the individual deviations:

$$b_{YX} \cdot b_{XY} = \frac{\sum_{i=1}^{N} (X - \bar{X}) \cdot (Y - \bar{Y})}{\sum_{i=1}^{N} (X - \bar{X}) \cdot (X - \bar{X})} \cdot \frac{\sum_{i=1}^{N} (X - \bar{X}) \cdot (Y - \bar{Y})}{\sum_{i=1}^{N} (Y - \bar{Y}) \cdot (Y - \bar{Y})} = \frac{\left[\sum_{i=1}^{N} (X - \bar{X}) \cdot (Y - \bar{Y})\right]^2}{\sum_{i=1}^{N} (X - \bar{X})^2 \cdot \sum_{i=1}^{N} (Y - \bar{Y})^2}$$

and

$$r_{YX}^2 = \frac{\sum_{i=1}^{N} (X - \bar{X}) \cdot (Y - \bar{Y})}{\sqrt{\sum_{i=1}^{N} (X - \bar{X})^2} \cdot \sqrt{\sum_{i=1}^{N} (Y - \bar{Y})^2}} \cdot \frac{\sum_{i=1}^{N} (X - \bar{X}) \cdot (Y - \bar{Y})}{\sqrt{\sum_{i=1}^{N} (X - \bar{X})^2} \cdot \sqrt{\sum_{i=1}^{N} (Y - \bar{Y})^2}}$$

$$= \frac{\left[\sum_{i=1}^{N} (X - \bar{X}) \cdot (Y - \bar{Y})\right]^2}{\sum_{i=1}^{N} (X - \bar{X})^2 \cdot \sum_{i=1}^{N} (Y - \bar{Y})^2}$$

Thus,

$$\text{Total SS} = \text{Explained SS} + \text{Unexplained SS}$$

$$\text{proportion of variance explained} = \frac{\text{Explained SS}}{\text{Total SS}}$$

Naturally, we can calculate the total sum of squares $\sum_{i=1}^{N} (Y_i - \bar{Y})^2$, the explained sum of squares $\sum_{i=1}^{N} (\hat{Y} - \bar{Y})^2$, and the unexplained sum of squares $\sum_{i=1}^{N} (Y_i - \hat{Y})^2$. However, it is bothersome to carry out all the steps of subtracting each estimated score from each actual score. Fortunately, we do not need to do so, because we can use the square of the correlation coefficient to represent the proportion of variance explained. Knowing that r_{YX}^2 represents the proportion of variance in Y that is explained by its linear relationship with X, we know that it is equal to the Explained SS divided by the Total SS. Furthermore, the proportion of variance in Y unexplained by X is equal to $(1.00 - r_{YX}^2)$, thus

$$r_{YX}^2 = \frac{\text{Explained SS}}{\text{Total SS}} = \text{the proportion of variance explained}$$

and

$$(1.00 - r_{YX}^2) = \frac{\text{Unexplained SS}}{\text{Total SS}} = \text{the proportion of variance unexplained}$$

One way in which two-variable correlations can be used is by comparing a set of coefficients for different groups. For example, Coleman presents the correlations between freshman grades and IQ for all the schools studied in *The Adolescent Society*, as is shown in Table 8.4.1.

Table 8.4.1   Correlation within a class between freshman grades and IQ, averaged over the four classes in schools

School	Boys Correlation	Girls Correlation
Farmdale	0.603	0.607
Marketville	0.591	0.504
Elmtown	0.516	0.645
Maple Grove	0.620	0.579
Green Junction	0.559	0.583
St. John's	0.404	
Newlawn	0.615	0.603
Millburg	0.493	0.446
Midcity	0.611	0.571
Executive Heights	0.540	0.682

Source: Adapted from James S. Coleman, *The Adolescent Society*. (New York: The Free Press, 1961), p. 261, Table 58.

To interpret any one of these coefficients, all we do is square it, and then we know the proportion of variance in one which is explained in the other. For example, for boys in Farmdale $r_{YX}^2 = (0.60)(0.60) = 0.36$. This indicates that 36 percent of the variance in boys' freshman grades is explained by its linear relationship with IQ.

Another way to use the zero-order product-moment correlation coefficient is to calculate it for each of the variables in a set just as we did for X and Y. Luckily, since $r_{YX} = r_{XY}$, we do not have to calculate two coefficient values, as we did with beta coefficients. For example, we could summarize the three possible zero-order correlation coefficients r_{YX}, r_{YT}, and r_{XT} for the eight students in the matrix shown in Table 8.4.2.

Table 8.4.2 Matrix of zero-order product moment correlation coefficients for variables *Y*, *X*, and *T* (*N* = 8)

	Y	X	T
Y	1.00	0.69	0.92
X	0.69	1.00	0.62
T	0.92	0.62	1.00

In our example, we know that the value of

$$r_{YX} = \frac{s_{XY}}{s_X s_Y} = \frac{1.81}{2.61} = +0.69$$

And, when we square it, $r_{XY}^2 = (+0.69)^2 = 0.47$.

The two methods give identical values for r_{XY}^2. We can put the value in words: "Forty-seven percent of the variance in grade point average (Y) is explained by its linear relationship with the number of hours spent studying (X)."

EXERCISES

1. In your own words, define, describe, or discuss the following terms and give an hypothetical example of each.
   correlation
   proportion of variance explained
   total variance
   total covariance
   r_{XY}
   Pearson's product moment correlation coefficient
   total sum of squares
   squared deviation

True or False Questions:

2. A. r_{XY} tells us the proportion of variance explained.
    B. r_{XY}^2 is the value of the explained sum of squares.
    C. The mean sum of the squared differences is the total variance.
    D. The b coefficient is similar to the correlation coefficient in that both use the covariance in the denominator.
    E. Pearson's Product Moment r is a coefficient for ordinal data.

3. Using the data presented in problem 6 in Chapter 7 calculate:
    A. r_{LM}
    B. r_{ML}

4. Using the data from problem 9 in Chapter 7 answer the questions below:
    A. What is the correlation between students' academic performance and the influence of T.V.?
    B. What proportion of the variance in student performance is explained by its linear relationship with T.V. influence?

5. Using the data presented in problem 4 in Chapter 7, calculate r_{YX} and r_{YT}.

6. Using the data in problem 5 in Chapter 7:
    A. Calculate r_{YX} and r_{XY}.
    B. What can be concluded about the relationship between age and academic aspirations?
    C. What proportion of the variance in academic aspirations is explained by its linear relationship with age?

7. Noting the trend of older students returning to school, you wonder if there is any correlation between age and grade point average. You poll the 9 people in your statistics class, and obtain the data presented below.

Name	Age (T)	Grade Point Average (X)	Statistics Examination Grade (Y)
Adrian	21	2.5	85.0
Carrie	19	2.5	81.0
Hal	20	3.5	87.5
Mike	21	3.0	92.0
Peter	19	3.5	85.0
Rose	20	2.5	78.0
Sarah	19	3.0	82.5
Wilbur	21	3.5	89.0
Zelda	20	3.0	85.0

   A. What are the correlations between age (T) and statistics examination grade (Y), between age (T) and grade point average (X), and between grade point average (X) and statistics examination grade (Y)?
   B. What proportion of the variance in statistics examination grade (Y) is explained by its linear relationship first with age (T) and then with grade point average (X)?

8. In a study of state employment agencies, Blau (1955) presents the following data. Competitiveness (X) is based on the degree to which the interviewer

185   *Explanation and Correlation with Two Interval Variables*

referred more than the expected number of applicants to job openings which he personally had reviewed. Productivity (Y) is measured by the percentage of openings per interviewer through which the applicant was actually hired. Table 1 presents statistics from Agency A where the group favors competition in the top of the table and statistics from Agency B where the group favors cooperation in the bottom of the table.

Table 1   Competitiveness and productivity in sections A and B

	Interviewer	Competi-tiveness† (4)	Produc-tivity‡ (5)
A	Adams	3.9	0.70
	Ahman	3.1	.49
	Ajax	4.9	.97
	Akers	3.2	.71
	Ambros	1.8	.45
	Atzenberg	2.9	.61
	Auble	2.1	.39
B	Babcock	2.2	.53
	Beers	1.6	.71
	Bing	1.5	.75
	Borden	2.1	.55
	Bush	2.1	.97

†Competitiveness index: The proportion of job openings received to which the recipient made a referral times the number of members of the section. (This represents the observed divided by the expected frequency of referrals made by the recipient of a job opening.)

‡Productivity index: The number of placements made divided by the number of job openings available, that is, the number of openings in the section per interviewer, which was 143 for Section A and 87 for Section B.

Source: Adapted from Peter M. Blau, *The Dynamics of Bureaucracy*, revised edition (Chicago: University of Chicago Press, 1963), p. 61.

A. Plot the figures on 2 graphs, and draw an estimated line which represents the pattern.
B. Calculate the regression figures for these data, and compare the results with your freehand estimate. What differences do you see?

9. Four groups of 120 employees each were asked the same question. The following table presents the number of "yes" and "no" answers.

Responses to Question	Group 1	Group 2	Group 3	Group 4
Yes	36	18	65	40
No	84	102	55	80
Total	120	120	120	120

A. Determine the proportion of affirmative responses for each of the four groups of employees.
B. The way in which groups are combined for analysis is an important matter. Discuss the meaning of and importance of the difference in the proportion of affirmative responses if the following groups were combined:

1 and 2	2 and 3
1 and 3	2 and 4
1 and 4	3 and 4

10. Explain this statement: "In no way do our statistics *prove* causation, prediction, or explanation."

part IV

Multivariate Descriptive Statistics

9

Partial Association for Nominal or Ordinal Variables

Earlier we learned that we can display cross-classified relationships by holding constant a third variable. This enabled us to classify statistical findings in tabular form, showing frequencies and percents according to relevant cross-classifications. We have also learned several ways to measure the relationship between two variables. Generally, we have considered the two variables to be a dependent variable Y and an independent variable X, and our concern has been to measure the form and strength of the relationship between the two. Now, our concern is with developing ways to measure cross-classified relationships holding constant one or more control or test variables.

Before we extend two-variable measures, we will resurrect the concept of "holding constant" or "controlling for" one additional independent variable. If you understand the three-variable case, that is, *one* dependent, *one* independent, and *one* control (another independent) variable, multivariate statistics are easy to understand. We merely extend the logic. The concept of introducing a test variable is not new. There are many instances in which we are concerned with the way in which two of the variables are influenced by a third one. We begin by learning about the process commonly called elaboration.

9.1 ELABORATING THE ASSOCIATION BETWEEN TWO VARIABLES

We can attempt to "explain" the relationship between two variables by controlling one or more additional variables. We have learned that in the cross-classification of two dichotomous variables, the inner cells of a fourfold table may take on a wide range of values depending upon the marginal limitations. Of course, the actual value in each cell is determined by the attributes of the respondents, and it is the *possibilities* for the observed values that are dependent upon the marginals. Adding a third dichotomous variable as a test or control variable helps clarify the concept of explanation. In this three-variable framework, explanation is part of the process commonly called *elaboration*.[1]

The basic idea of elaboration is that we analyze the relationship between dependent (Y) and independent (X) variables holding constant a test (T) variable. Starting with a fourfold (2×2) table cross-classifying variables X and Y, we can construct an eightfold ($2 \times 2 \times 2$) table by cross-classifying variable T with the other two simultaneously as shown in Table 9.1.1.

When we cross-tabulate three dichotomous variables, we get $2^3 = 8$ inner cells which form an eightfold table. For the sake of consistency and ease of expansion, and since we are concerned with examining the association between X and Y, we continue to designate the inner cells a, b, c, and d. However, to show their value on the third variable T, we use $T+$ and $T-$ combined with the cell letters. For example, all b's possess $+$ on X and $+$ on Y, some b's possess $+$ on T and some possess $-$ on T, and they are designated $b+$ and $b-$ respectively. Table 9.1.2 presents the abstract layout for a $2 \times 2 \times 2$, eightfold table.

There are three ways we can calculate first-order versions of Q_{XY}. First, we can look at the eightfold table as if it were two fourfold tables, one on top of the other. The top one shows the relationship between X and Y on the constant condition that T is high, and the bottom one shows the relationship between X and Y on the constant condition that T is low. These are called *conditional tables* and are defined by the classes or categories of the control variable. Put differently, on the condition that T is high or the condition that T is low, we can examine the X and Y relationship.

From this viewpoint, we can compute what are analogous to two zero-order Q's, one for $T+$ and one for $T-$. Such values can be noted as

---

[1]Elaboration of a three-variable cross-classification system was originally developed by Paul F. Lazarsfeld and later formalized by Patricia L. Kendall and Paul F. Lazarsfeld. Patricia L. Kendall and Paul F. Lazarsfeld, "Problems of Survey Analysis," in *Continuities in Social Research*, eds. R. K. Merton and Paul F. Lazarsfeld (New York: The Free Press, 1950), pp. 147–67; and Paul F. Lazarsfeld, "The Algebra of Dichotomous Systems" in *Studies in Item Analysis and Prediction*, ed. Herbert Solomon (Stanford, California: Stanford University Press, 1961), pp. 111–57.

Table 9.1.1.   Notation for two and three cross-classified dichotomous variables

2 × 2 × 2 Table

T	X	Low	High
			Y	
High	High	High T / High X / Low Y	High T / High X / High Y
	Low	High T / Low X / Low Y	High T / Low X / High Y
Low	High	Low T / High X / Low Y	Low T / High X / High Y
	Low	Low T / Low X / Low Y	Low T / Low X / High Y

2 × 2 Table

X	Low	High
		Y	
High	High X / Low Y	High X / High Y
Low	Low X / Low Y	Low X / High Y

Table 9.1.2.   Notation for a 2 × 2 × 2 eightfold table

Variable T	Variable X	Variable Y		Sum
		−	+	
+	+	$a+$	$b+$	$T+ X+$
	−	$c+$	$d+$	$T+ X-$
	Sum	$T+ Y-$	$T+ Y+$	$T+$
−	+	$a-$	$b-$	$T- X+$
	−	$c-$	$d-$	$T- X-$
	Sum	$T- Y-$	$T- Y+$	$T-$

$Q_{XY \cdot COND\ T+}$ and $Q_{XY \cdot COND\ T-}$, and they are called the conditional Q association values for X and Y, depending upon the "condition" of T or when T is held constant. They can be expressed symbolically as:

$$Q_{XY \cdot COND\ T+} = \frac{[(b+)(c+)] - [(a+)(d+)]}{[(b+)(c+)] + [(a+)(d+)]}$$

$$Q_{XY \cdot COND\ T-} = \frac{[(b-)(c-)] - [(a-)(d-)]}{[(b-)(c-)] + [(a-)(d-)]}$$

Second, we can put together the cross products of the two conditional tables that are "tied" on T. By doing this we can examine that part of the total zero-order associations between X and Y that remains after controlling for T. This Q value is called the first-order partial association. The number of test variables determines the order of the partial association value. That is, one test variable produces a first-order partial, two test variables produce second-order partials, three test variables produce third-order partials, etc. For now, we can limit ourselves to the first-order Q values.

The partial Q_{XY} is computed using the cross products of X and Y by adjusting the zero-order value controlling for both categories of T after it has been held constant (also called "tied on T"). To compute the partial Q value, we add together the cross products of X and Y when T is controlled, and then compute Q as usual. The formula for the first-order partial can be expressed symbolically:

$$Q_{XY \cdot PART\ T} = \frac{\overset{\text{consistent cross products with } T \text{ tied}}{([(b+)(c+)] + [(b-)(c-)])} - \overset{\text{inconsistent cross products with } T \text{ tied}}{([(a+)(d+)] + [(a-)(d-)])}}{([(b+)(c+)] + [(b-)(c-)]) + ([(a+)(d+)] + [(a-)(d-)])}$$

The partial Q between X and Y, controlling for T, therefore, is derived by summing the cross products of the consistent and inconsistent cells for the two conditional tables and then calculating Q focusing on consistent and inconsistent cross products. The partial Q formula takes into account not just consistency and inconsistency between X and Y, but also that of T as well. Specifically, we multiply $(b+)$ times $(c+)$, $(b-)$ times $(c-)$, $(a+)$ times $(d+)$, and $(a-)$ times $(d-)$. In other words, we deal with cross products that are "tied" on T. This leads to the third way we can calculate a version of Q_{XY} based on an eightfold table.

Third, we can put together the cross products of the two conditional tables that are "not tied" on T. By doing this, we can examine the difference between the zero-order association between X and Y and that which remains after controlling for T. This Q value is called the *first-order differential association*, and it, like the partial, takes into account the fact that there are two categories of T, but it focuses on those cross products that occur when the X and Y values are computed for *differing states* of T after it has been held constant (also called "differing" on T). The formula for the first-order differential can be expressed symbolically:

$$Q_{XY \cdot DIFF\ T} = \frac{\overset{\text{consistent cross products with } T \text{ differing}}{([(b+)(c-)] + [(b-)(c+)])} - \overset{\text{inconsistent cross products with } T \text{ differing}}{([(a+)(d-)] + [(a-)(d+)])}}{([(b+)(c-)] + [(b-)(c+)]) + ([(a+)(d-)] + [(a-)(d+)])}$$

The zero-order, conditional, partial, and differential Q's cover all the possible combinations of the X and Y associations. However, we must consider one last matter involving a weighting techique to average the partial and differential Q values. This technique will enable us to put the two together to equal the total zero-order Q_{XY}. As in earlier averaging procedures, we want to know the weight of the partial and the weight of the differential. Put differently, we are seeking the weights for the proportions of cross products that are (1) tied on T and (2) differing on T. The *partial weight* is nothing more than the proportion of cross products on X and Y tied on T to the cross products on X and Y not controlling for T. It can be expressed symbolically:

partial weight $= PW$

$$= \frac{([(b+)(c+)] + [(b-)(c-)]) + ([(a+)(d+)] + [(a-)(d-)])}{([(b+) + (b-)][(c+) + (c-)]) + ([(a+) + (a-)][(d+) + (d-)]}$$

or

$$PW = \frac{\text{the denominator of } Q_{XY \cdot PART\ T}}{\text{the denominator of } Q_{XY}}$$

The formula for the *differential weight* uses the analogous cross products and is the proportion of cross products on X and Y differing on T to the cross products of X and Y not controlling for T. It can be expressed symbolically:

differential weight $= DW$

$$- \frac{([(b+)(c-)] + [(b-)(c+)]) + ([(a+)(d-)] + [(a-)(d+)])}{([(b+) + (b-)][(c+) + (c-)]) + ([(a+) + (a-)][(d+) + (d-)])}$$

or

$$DW = \frac{\text{the denominator of } Q_{XY \cdot DIFF\ T}}{\text{the denominator of } Q_{XY}}$$

By putting together the partial and differential Q's and their appropriate weights, we can show that

$$Q_{XY} = [(Q_{XY \cdot PART\ T}) \cdot (PW)] + [(Q_{XY \cdot DIFF\ T}) \cdot (DW)]$$

From this formula for three dichotomous variables, we can say that Q_{XY} is a weighted average of the partial and differential Q values, where the weights for each are the proportions of cross products tied on T and differing on T. This mathematical necessity means that the zero-order, partial, and differential Q values have the following characteristics:[2]

Principle 1.   If any two are equal in sign and size, the third must have the same sign and size.

[2]James A. Davis, *Elementary Survey Analysis* (Englewood Cliffs, New Jersey: Prentice-Hall, Inc., 1971), pp. 83–86.

Principle 2. If the partial is less than the zero-order, the differential must be larger than either.

Principle 3. If the partial is greater than the zero-order, the differential must be smaller than either.

Principle 4. Whatever the value of the zero-order, the partial can have any value between $+1.00$ and -1.00, provided that the weights and the value of the differential are appropriate.

9.2 ELABORATION OUTCOMES

In order to carry out elaboration, we examine the two-variable associations controlling for one or more test variables. We introduce these variables to *test* the properties of the original association by recalculating the coefficient among subgroups of cases that are similar in their category of the test variable.

In multivariate survey analysis, we introduce one or more test variables to *test* the properties of the original Q_{XY} by recalculating the Q coefficient among subgroups of cases that are similar in their category of the test variable. This statistic is the partial association value $Q_{XY \cdot PART\ T}$, and we may compare it with the zero-order Q_{XY} value to assess the amount of change or explanatory quality of the test variable(s).[3]

We derive partial coefficients for Q by cross-classifying the cases according to the simultaneous possession of several variables. What this means, then, is that the partial Q's are obtained by cross-tabulating the data and then calculating the partials from the 2^K-fold tables, with K being the number of variables.

The first-order partial Q values can be interpreted in relation to the zero-order Q value. There are many ways in which the cases can be distributed on X and Y controlling for T, and the system of elaboration gives a useful interpretation to some of the most important possibilities.[4] If the conditional relationships are considerably different from each other, then we say that the control variable *specifies* the conditions under which the relationship exists. If the partial relationship is approximately equal to the zero-order relationship, then we say that the control variable has *no effect* on the original relationship. If the partial relationship is reduced to a negligible one (vanishes), then we say that the control variable *explains* the original relationship. If the partial relationship is elevated well above the zero-order relationship, then we say that the control variable *suppresses* the original relationship. Table 9.2.1 summarizes the way we can interpret these four outcomes.

These interpretations are displayed in Figure 9.2.1. The following interpretive rules guide our analysis of elaboration outcomes.[5]

[3] Davis, *ibid.*, uses the symbol $Q_{XY \cdot TIED\ T}$ to express the first-order partial coefficient.

[4] Davis, *ibid*, p. 86.

[5] Davis, *ibid.*, pp. 87–105.

Table 9.2.1.   Interpretation of possible outcomes in a three variable elaboration

IF	THEN
If the two conditionals are considerably different from each other	*Then we say that T "specifies" the condition under which the relationship exists*
If the partial relationship between X and Y is ...	Then we say that T ...
Equal to the zero-order	Has "no effect" on the relationship REGION A and REGION C
Negligible and lower than the zero-order	"Explains" the relationship REGION B
Nonnegligible and lower than the zero-order	"Somewhat explains" the relationship, but there is no neat label for this outcome REGION E
Elevated above the zero-order	"Suppresses" the relationship REGION D

Source: Adapted from James A. Davis, *Elementary Survey Analysis*. Englewood Cliffs, New Jersey: Prentice-Hall, 1971, pp. 87–104; and Kendall and Lazarsfeld, *op. cit.*

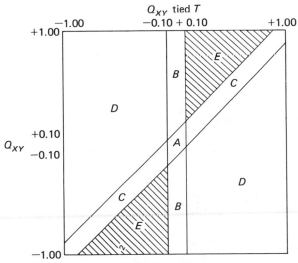

Figure 9.2.1. Possible outcomes in three-variable analysis. (From *James A. Davis,* Elementary Survey Analysis. *Englewood Cliffs, New Jersey: Prentice-Hall, 1971, p. 87.*)

Rule 1

If we find differing conditionals, before claiming to have found a specification, make sure that (1) the conditional Q's differ by 10 Q units and (2) both conditional tables meet the standard for expected cell frequencies of 5 or more expected cases in each cell. Even then be wary of the finding if T is highly skewed.

Rule 2

To interpret a three-way specification, arrange the data in a percentage table and seek a plausible statement in terms of (1) column differences in a row association, (2) row differences in a column association, or (3) a particular cell with an especially high or low percentage.

Rule 3

If we find a nonnegligible zero-order association between X and Y and if $Q_{XY \cdot PART\ T}$ is negligible or zero, then we say that T *explains* the association.

Rule 4

If we find a nonnegligible zero-order association between X and Y and if a test variable reduces the zero-order association but the partial is still nonnegligible, we say that T *partly explains* the association. In such cases we should test to see whether a partial calculated among finer groupings of T will lead to a negligible partial.

Rule 5

If the difference between the partial and the differential is less than 10 points on the Q scale, treat the data as falling in regions A or C, depending on the value of the zero-order.

If there is a difference of 10 points or more:
  a) If the partial is negligible and the zero-order is nonnegligible, you may say that T explains XY, thus, assign the results to Region B.
  b) Otherwise, assign the results to region D or E depending on the values of the coefficients.

Rule 6

To predict the sign of "D minus P" (the value of the differential minus the value of the partial), multiply the signs of TX and TY.

6.a) *Sign Rule for Explanation*: A test variable T that explains Q_{XY} must have a zero-order association with both X and Y.
If Q_{XY} is positive, Q_{TX} and Q_{TY} must have the same sign (both positive or both negative).
If Q_{XY} is negative, Q_{TX} and Q_{TY} must have opposite signs (one negative and one positive).

6.b) *Sign Rule of Thumb for Explanation:* When seeking a test variable to explain Q_{XY}, the zero-order associations should not only fit the sign rule, but, generally speaking, Q_{TX} and Q_{TY} should be stronger than Q_{XY}.

6.c) *Sign Rule for a Suppressor Variable:* When T raises a positive XY correlation, it must have opposite-sign zero-order correlations with X and Y. When T strengthens a negative XY correlation, it must have same-sign (both positive or both negative) zero-order correlations with X and Y. When T reverses the sign of a correlation, the sign rule conditions for explanations must hold.

6.d) *Sign Rule of Thumb for a Suppressor Variable:* If you "want to make" the XY correlation more positive (or reverse a negative correlation) seek a test variable where TX and TY have opposite signs and both are of greater magnitude than XY. If you "want to make" the XY correlation more negative (or reverse a positive correlation), seek a test variable where TX and TY have the same sign (both positive or both negative) and both are of greater magnitude than XY.

Suppose we are interested in elaborating a low negative zero-order association between age (low age being defined as under 40 and high age being defined as 40 or over) and participation in voluntary organizations (the number of voluntary organizations to which the respondents belong). Table 9.2.2 shows that a hypothetical sample of 881 respondents has a zero-order Q_{XY} of -0.241. This tells us that differences in participation are associated with differences in age.

Also suppose we decide that it seems reasonable that education might be a test variable which can account for the association between age and participation. Table 9.2.3 presents the hypothetical data of the three variable (eightfold) association.

Using the appropriate formula we can show the elaborated version of Q_{XY} symbolically as:

$$Q_{XY} = [(Q_{XY \cdot PART\ T}) \cdot (PW)] + [(Q_{XY \cdot DIFF\ T}) \cdot (DW)]$$
$$-0.241 = [(-0.002) \cdot (0.498)] + [(-0.480) \cdot (0.502)]$$

Evaluating the partial Q coefficient of -0.002 is all that remains. Using Rule 3, which defines Region B, since the zero-order is nonnegligible and the

Table 9.2.2. Age and participation

	Participation		
Age	Low	High	Total
High	262	151	413
Low	241	227	468
Total	503	378	881

$$Q_{XY} = \frac{bc - ad}{bc + ad}$$

$$Q_{XY} = \frac{(151 \cdot 241) - (262 \cdot 227)}{(151 \cdot 241) + (262 \cdot 227)} = \frac{36{,}391 - 59{,}474}{36{,}391 + 59{,}474}$$

$$Q_{XY} = \frac{-23{,}083}{95{,}865} = \underline{-0.241}$$

Table 9.2.3. Education, age, and participation

		Participation		
Education	Age	Low	High	Total
High	High	102	125	227
	Low	178	216	394
Low	High	160	26	186
	Low	63	11	74
	Total	503	378	881

$$Q_{XY \cdot PART\ T} = \frac{([(b+)(c+)] + [(b-)(c-)]) - ([(a+)(d+)] + [(a-)(d-)])}{([(b+)(c+)] + [(b-)(c-)]) + ([(a+)(d+)] + [(a-)(d-)])}$$

$$Q_{XY \cdot PART\ T} = \frac{[(125 \cdot 178) + (26 \cdot 63)] - [(102 \cdot 216) + (160 \cdot 11)]}{[(125 \cdot 178) + (26 \cdot 63)] + [(102 \cdot 216) + (160 \cdot 11)]}$$

$$Q_{XY \cdot PART\ T} = \frac{(22{,}250 + 1{,}638) - (22{,}032 + 1{,}760)}{(22{,}250 + 1{,}638) + (22{,}032 + 1{,}760)}$$

$$Q_{XY \cdot PART\ T} = \frac{23{,}888 - 23{,}792}{23{,}888 + 23{,}792} = \frac{96}{47{,}680} = \underline{0.002}$$

first-order partial is negligible, we can say that T explains the association between X and Y. Put differently, we may conclude that for our sample education accounts for the low negative association between age and participation.

9.3 COMPUTATIONS FOR COEFFICIENTS OF FOUR NOMINAL OR ORDINAL VARIABLES

We can compute higher order partial Q values that can be similarly interpreted as we elaborate the X and Y association. Table 9.3.1 shows the notation scheme for four dichotomous variables in a $2 \times 2 \times 2 \times 2$ sixteenfold table.

Table 9.3.1. Notation for a 2 × 2 × 2 × 2 sixteenfold table

				Y	
S	T	X		−	+
+	+	+		$a++$	$b++$
+	+	−		$c++$	$d++$
+	−	+		$a-+$	$b-+$
+	−	−		$c-+$	$d-+$
−	+	+		$a+-$	$b+-$
−	+	−		$c+-$	$d+-$
−	−	+		$a--$	$b--$
−	−	−		$c--$	$d--$

In order to compute the second-order partial and differential Q values, we extend the logic of the first-order values further elaborating the X and Y association, now controlling for T and S.[6] Four second-order Q values result from the appropriate formulas. The names for these values can be seen in Table 9.3.2 and the formulas in Table 9.3.3.

Table 9.3.2. Second-order coefficients

		T	
		Differing	Tied
	Tied	Partial differential	Second-order partial
S			
	Differing	Second-order differential	Differential partial

Source: Adapted from James A. Davis, *Elementary Survey Analysis* (Englewood Cliffs, New Jersey: Prentice-Hall, Inc., 1971), Table 6.6, p. 137.

[6]Davis, *ibid.*, p. 137.

Table 9.3.3. Formulas for second-order Q coefficients

$$\text{PART S,T} = \frac{[(b++)(c++) + (b-+)(c-+) + (b+-)(c+-) + (b--)(c--)]}{-\,[(a++)(d++) + (a-+)(d-+) + (a+-)(d+-) + (a--)(d--)]}$$

$$\begin{aligned}\text{PART S,}\\ \text{DIFF T}\end{aligned} = \frac{[(b++)(c-+) + (b-+)(c++) + (b+-)(c--) + (b--)(c+-)]}{-\,[(a++)(d-+) + (a-+)(d++) + (a+-)(d--) + (a--)(d+-)]}$$

$$\begin{aligned}\text{DIFF S,}\\ \text{PART T}\end{aligned} = \frac{[(b++)(c+-) + (b-+)(c--) + (b+-)(c++) + (b--)(c-+)]}{-\,[(a++)(d+-) + (a-+)(d--) + (a+-)(d++) + (a--)(d-+)]}$$

$$\text{DIFF S,T} = \frac{[(b++)(c--) + (b-+)(c+-) + (b+-)(c-+) + (b--)(c++)]}{-\,[(a++)(d--) + (a-+)(d+-) + (a+-)(d-+) + (a--)(d++)]}$$

To compute the appropriate weights, divide the denominator of the coefficient by the denominator for Q_{xy}.

We can now expand the formula for the zero-order Q_{XY} and show the equivalent terms for the second-order partial Q values.

$$Q_{XY} = [(Q_{XY \cdot PART\ S,T})(P_{ST}W)] + [(Q_{XY \cdot PART\ S,\ DIFF\ T})(P_S D_T W)]$$
$$+ [(Q_{XY \cdot DIFF\ S,\ PART\ T})D_S P_T W)] + [(Q_{XY \cdot DIFF\ S,T})(D_{ST}W)]$$

We can combine this equation in a fourfold table and show the zero-order, first-order, and second-order versions of Q_{XY} for the possible total, partial, and differential expressions, as shown in Table 9.3.4.

Table 9.3.4.   Zero-order, three-variable, and four-variable coefficients for *XY*

		T		
		Differing	Tied	Weighted Sum
S	Tied	$Q_{XY \cdot PART\ S,\ DIFF\ T}$	$Q_{XY \cdot PART\ S,T}$	$Q_{XY \cdot PART\ S}$
	Differing	$Q_{XY \cdot DIFF\ S,T}$	$Q_{XY \cdot DIFF\ S,\ PART\ T}$	$Q_{XY \cdot DIFF\ S}$
Weighted Sum		$Q_{XY \cdot DIFF\ T}$	$Q_{XY \cdot PART\ T}$	Q_{XY}

1. A first-order partial is the weighted sum of the second-order partial and the appropriate partial differential.
2. A first-order differential is the weighted sum of the second-order differential and the "other" partial differential.
3. The zero-order may be seen alternatively as:
   a) The weighted average of the four second-order coefficients.
   b) The weighted average of either pair of first-order coefficients.
   c) The weighted average of the four-variable coefficients in a particular row (column) and the three-variable coefficient for the other row (column).

Source: Adapted from James A. Davis, *Elementary Survey Analysis* (Englewood Cliffs, New Jersey: Prentice-Hall, Inc., 1971), p. 138, Table 6.7, and p. 139.

9.4 FOUR VARIABLE ELABORATION

Using data from a National Opinion Research Center study,[7] we will examine a sample of 4284 respondents in terms of four status variables. For present purposes, we cross-tabulate the variables as dichotomous ordinal ones; thus

[7]Briefly summarized, the All-American Sample consists of 10,557 males from the following groups: Civilian veterans and nonveterans, Army enlisted men and officers, Navy enlisted men and officers, Air Force enlisted men and officers, and Marine Corps enlisted men and officers. The samples from which the present data were gathered were prepared by Current Population Surveys and by the research units of the various branches of the military service in October, 1964.

we do not need to make any assumptions about measured intervals as is necessary in forms of interval variable analyses.

V: *Father's Education* is dichotomized high $(+)$ if the father had graduated from high school and low $(-)$ if the father had not graduated from high school or less.

X: *Father's Occupation* is a dichotomous variable, dividing the sample into high $(+)$, those whose fathers were white-collar workers, managers, proprietors, officials, and professionals, and low $(-)$, those whose fathers were blue-collar workers or farmers.

U: *Respondent's Education* is dichotomized into those who had had some college or more, high $(+)$, and those who had graduated from high school or less, low $(-)$.

W: *Respondent's First Job* is dichotomized into high $(+)$, those whose first job was as a white-collar worker, manager, proprietor, official, or professional, and low $(-)$, those whose first job was as a blue-collar worker or farmer.

Table 9.4.1 presents the $2 \times 2 \times 2 \times 2$ sixteenfold table for the four status variables.

Table 9.4.1.   All-American sample
Raw data for four status variables cross-tabulated in temporal order

Father's Education V	Father's Occupation X	Respondent's Education U	Respondent's First Job W $-$	$+$
$+$	$+$	$+$	99	332
$+$	$+$	$-$	95	45
$+$	$-$	$+$	138	141
$+$	$-$	$-$	266	72
$-$	$+$	$+$	61	151
$-$	$+$	$-$	158	58
$-$	$-$	$+$	272	289
$-$	$-$	$-$	1806	301
				$N = 4284$

Table 9.4.2 presents all the possible zero-order, first-order, and second-order Q values that can result from a four-variable, $2 \times 2 \times 2 \times 2$, sixteenfold table. By testing the difference between various Q values, we can compare them to determine the single effects of one or the other test variables or the joint effects of the two together on the original association.[8]

The underlined Q values are the appropriate ones to use for comparative purposes. For example, the Q_{VX} of 0.705 is the zero-order association between

[8] For a fuller discussion see Davis, *Elementary Survey Analysis, op. cit.*, Chapter 6.

Table 9.4.2. Decomposition of Q
zero-order, three-variable, and four-variable coefficients, variables V, X, U, and W

Relationship		Diff : X : Tied		1st Order
$UW \cdot XV$	Tied	0.797	0.721	0.751
	V			
	Diff	0.854	0.702	0.795
	1st Order	0.832	0.713	0.773

Relationship		Diff : U : Tied		1st Order
$XW \cdot UV$	Tied	0.638	0.406	0.55
	V			
	Diff	0.774	0.458	0.687
	1st Order	0.722	0.432	0.629

Relationship		Diff : U : Tied		1st Order
$VW \cdot UX$	Tied	0.408	0.165	0.306
	X			
	Diff	0.717	0.344	0.613
	1st Order	0.597	0.247	0.479

Relationship		Diff : W : Tied		1st Order
$XU \cdot WV$	Tied	0.653	0.444	0.574
	V			
	Diff	0.812	0.633	0.752
	1st Order	0.751	0.551	0.68

Relationship		Diff : W : Tied		1st Order
$VU \cdot WX$	Tied	0.515	0.511	0.513
	X			
	Diff	0.773	0.648	0.731
	1st Order	0.674	0.574	0.634

Relationship		Diff : W : Tied		1st Order
$VX \cdot WU$	Tied	0.605	0.605	0.605
	U			
	Diff	0.799	0.695	0.764
	1st Order	0.745	0.649	0.705

father's education and father's occupation, and it is not examined at a higher order. The Q_{VU} and Q_{UX} values of 0.513 and 0.574 are first-order partial values, both of which are lower than the zero-order and first-order differentials. The $Q_{VW} = 0.165$, $Q_{XW} = 0.406$, and $Q_{UW} = 0.721$ are second-order partial values, and they all are lower than both their zero-order and differential values.

To read the first-order partial and differential Q's, we look only in the causal direction. The statistic's value does not really tell us the difference between directions, but our theory should take this into account. For example, we can look at the partial VU relationship in two ways. First, as we did above, we can control for X, the association between father's education and respondent's education controlling for father's occupation. Second, we could look at the same association, controlling for respondent's first job. From a theoretical perspective, the second view would make little sense.

It is possible to continue expanding the number of variables under consideration. Each additional nominal or ordinal variable means that we must cross-tabulate all of the variables in order to compute the appropriate higher order coefficients. Here, of course, is where we may run into difficulty because of the number of cells resulting from extensive cross-tabulations. Even if we use only dichotomous variables in cross-tabulated tables with 7 variables, we progress from 4 inner cells to 8 to 16 to 32 to 64 to 128. Obviously, unless we have a very large sample to begin with, we may well run out of sufficient expected cases to calculate meaningful values for higher-order versions of Yule's Q. However, we need not be too disheartened by these limitations, since we can very effectively study the structure of relationships among a system of several dichotomous nominal or ordinal variables. For many sociological purposes this will be of great benefit, as we will learn in Chapter 18.

EXERCISES

1. In your own words define, describe, or discuss the following terms and give an hypothetical example of each:
   "explain" the relationship
   elaboration
   conditional table
   first-order differential association
   partial weight
   "suppresses" the relationship
   "somewhat explains" the relationship
   conditional Q association values
   multivariate statistics
   consistent cross products
   differential weight

2. Explain the similarities and differences between a partial Q and a differential Q.

3. Describe in a few brief paragraphs the "workings" of the elaboration model and its value in analyzing data.

4. Show arithmetically how to "expand the formula" for Q_{XY} to show the "equivalent terms for the second-order partial Q values."

5. In question 5, Chapter 6, we addressed Davis' question of "whether new groups have trouble, or new individuals tend to drop out, or both" in terms of the group's age and the individual's longevity. Now, let us consider the additional aspect of the drop-out rate by examining the cross-tabulation of the three variables as shown in Table A. Use these data for problems 5-10.

Table A. Age, longevity, and drop-out rate

Age of the Reading Group (T)	Individual Longevity in Reading Program (X)	Drop-Out Rate (Y)		Total
		Low	High	
High	High	201	541	742
	Low	64	155	219
Low	High	44	46	90
	Low	273	283	556
Total		582	1025	1607

Source: Adapted from James A. Davis, *Great Books and Small Groups* (New York: The Free Press of Glencoe, Inc., 1961), pp. 100–104 and Chart 4.1, p. 224.

A. Combine the appropriate cells in Table A and compute the three zero-order values Q_{XY}, Q_{TY}, and Q_{TX}.

B. Discuss and interpret the findings for the three bivariate associations.

6. Compute the conditionals $Q_{XY \cdot COND \, T+}$ and $Q_{XY \cdot COND \, T-}$ and describe and discuss the differences between the conditionals and the zero-order Q_{XY} in problem 5.

7. Compute the first-order partial $Q_{XY \cdot PART \, T}$ and discuss the difference between it and the zero-order Q_{XY} in problem 5.

8. Compute the first-order differential $Q_{XY \cdot DIFF \, T}$ and discuss the difference between it and the zero-order Q_{XY} in problem 5.

9. Compute the appropriate weights for the partial (PW) and the differential (DW). Combine all the elements of the first-order coefficients into the formula for the zero-order Q_{XY} and compute its value in problem 5.

10. Summarize, discuss, and interpret the elaborated findings which emerge from problems 5 through 9.

10

Partial and Multiple
Linear Regression

In social statistics, we are often interested in measuring, estimating, and predicting relationships between one dependent interval variable and *more than one* independent variables. Two-variable linear regression is merely a special case in which we have *one* dependent variable and *one* independent variable, which we use for prediction purposes. To conceptualize what the regression formula amounts to is not too complex for *two* variables, because most of us are used to thinking about two things at a time, that is, in two dimensions. To conceptualize what is going on with three or more variables is a more difficult job; however, it too fits in with the way we think about problems in everyday life. A little effort and some thought ought to enable you to grasp what we are getting at in the next few pages. It is important not to give this section a once-over-lightly treatment, because a clear understanding is essential for what comes later.

10.1 ONE CONTROL VARIABLE

Suppose that we want to examine family income (Y) given education (X) and holding race (T) constant. Table 10.1.1 shows the percent distribution of family income level and years of school completed holding race constant. The similarities and differences can be analyzed in many ways: We could examine the number of families, the percent distribution, or the median income.

We now want to develop measures to help us measure such findings as these. Let us begin with the case of one control variable.

Table 10.1.2 shows information about three variables, the weekly hours worked (T), the hourly earnings (X), and the weekly earnings (Y), for private nonagricultural industries in 1970. Clearly, there are notable differences among the various industries. Despite the fact that there are fairly sizable differences in the hours worked, our main concern now is in the relationship between hourly earnings and weekly earnings.

Figure 10.1.1 shows the joint locations of hourly earnings and weekly earnings connected by a line. There is clearly not a perfect straight-line relationship between the two variables X and Y. Furthermore, none of the formulas presented thus far would enable us to make exact predictions about weekly earnings (Y) from the information we have about hourly earnings (X) alone. Of course, if they are linearly related and we use information about both variables X and T simultaneously, we could predict Y more precisely. In this case, if we multiply X and T, we are able to get Y exactly.

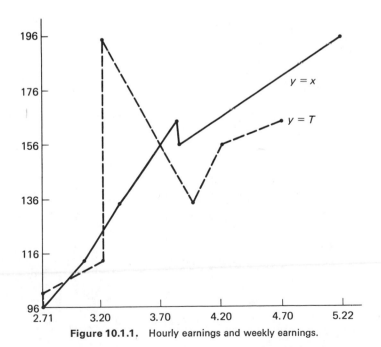

Figure 10.1.1.   Hourly earnings and weekly earnings.

Table 10.1.1. Money income—percent distribution of families, by income level, by years of school completed and race of head: 1961 and 1969

Race of Head and Years of School Completed	Number of Families (1,000)	Percent Distribution by Income Level						Median income
		Under $3,000	$3,000–$4,999	$5,000–$6,999	$7,000–$9,999	$10,000–$14,999	$15,000 and Over	
1961								
White families	39,630	18.3	18.8	22.7	22.4	12.6	5.4	$6,100
Elementary school	13,525	32.9	24.2	19.3	15.0	6.7	1.9	4,419
High school	17,645	12.3	19.2	27.3	25.6	12.1	3.4	6,344
1–3 years	7,103	15.2	20.1	27.4	24.2	10.5	2.6	6,036
4 years	10,542	10.4	18.5	27.2	26.6	13.2	3.9	6,548
College	8,460	6.9	9.3	18.4	27.3	22.8	15.3	8,560
1–3 years	3,773	9.9	11.9	21.4	28.4	17.8	10.5	7,586
4 years or more	4,687	4.3	7.2	16.2	26.4	26.9	19.0	9,503
Black and other families	4,190	45.7	24.7	14.4	9.3	4.8	1.3	3,340
Elementary school	2,416	57.4	24.2	10.7	5.1	2.0	0.7	2,593
High school	1,414	33.7	27.8	18.3	11.8	6.3	2.0	4,115
1–3 years	804	39.6	28.9	17.7	6.8	4.9	2.1	3,711
4 years	610	26.0	26.5	19.3	18.2	8.1	1.9	4,773
College	360	16.7	14.3	23.1	26.5	17.0	2.3	6,593
1–3 years	201	23.6	16.8	24.1	22.8	11.4	1.3	(B)
4 years or more	159	7.8	11.3	21.8	31.3	24.3	3.4	(B)

Table 10.1.1. Cont'd.

1969

White families	42,967	7.7	9.3	11.1	21.4	28.8	21.9	10,089
Elementary school	10,852	17.4	18.6	15.6	21.4	18.6	8.3	6,769
Less than 8 years	5,207	23.1	20.0	16.2	19.1	15.1	6.5	5,799
8 years	5,645	12.3	17.2	15.1	23.5	22.0	9.8	7,651
High school	20,984	5.3	7.3	11.6	24.5	32.4	19.1	10,181
1–3 years	7,026	7.6	9.8	13.3	24.4	29.4	15.7	9,342
4 years	13,958	4.2	6.0	10.6	24.4	34.1	20.7	10,563
College	11,131	2.9	3.6	5.6	15.8	31.7	40.5	13,426
1–3 years	4,914	3.8	4.6	7.5	19.4	33.8	30.8	11,949
4 years or more	6,216	2.3	2.7	4.1	12.9	30.0	48.1	14,685
Black and other families	4,746	19.5	18.9	17.3	19.7	15.9	9.0	6,340
Elementary school	1,988	29.1	23.5	17.8	16.0	9.9	3.7	4,754
Less than 8 years	1,485	32.1	24.6	16.9	14.5	8.5	3.5	4,351
8 years	503	20.3	20.1	20.6	20.5	14.2	4.4	5,927
High school	2,178	14.6	16.8	18.6	22.7	19.3	8.1	7,002
1–3 years	1,078	19.6	20.2	17.9	22.0	14.8	5.6	6,217
4 years	1,101	9.9	13.6	19.2	23.2	23.7	10.6	7,875
College	580	4.3	11.0	10.0	21.0	23.7	30.0	10,555
1–3 years	306	6.3	14.3	13.5	25.7	23.0	17.0	9,194
4 years or more	274	2.1	7.4	5.9	15.6	24.5	44.4	13,682

B Base less than 200,000.

Source: Dept. of Commerce, Bureau of the Census; *Current Population Reports*, Series P–60, and unpublished data. U.S. Bureau of the Census, *Statistical Abstract of the United States: 1971* (92nd edition). Washington, D.C., 1971, p. 319, No. 508.

Table 10.1.2.   Hourly earnings, weekly hours, and weekly earnings (1970)

Industry	Weekly Hours T	Hourly Earnings X	Weekly Earnings Y
Mining	42.7	$3.84	$163.97
Contract Construction	37.4	5.22	195.23
Manufacturing	39.8	3.36	133.73
Transportation and Public Utilities	40.5	3.85	155.93
Wholesale and Retail Trade	35.3	2.71	95.66
Financial, Insurance, Real Estate	36.8	3.07	112.98
Services	34.5	2.84	97.98
Total (all industries)	37.2	$3.23	$120.16

Weekly Earning = Hourly Earnings·Weekly Hours

$$Y = X \cdot T$$

Thus far we have focused on just two variables and their total variances and covariances. It is important to understand that we can calculate the variance of a number of other variables and then calculate the covariances with variables X and Y. For example, let us examine the eight students according to their responses to the question "What was your overall high school average?" called variable T. Table 10.1.3 carries out the calculation for the total variance of T, which we find to be 20.75.

Table 10.1.3.   Computations for the mean and variance of T

Student	i	High School Average (T)	$\bar{T}$	$(T_i - \bar{T})$	$(T_i - \bar{T})^2$
Alan	1	95.00	87.50	7.50	56.25
Arthur	2	85.00	87.50	−2.50	6.25
Betty	3	90.00	87.50	2.50	6.25
Candy	4	87.00	87.50	−0.50	0.25
Helen	5	88.00	87.50	0.50	0.25
Maxine	6	83.00	87.50	−4.50	20.25
Roland	7	80.00	87.50	−7.50	56.25
Sam	8	92.00	87.50	4.50	20.25

$$\sum_{i=1}^{N} T_i = 700.00 \qquad \sum_{i=1}^{N} (T_i - \bar{T})^2 = 166.00$$

$$\bar{T} = \frac{700}{8} = 87.50 \qquad \text{var } T = s_T^2 = 20.75$$

We now must consider variable T in combination first with variable Y and then with variable X. The steps are identical to those carried out in the

XY case, and are summarized in Tables 10.1.4 and 10.1.5. We can see that the covariance of *T* and *Y* is $s_{TY} = 4.81$, and the covariance of *T* and *X* is $s_{TX} = 6.50$.

We now have the total variances for variables *Y*, *X*, and *T*. We also have covariances for the pairs of variables *Y* and *X*, *Y* and *T*, and *X* and *T*. Since all these values are interrelated, they are commonly displayed in a covariance matrix.

For our purposes we treat the covariance matrix as a square one. This means that it displays all of the variables in two places, once across the top and once down the left-hand side. Since we merely square the number of variables to determine the number of cells in a square matrix, a three-variable matrix is composed of 9 cells, a four-variable matrix is composed of 16 cells, a five-variable matrix is composed of 25 cells, and so forth.

Table 10.1.4.   Computations for the covariance of *T* and *Y*

Student	i	T_i	$\bar{T}$	$(T_i - \bar{T})$	Y_i	$\bar{Y}$	$(Y_i - \bar{Y})$	$(T_i - \bar{T})(Y_i - \bar{Y})$
Alan	1	95.00	87.50	(7.50)	4.0	2.25	(1.75)	13.1250
Arthur	2	85.00	87.50	(−2.50)	1.5	2.25	(−0.75)	1.8750
Betty	3	90.00	87.50	(2.50)	3.5	2.25	(1.25)	3.1250
Candy	4	87.00	87.50	(−0.50)	2.5	2.25	(0.25)	−0.1250
Helen	5	88.00	87.50	(0.50)	2.0	2.25	(−0.25)	−0.1250
Maxine	6	83.00	87.50	(−4.50)	0.5	2.25	(−1.75)	7.8750
Roland	7	80.00	87.50	(−7.50)	1.0	2.25	(−1.25)	9.3750
Sam	8	92.00	87.50	(4.50)	3.0	2.25	(0.75)	3.3750

$$\sum_{i=1}^{N} (T_i - \bar{T})(Y_i - \bar{Y}) = 38.5000$$

$$\text{cov } TY = s_{TY} = 4.81$$

Table 10.1.5.   Computations for the covariance of *T* and *X*

Student	i	T_i	$\bar{T}$	$(T_i - \bar{T})$	X_i	$\bar{X}$	$(X_i - \bar{X})$	$(T_i - \bar{T})(X_i - \bar{X})$
Alan	1	95.00	87.50	(7.50)	4	4.5	(−0.5)	− 3.75
Arthur	2	85.00	87.50	(−2.50)	2	4.5	(−2.5)	6.25
Betty	3	90.00	87.50	(2.50)	7	4.5	(2.5)	6.25
Candy	4	87.00	87.50	(−0.50)	5	4.5	(0.5)	− 0.25
Helen	5	88.00	87.50	(0.50)	6	4.5	(1.5)	0.75
Maxine	6	83.00	87.50	(−4.50)	1	4.5	(−3.5)	15.75
Roland	7	80.00	87.50	(−7.50)	3	4.5	(−1.5)	11.25
Sam	8	92.00	87.50	(4.50)	8	4.5	(3.5)	15.75

$$\sum_{i=1}^{N} (T_i - \bar{T})(X_i - \bar{X}) = 52.00$$

$$\text{cov } TX = s_{TX} = 6.50$$

We can summarize the information provided by Tables 10.1.3, 10.1.4, and 10.1.5, in a three-variable covariance matrix for variables Y, X, and T.

The means and covariance matrix for variables Y, X, and T
($N = 8$)

Mean		Y	X	T
2.25	Y	1.31	1.81	4.81
4.50	X	1.81	5.25	6.50
87.50	T	4.81	6.50	20.75

Moving across the first variable line of the matrix, we find the mean score of the variable grade point average $\bar{Y} = 2.25$. In the body of the matrix are the numbers 1.31, 1.81, and 4.81. The first number is the total variance of Y, $s_Y^2 = 1.31$. To see how the variance fits into the covariance matrix, you can think of it as the "covariance" of Y with itself, or $s_Y^2 = 1.31$. The next value in the top row represents the covariance between variable Y and variable X, $s_{YX} = 1.81$. Verbally it is the covariance between grade point average and the number of hours spent studying. The covariance of each of the other sets of variables is shown in the cell whose margins carry the appropriate letter abbreviations. The diagonal lines isolate the total *variance* of each variable; however, each of these may also be thought of as the "*covariance*" of the variable with itself.

Notice that the entries below the diagonal are the mirror image of those above it: Moving down column 1, for example, the numbers 1.31, 1.81, and 4.81 correspond with the numbers across row 1. Because of this duplication, some statistical researchers present a matrix showing only the diagonal and either the top or the bottom half of the matrix. We prefer to see a full square matrix, however, for it is often handy to glance down a column or across a row to check all the covariances for a given variable.

Now let us put this sort of interest in terms of linear regression and specific techniques to carry it out. Suppose that we have variables Y, X, and T, and that we have a set of Y values in linear combination with X and T values, that is, a set of $\hat{Y}$ values. If we start at one level of T and estimate Y given X through all levels of T, there will be a unique set of values represented by the $\hat{Y}$ regression estimates within each level of T. Now, if we hold T constant at a specific level, we can find a unique set of estimated values of the regression of Y on X. "Holding constant," or "controlling for," additional independent variables are the statistician's terms that mean what we might say in everyday language as "if the test variable were equal for all cases." Technically, the difference between "holding constant" and "controlling for" is that the former is an examination of a two-variable relationship at constant levels of the test variable while the latter is an adjustment of the values of the zero-order relationship according to the level of the control variable.[1]

You have already seen how this concept is put into practice. For a given coefficient, all the variables being "controlled" appear to the right of the dot, "·". Thus, $a_{Y \cdot XT}$ is the *a* coefficient of *Y* controlling for *X* and *T*, $b_{YX \cdot T}$ is the *b* coefficient of *Y* on *X* controlling for *T*, and so forth.

We call the multiple-linear-regression *b* coefficients *partial b coefficients*. This is because they represent that *part* of the regression between two variables that remains after controlling for the other independent variables in the equation. When there is one control variable we call it the *first-order* partial, when there are two control variables it is the *second-order* partial, and so forth. In the two-variable zero-order case b_{YX} we in essence ignored any other independent variables; that is, we had zero control variables. This means that the multiple-linear-regression-partial *b* coefficient $b_{YX \cdot T}$ represents only part of the overall total regression of variable *Y* on *X*. Specifically, it represents that part of the regression that is left after subtracting the joint regressions of the two other variables.

We stated the two-variable-linear-regression equation as:

$$\hat{Y} = a_{Y \cdot x} + (b_{YX} \cdot X_i)$$

Since additional variables may provide additional useful predictive information, let us expand the above equation to take into account any number of independent variables. Even though it may look complex, let us state the three-variable-linear-regression equation symbolically:

$$\hat{Y} = a_{Y \cdot XT} + (b_{YX \cdot T} \cdot X_i) + (b_{YT \cdot X} \cdot T_i)$$

One of the basic elements of multiple linear regression is that the variables must vary. Namely, we must have a distribution of Y_i values, X_i values, and T_i values, the sums of squares of which do not equal zero, that is,

$$\text{Total SS} = \sum_{i=1}^{N} (Y_i - \bar{Y})^2 \neq 0$$

$$\text{Total SS} = \sum_{i=1}^{N} (X_i - \bar{X})^2 \neq 0$$

$$\text{Total SS} = \sum_{i=1}^{N} (T - \bar{T})^2 \neq 0$$

[1]For a more detailed discussion see Hubert M. Blalock, Jr., *Social Statistics*, 2nd ed. (New York: Holt, Rinehart and Winston, Inc., 1969), pp. 303–336, and 433–37.

This means that part of our concern is with how each variable varies, as measured by the total sums of squares or the variances. We are also interested in the ways in which the variables covary with each other, as measured by the sum of the cross products, or the covariance. Finally, we are interested in how *combinations* of variables covary, as measured by the relationships between variables. The third variable in this set is called the *test* or *control variable*.

We must make the same basic assumptions that we made for two-variable linear regression. Summarized briefly, these assumptions are:

1. We assume that the *a* and *b* coefficients are constant values and that the values of all the variables are randomly distributed.
2. We assume that the relationships between the variables are linear.
3. We assume that the independent variables provide additional estimative information above and beyond that provided by the means of the dependent and independent variables.
4. We assume that the distribution of *Y* values for fixed values of *X* and *T* (or all combinations of variables) is normal and that the standard deviation of each *Y* distribution is equal.

It is a simple matter to add a fourth variable, because it is merely an extension of the above logic. With this in mind, let us express the four-variable-multiple-linear-regression equation symbolically:

$$\hat{Y} = a_{Y.XTV} + (b_{YX.TV} \cdot X_i) + (b_{YT.XV} \cdot T_i) + (b_{YV.XT} \cdot V_i)$$

This equation states that we predict values of *Y* from knowledge we have of the independent variables *X*, *T*, and *V*. Thus what we do is predict the value of a given Y_i with $\hat{Y}$ based on the simultaneous possession of scores on all the variables in our multiple linear model. Among the crucial aspects of this equation is that the values of $a_{Y.XTV}$ and the partial *b* coefficients ($b_{YX.TV}$, etc.) are constants that can be used in combination with a varying value for each variable.

10.2 PARTIAL REGRESSION COEFFICIENTS

Our concern now is to determine *partial* coefficients between one dependent variable and one independent variable controlling for one test (an additional independent) variable. Actually it is rather easy, for we can use the zero-order total *b* coefficients between all the pairs of variables in our multiple-linear-regression system. We want to know how much (what part) of the total *b* coefficient of *Y* given one variable remains after we control for the third variable.

Since we are interested in predicting values for the dependent variable *Y* as it is influenced by the variables X and T, we are seeking the first-order partial regression coefficient of Y on X controlling for T, that is, $b_{YX \cdot T}$, and the first-order partial regression coefficient of Y on T controlling for X, that is, $b_{YT \cdot X}$. The zero-order b_{YX} tells us the total regression of Y on X, or the amount of change in Y that would be produced by a given change in X. The zero-order b_{YT} tells us the total regression of Y on T, or the amount of change in Y that would be produced by a given change in T.

We start with the total regression between the dependent variable and one independent variable. Then we subtract from it the impact of the additional independent or control variable. For practical purposes, we use a formula that enables us to calculate the first-order partial b coefficient of the dependent variable Y on the independent variable X controlling for one additional independent variable T. This formula uses the zero-order coefficients b_{YX}, b_{YT}, b_{TX}, and b_{XT}.[2] Pay close attention to the order of the subscripts; it is very important. The equation for the first-order partial b coefficient of Y on X controlling for T can be expressed symbolically:

$$b_{YX \cdot T} = \frac{b_{YX} - (b_{YT}) \cdot (b_{TX})}{1.00 - (b_{XT}) \cdot (b_{TX})}$$

[2]The way this comes about follows our earlier treatment of variances and covariances. The first-order partial b coefficient can be expressed in terms of the variances and covariances between the variables. The formula for $b_{YX \cdot T}$ can be expressed symbolically:

$$b_{YX \cdot T} = \frac{\dfrac{s_{XY}}{s_X^2} - \dfrac{s_{TY}}{s_T^2} \cdot \dfrac{s_{TX}}{s_X^2}}{1.00 - \dfrac{s_{TX}}{s_T^2} \cdot \dfrac{s_{TX}}{s_X^2}} = \frac{\dfrac{s_{XY}}{s_X^2} - \dfrac{s_{TY} \cdot s_{TX}}{s_T^2 \cdot s_X^2}}{1.00 - \dfrac{s_{TX}^2}{s_T^2 \cdot s_X^2}}$$

The term in the numerator of the numerator, $s_{TY} \cdot s_{TX}$, is the product of two covariances, providing a sort of "co-covariance" between the control variable and both the dependent and the independent variables. The numerator of the denominator, s_{TX}^2, is the covariance squared, providing a positive square of the covariance of the control and independent variables. The denominators are identical, $s_T^2 \cdot s_X^2$.

The term in the numerator is the ratio of the product of the two covariances to the product of the two variances, $\dfrac{s_{TY} \cdot s_{TX}}{s_T^2 \cdot s_X^2}$. As such, it is the ratio of the combined actual covariances of the control variable (T) with the dependent (Y) and the independent (X) variables, $s_{TY} \cdot s_{TX}$, to the maximum possible covariance between the control and the independent variables, $s_T^2 \cdot s_X^2$. In this sense, it serves as a factor to correct for the possibility that T and X may combine to influence Y. This factor is subtracted from the ratio of the covariance of X and Y to the variance of X, $\dfrac{s_{XY}}{s_X^2}$. This step gets at the influence of T on X and Y.

The denominator is made up of the ratio of the covariance between X and T squared to the product of the two variances, $\dfrac{s_{TX}^2}{s_T^2 \cdot s_Y^2}$. The term s_{TX}^2 is a measure of the actual covariance between the two variables, while the term $s_T^2 \cdot s_X^2$ is a measure of the maximum possible covariance between them. Thus, the quotient of the two terms is always a fraction which when subtracted from 1.00 corrects for the fact that X and T may leave some of the variance in one unaccounted for by the other. The closer $\dfrac{s_{TX}^2}{s_T^2 \cdot s_X^2}$ is to 1.00, the closer the denominator is to 0, and this influences the first-order partial b coefficient.

Put verbally, in order to obtain the first-order partial regression of Y on X holding T constant, we start with the zero-order total regression of Y on X. From it we subtract the product of the regression of Y on T times the regression of T on X. We then divide this difference by the difference between 1.00 less the product of the regression of X on T times the regression of T on X.

We can sum this up as follows. The zero-order b coefficient measures both the direct and the indirect influence of X on Y, while the partial b coefficient measures only the direct influence of X on Y controlling for T. Thus the *difference* between the zero-order b and the first-order partial b may be thought of as the indirect influence of X on Y as it operates through T.

These steps tell us that since we already know b_{YX}, we need a correction factor that first shows how the control variable T affects the dependent variable Y, and that then shows how the control variable T is affected by the independent variable X. We can get such a factor by multiplying the zero-order b coefficient of Y on T times the zero-order b coefficient of T on X, thus, $(b_{YT}) \cdot (b_{TX})$.

We can put the zero-order b coefficient and the correction factor together in one term, which measures the amount of change in Y given X less the product of the amount of change in Y given T times the amount of change in T given X. Thus the numerator may be stated as $b_{YX} - (b_{YT}) \cdot (b_{TX})$.

We must also take into account the amount of change in the independent variable X given the control variable T. We can do so with a factor that gets at the joint impact of the two variables, that is, the effects of X on T and T on X. To do this we multiply the zero-order b coefficient of X given T times the zero-order b coefficient of T given X, that is, $(b_{XT}) \cdot (b_{TX})$.

Since the product of the two zero-order regression coefficients for a given pair of variables can never exceed 1.00, the term $(b_{XT}) \cdot (b_{TX})$ must be equal to or less than 1.00. Because we are considering the regression of Y on X controlling for T, we want to know what part of the variance in X is not affected by T. Our interest is in that proportion of the maximum *possible* regression between X and T that is not accounted for by the linear relationship between them. To find this proportion, we subtract the term $(b_{XT}) \cdot (b_{TX})$ from 1.00. Thus the denominator may be stated as $1.00 - (b_{XT}) \cdot (b_{TX})$. This factor enables us to take into account the proportionate joint impact of the two variables on each other: that is, it is the proportion of variance in X that is not explained by its linear relationship with T.

The point of these steps is to derive a coefficient that indicates *part* of the regression of Y on X, that is, a *partial b* coefficient. In the above case we are interested in that part of the regression of Y on X that remains after we have controlled for knowledge of T, and the first-order partial coefficient $b_{YX \cdot T}$ tells us just that.

The first-order partial b of Y on T controlling for X is computed in an identical manner with the appropriate zero-order coefficients entered into the formula. It can be expressed symbolically:

$$b_{YT \cdot X} = \frac{b_{YT} - (b_{YX}) \cdot (b_{XT})}{1.00 - (b_{TX}) \cdot (b_{XT})}$$

As yet, we have not learned how to calculate the coefficient $a_{Y \cdot XT}$, but it is a straightforward extension of our earlier treatment of the coefficients $a_{Y \cdot X}$ and $a_{Y \cdot T}$. You may recall $a_{Y \cdot X}$ measures the average influence on Y of the average value of X, and to calculate it we subtract $(b_{YX} \cdot \bar{X})$ from the mean $\bar{Y}$, that is, $a_{Y \cdot X} = \bar{Y} - (b_{YX} \cdot \bar{X})$. Similarly, $a_{Y \cdot T} = \bar{Y} - (b_{YT} \cdot \bar{T})$. For the three-variable $a_{Y \cdot XT}$, we just combine the two average influences. To do this, we subtract first the average X value and then the average T value. It can be expressed symbolically as:

$$a_{Y \cdot XT} = \bar{Y} - (b_{YX \cdot T} \cdot \bar{X}) - (b_{YT \cdot X} \cdot \bar{T})$$

10.3 CALCULATING AND INTERPRETING PARTIAL REGRESSION COEFFICIENTS

In Chapter 7, we learned about two-variable linear regression, examining the regression slopes of grade point average (Y) on academic effort (X) and past academic performance (T). Table 10.3.1 again presents the matrix of zero-order total b coefficients for the three variables.

Table 10.3.1.   Matrix of zero-order b coefficients for variables Y, X, and T

Means		Y	X	T
2.25	Y	1.00	0.34	0.23
4.50	X	1.38	1.00	0.31
87.50	T	3.67	1.24	1.00

Remember, the two total b coefficients for a given pair of variables are not necessarily equal. Therefore, in a matrix of regression coefficients you should read the left-hand column variable abbreviation as the first subscript and the row variable as the second subscript, e.g., $b_{YX} = 0.34$ and $b_{XY} = 1.38$, etc. Using the zero-order total b coefficients of X and Y, the partial $b_{YX \cdot T}$ now can be calculated:

$$b_{YX \cdot T} = \frac{b_{YX} - (b_{YT}) \cdot (b_{TX})}{1.00 - (b_{XT}) \cdot (b_{TX})}$$

$$b_{YX \cdot T} = \frac{0.34 - (0.23) \cdot (1.24)}{1.00 - (0.31) \cdot (1.24)} = \frac{0.34 - 0.29}{1.00 - 0.38}$$

$$b_{YX \cdot T} = \frac{0.05}{0.62} = \underline{\underline{0.08}}$$

The partial coefficient $b_{YX \cdot T}$ can be interpreted as indicating that a change of one unit of academic effort (X) produces a change of 0.08 grade point units (Y) when past academic performance (T) is held constant. In other words, compared to looking at X and Y alone, if we control for past performance we reduce the amount of change in grade point average that is produced by a change in academic effort. The zero-order b of Y given X is $b_{YX} = 0.34$, and the first-order partial is $b_{YX \cdot T} = 0.08$.

This can also be interpreted as indicating that the direct influence of X on the slope of Y is 0.08 of a grade point unit, whereas the indirect influence of X on the slope of Y is 0.26 of a grade point unit, that is,

$$0.34 - 0.08 = 0.26.$$

Thus, the total (direct and indirect) influence of X on Y is

$$0.26 + 0.08 = 0.34 = b_{YX},$$

the zero-order b coefficient. This can be stated diagrammatically, as shown in Figure 10.3.1.

We could do the same thing with the first-order partial coefficients $b_{XY \cdot T}$, $b_{YT \cdot X}$, and $b_{TY \cdot X}$, as can be seen in Table 10.3.2.

The partial coefficient $b_{YT \cdot X}$ can be interpreted as indicating that a change of one unit of past academic performance (T) produces a change of 0.19 of a unit of grade point average (Y) controlling for academic effort (X). Put

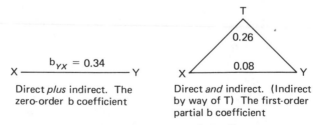

Direct *plus* indirect. The zero-order b coefficient

Direct *and* indirect. (Indirect by way of T) The first-order partial b coefficient

Figure 10.3.1.   Direct and indirect effects.

Table 10.3.2. Computations for first-order regression coefficients

$$b_{XY \cdot T} = \frac{b_{XY} - (b_{XT}) \cdot (b_{TY})}{1.00 - (b_{YT}) \cdot (b_{TY})}$$

$$b_{XY \cdot T} = \frac{1.38 - (0.31) \cdot (3.67)}{1.00 - (0.23) \cdot (3.67)} = \frac{1.38 - 1.14}{1.00 - 0.84}$$

$$b_{XY \cdot T} = \frac{0.24}{0.16} = \underline{\underline{1.50}}$$

$$b_{YT \cdot X} = \frac{b_{YT} - (b_{YX}) \cdot (b_{XT})}{1.00 - (b_{TX}) \cdot (b_{XT})}$$

$$b_{YT \cdot X} = \frac{0.23 - (0.34) \cdot (0.31)}{1.00 - (1.24) \cdot (0.31)} = \frac{0.23 - 0.11}{1.00 - 0.38}$$

$$b_{YT \cdot X} = \frac{0.12}{0.62} = \underline{\underline{0.19}}$$

$$b_{TY \cdot X} = \frac{b_{TY} - (b_{TX}) \cdot (b_{XY})}{1.00 - (b_{YX}) \cdot (b_{XY})}$$

$$b_{TY \cdot X} = \frac{3.67 - (1.24) \cdot (1.38)}{1.00 - (0.34) \cdot (1.38)} = \frac{3.67 - 1.71}{1.00 - 0.47}$$

$$b_{TY \cdot X} = \frac{1.96}{0.53} = \underline{\underline{3.70}}$$

differently, a one-unit change in past performance produces nearly as much of a change in grade point average when we control for academic effort as is produced by T alone. The zero-order $b_{YT} = 0.23$, and the first-order partial $b_{YT \cdot X} = 0.19$.

Just as we did with the first-order partial regression of Y on X controlling for T, we can interpret these two partials in terms of direct and indirect influences. Since the zero-order $b_{YT} = 0.23$, and the first-order partial $b_{YT \cdot X} = 0.19$, we can say that the direct influence of T on Y is 0.19 of a unit while the indirect influence is 0.04 of a unit. Thus the zero-order $b_{YX} = 0.23$ indicates the total (direct and indirect) influence of X on Y.

We learned that the product of the zero-order coefficients, $b_{YX} \cdot b_{XY}$, cannot exceed 1.00. This important feature is also true of the partial b coefficients, and the product of the two first-order partial regression coefficients for a given pair of variables must be equal to or less than 1.00. This means that when we multiply the two first-order partial coefficients of X and Y by each other, the product is a proportion. For example,

$$b_{YX \cdot T} \cdot b_{XY \cdot T} = (0.08)(1.50) = 0.12$$

And,

$$b_{YT \cdot X} \cdot b_{TY \cdot X} = (0.19)(3.70) = 0.70$$

Notice that even though we use the first-order partial b coefficients, the coefficient $a_{Y \cdot XT}$ is a second-order one. This is because we have two control variables, and because there is only *one* a for each variable for any number of other variables taken simultaneously: Furthermore, $a_{Y \cdot XT} = a_{Y \cdot TX}$. The a coefficient measures the average constant amount of change in Y given X and T. In our example, the a coefficient can be computed as follows:

$$a_{Y \cdot XT} = \bar{Y} - (b_{YX \cdot T} \cdot \bar{X}) - (b_{YT \cdot X} \cdot \bar{T})$$
$$a_{Y \cdot XT} = 2.25 - (0.08 \cdot 4.50) - (0.19 \cdot 87.50)$$
$$a_{Y \cdot XT} = 2.25 - 0.36 - 16.63$$
$$a_{Y \cdot XT} = \underline{\underline{-14.74}}$$

Even if the steps are clear, what the a coefficient indicates may not be so clear. The coefficient $a_{Y \cdot XT}$ is the value of $\hat{Y}$ when X and T are equal to zero. Thus, $a_{Y \cdot XT} = -14.74$ means that for someone who has a high school average of 0 and who studies 0 hours each week, we would predict a grade point average of -14.74. Obviously this is impossible, since GPA does not go below 0.00.

The problem is in our measurement scales. For grade point average we use nine values that range from 0.0 to 4.0, that is, 0.0, 0.5, 1.0, 1.5, and so forth. Similarly, for hours spent studying we use nine values that range from 0 to 8. In both cases we represent the students with actual and discrete values. On the other hand, high school average is measured on a scale that locates the respondents on values that range from 0 to 100.

Even though we have these different scales, when we calculate the coefficient $a_{Y \cdot XT}$ we deal with the means of all three variables. While it is possible to study 0 hours, it is inconceivable that someone could have a high school average of 0. Thus the mean of 87.50 for variable T, high school average, will modify the minus grade point average of $a_{Y \cdot XT}$. In other words, when the a coefficient is used in the multiple-linear-regression formula, its negative value will be corrected by the fact that the term for high school average, $(b_{YT \cdot X} \cdot T)$, will elevate the predicted score $\hat{Y}$.

10.4 LINEAR REGRESSION WITH ONE CONTROL VARIABLE

The general idea of linear regression makes a great deal of common as well as statistical sense. For example, from the common-sense perspective, whenever we cross a street we "semiconsciously" judge such things as the width of the street, the direction and speed of oncoming traffic, the weather, road, and visibility conditions, our own walking speed, and so forth. Then we use a pragmatic form of multiple linear regression to "predict" whether to cross or not. An imperfect prediction can be devastating, and small children receive considerable training in street crossing because of the danger. The point of such training is to develop a built-in method of handling everyday problems with a sound prediction system. The version of multiple linear regression we are learning is essentially this sort of tool for statistical problems.

From the statistical perspective, Blau and Duncan put it this way:

> Actually, the principal motivation for regression techniques is not their convenience as a summary of the two-variable relationship but their power in expressing the essential facts of multi-variable relationships. If the requisite assumptions hold, all the information needed for an adequate description of a *k* variable system is contained in the $k(k - 1)/2$ distinct two-variable relationships included in the set.
>
> From this standpoint, the motivation for multiple regression is just like that for multiple classification. Moreover, the essential assumption . . . is also the same, namely, that net relationships are additive.[3]

Since we are familiar with the relationship among present academic performance, past academic performance, and academic effort, let us examine the details of linear regression with one control variable. We will utilize a minimum high school average of 78 for admission and the minimum possible hours spent studying of 0. Substituting these values into the multiple-linear-regression equation, we find

$$\hat{Y} = a_{Y \cdot XT} + (b_{YX \cdot T} \cdot X_i) + (b_{YT \cdot X} \cdot T_i)$$
$$\hat{Y} = -14.74 + (0.08 \cdot 0) + (0.19 \cdot 78)$$
$$\hat{Y} = -14.74 + 0 + 14.83$$
$$\hat{Y} = \underline{0.09}$$

This indicates that for a student whose high school average was 78 and who spends 0 hours studying each week, we would predict a grade point average of 0.09 based on our regression values. Incidentally, since we used the mean $\bar{T}$ to calculate the regression coefficients, if the distribution of high school averages had been around a lower mean the *a* coefficient would not have been such a sizable negative value.

Now that we have all the terms for our equation, it is a simple matter to enter them in the multiple-linear-regression equation, thereby enabling us to predict the value of *Y* given any (hypothetical or observed) values of *X* and *T*. We can do this by using the formula for $\hat{Y}$, as can be seen in Table 10.4.1.

Earlier we learned that when we have one dependent variable and one

[3]Peter M. Blau and Otis Dudley Duncan, *The American Occupational Structure* (New York: John Wiley & Sons, Inc., 1967), p. 146.

Table 10.4.1.   Computations for predicted Y scores

Student	i	X_i	T_i	$a_{Y \cdot XT}$ -14.74	$+$ $+$	$(b_{YX \cdot T} \cdot X_i)$ $(0.08 \cdot X_i)$	$+$ $+$	$(b_{YT \cdot X} \cdot T_i)$ $(0.19 \cdot T_i)$	$=$ $=$	$\hat{Y}$ $\hat{Y}$
Alan	1	4	95.00	-14.74	$+$	0.32	$+$	18.05	$=$	3.63
Arthur	2	2	85.00	-14.74	$+$	0.16	$+$	16.15	$=$	1.57
Betty	3	7	90.00	-14.74	$+$	0.56	$+$	17.10	$=$	2.92
Candy	4	5	87.00	-14.74	$+$	0.40	$+$	16.53	$=$	2.19
Helen	5	6	88.00	-14.74	$+$	0.48	$+$	16.72	$=$	2.46
Maxine	6	1	83.00	-14.74	$+$	0.08	$+$	15.77	$=$	1.11
Roland	7	3	80.00	-14.74	$+$	0.24	$+$	15.20	$=$	0.70
Sam	8	8	92.00	-14.74	$+$	0.64	$+$	17.48	$=$	3.38

independent variable, we can plot individuals according to their scores on the two variables and draw a graph that represents the two variables and describes our observations. The two-variable-linear-regression equation enables us to describe the respondents with a unique straight line, which meets the least-squares criterion. Such a line is drawn on one dimension, but it represents the two dimensions provided by variables X and Y. This means that the regression line represents a linear combination of predicted values on two dimensions. Put differently, we can describe our observations with a set of one-dimensional predictions in two-dimensional space. By definition, this is a straight line, as is shown in Figure 10.4.1.

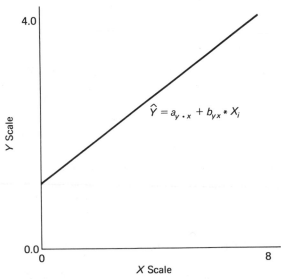

$$\hat{Y} = a_{y \cdot x} + b_{yx} * X_i$$

Figure 10.4.1.   Regression line (estimates) of Y given X.

It is easy to understand the two-variable example because it is not difficult to visualize and conceptualize. It is easy enough to visualize three variables simultaneously, but to conceptualize them in terms of a set of two-dimensional linear estimates in three dimensions is a bit more difficult. Even though we are capable of imagining many dimensions, beyond the three-variable case it is very difficult to describe graphically either the space or the linear combinations of predictions within it. For this reason, we learn about the graphic presentation for three variables. Knowing what is happening in three-dimensional space simplifies understanding the multidimensions of many variables —adding more variables merely means dealing with more *control* variables, but the principle is the same.

The multiple-linear-regression formula produces a linear combination of predictions in any number of dimensions. The total number of variables is what determines the dimension of the linear combination, and the predictions are always one dimension less than the total number of variables. Thus, an l dimensional space can provide a set of $l - 1$ dimensional predictions.

It is not too difficult to locate individuals in a three-dimensional space according to their scores on three variables: Persons in the labor force have joint values on occupational prestige, on education, and on age; students have joint scores on grade point average, on their high school average, and on the weekly number of hours spent studying; and so forth. Since we are interested in the way the combination of the independent and control variables influence the dependent variable, for each level of X and each level of T we assume that there will be a corresponding distribution of scores on Y. This is a straightforward extension of our earlier treatment of X and Y together with no control variable.

We can draw a cube to represent the combination of three variables. First we locate variable Y on the vertical axis and variable X on the horizontal axis. Then we place variable T on the depth axis moving backward, creating a *third* dimension, as can be seen in Figure 10.4.2. The variable Y is located from top-to-bottom creating the vertical dimension, which ranges from 0.00 to 4.0 in intervals of 0.5. The left-to-right surface creates the horizontal dimension representing variable X, which ranges from 0 to 8, in intervals of 1. Finally, variable T becomes the front-to-rear dimension, and it ranges from 78.5 to 100 in intervals of 2.5. The T scale is set at these intervals for ease of description and diagraming purposes.

It is a straightforward matter to locate the respondents in the cube. For example, we could describe Helen as a student whose grade point average Y is 2.0, who spends 6 hours each week studying (X), and whose high school average T was 85. Helen can then be located at the unique spot where $Y = 2.0$, $X = 6$, and $T = 85$, as can be seen in Figure 10.4.2.

If we locate all the respondents in the cube, we get a three-variable scatter-cube. However, as you can see from Figure 10.4.2, to locate even one person

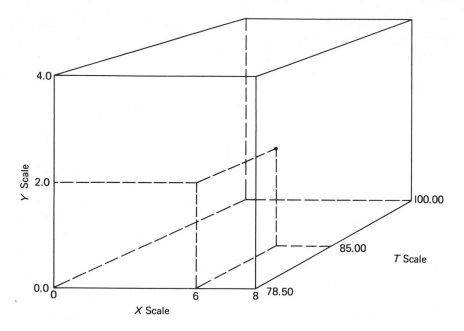

Figure 10.4.2. The three variable-location of Helen.

in a cube requires some rather detailed plotting. Since this is the case, it is convenient to describe the respondents with a unique summarizing measure such as the mean line or the regression line. As you may have guessed, the multiple-linear-regression equation predicted values $\hat{Y}$ provide us with just such a summary measure. In order to summarize three-variable observations, we use a set of *two*-dimensional linear predictions in the *three*-dimensional space provided by variables Y, X, and T, that is,

$$\hat{Y} = a_{Y.XT} + (b_{YX.T} \cdot X_i) + (b_{YT.X} \cdot T_i)$$

This set of predicted values may be represented by a flat surface *plane* in a cube rather than by a *line* on a graph. To make this clearer, imagine a cube or an empty box and follow the steps described below. First, we can draw variables Y, X, and T on the various surfaces of the cube in the same pattern as that in Figure 10.4.2. Y is the top-to-bottom dimension, X is the left-to-right dimension, and T is the front-to-rear dimension.

Since the first value of T that is plotted on the box is 78.00, the front surface of the box represents variables X and Y as if they were drawn on a graph with T held constant at 78.00. Now, suppose we could peel off the entire front surface at the point marked 80.00 on the T scale. The surface on the new front of the box would represent a graph of X and Y with T held constant at 80.00.

224

We could similarly treat X and Y at any level of T, proceeding to the back of the box. At that point the surface would represent a graph of variables X and Y with T held constant at 100.00, which is the highest level of T plotted on the box. Incidentally, these steps graphically describe what we mean by "holding T constant."

Using the values for the lowest and highest high school averages (80.00 for Roland and 95.00 for Alan), we will draw a cube in which we plot the linear predictions on the appropriate surfaces of the box. The predictions are calculated in Table 10.4.2. On the front we use the linear regression values of Y on X holding T constant at 80.00. Thus, given that $X_i = 0$ and $T_i = 80.00$, we would predict the value of Y to be $\hat{Y} = 0.46$, and given that $X_i = 1$ and $T_i = 80.00$, we would predict the value of Y to be $\hat{Y} = 1.10$. These linear regression predictions may be represented by a one-dimensional line on the

Table 10.4.2.   Estimated values of Y given X_i with T held constant at 80.00 and 95.00

	$\hat{Y} = a_{Y \cdot XT}$	$+$	$(b_{YX \cdot T} \cdot X_i)$	$+$	$(b_{YT \cdot X} \cdot T_i)$	$=$	$\hat{Y}$
T held constant at 80.00:							
if $X_i = 0$,	$\hat{Y} = -14.74$	$+$	$(0.08 \cdot 0)$	$+$	$(0.19 \cdot 80)$	$=$	0.46
if $X_i = 1$,	$\hat{Y} = -14.74$	$+$	$(0.08 \cdot 1)$	$+$	$(0.19 \cdot 80)$	$=$	0.54
if $X_i = 2$,	$\hat{Y} = -14.74$	$+$	$(0.08 \cdot 2)$	$+$	$(0.19 \cdot 80)$	$=$	0.62
if $X_i = 3$,	$\hat{Y} = -14.74$	$+$	$(0.08 \cdot 3)$	$+$	$(0.19 \cdot 80)$	$=$	0.70
if $X_i = 4$,	$\hat{Y} = -14.74$	$+$	$(0.08 \cdot 4)$	$+$	$(0.19 \cdot 80)$	$=$	0.78
if $X_i = 5$,	$\hat{Y} = -14.74$	$+$	$(0.08 \cdot 5)$	$+$	$(0.19 \cdot 80)$	$=$	0.86
if $X_i = 6$,	$\hat{Y} = -14.74$	$+$	$(0.08 \cdot 6)$	$+$	$(0.19 \cdot 80)$	$=$	0.94
if $X_i = 7$,	$\hat{Y} = -14.74$	$+$	$(0.08 \cdot 7)$	$+$	$(0.19 \cdot 80)$	$=$	1.02
if $X_i = 8$,	$\hat{Y} = -14.74$	$+$	$(0.08 \cdot 8)$	$+$	$(0.19 \cdot 80)$	$=$	1.10
T held constant at 95.00:							
if $X_i = 0$,	$\hat{Y} = -14.74$	$+$	$(0.08 \cdot 0)$	$+$	$(0.19 \cdot 95)$	$=$	3.31
if $X_i = 1$,	$\hat{Y} = -14.74$	$+$	$(0.08 \cdot 1)$	$+$	$(0.19 \cdot 95)$	$=$	3.39
if $X_i = 2$,	$\hat{Y} = -14.74$	$+$	$(0.08 \cdot 2)$	$+$	$(0.19 \cdot 95)$	$=$	3.47
if $X_i = 3$,	$\hat{Y} = -14.74$	$+$	$(0.08 \cdot 3)$	$+$	$(0.19 \cdot 95)$	$=$	3.55
if $X_i = 4$,	$\hat{Y} = -14.74$	$+$	$(0.08 \cdot 4)$	$+$	$(0.19 \cdot 95)$	$=$	3.63
if $X_i = 5$,	$\hat{Y} = -14.74$	$+$	$(0.08 \cdot 5)$	$+$	$(0.19 \cdot 95)$	$=$	3.71
if $X_i = 6$,	$\hat{Y} = -14.74$	$+$	$(0.08 \cdot 6)$	$+$	$(0.19 \cdot 95)$	$=$	3.79
if $X_i = 7$,	$\hat{Y} = -14.74$	$+$	$(0.08 \cdot 7)$	$+$	$(0.19 \cdot 95)$	$=$	3.87
if $X_i = 8$,	$\hat{Y} = -14.74$	$+$	$(0.08 \cdot 8)$	$+$	$(0.19 \cdot 95)$	$=$	3.95

front of the cube. We plot the $\hat{Y}$ line according to the values of X holding T constant at 80.00, as is shown by the line on the front of the cube in Figure 10.4.3.

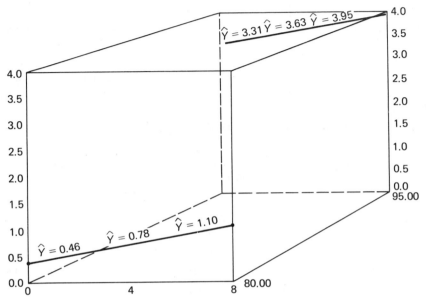

Figure 10.4.3. The partial regression slopes of Y on X holding T constant at 80.00 and 95.00.

We can also carry out the linear predictions and plot the regression line for Y on X holding T constant at 95.00 on the back of the cube. Thus given that $X_i = 0$ and $T_i = 95.00$, we would predict the value of Y to be $\hat{Y} = 3.31$, and given that $X_i = 1$ and $T_i = 95.00$, we would predict the value of $\hat{Y}$ to be $Y = 3.39$. As we did with the linear regression predictions with T held constant at 80.00, we plot the predicted values with T held constant at 95.00 on the back surface of the cube. This one-dimensional line represents the regression of Y on X with T held constant at 95.00, as is shown by the line on the back of the cube in Figure 10.4.3.

These two regression lines of $b_{YX \cdot T} \cdot X_i$ with T held constant represent the lower and upper limits of the linear predictions of Y given X holding T constant for the respondents. In other words, we plot the partial regression lines along the appropriate surface of the three-variable cube. Naturally, we could carry out similar calculations estimating Y given X holding T constant at any level of T.

We can also plot the regression lines of $b_{YT \cdot X} \cdot T_i$ with X held constant. These predictions are shown in Table 10.4.3. Given that $T_i = 80.00$ and

Table 10.4.3.   Estimated values of Y given T_i with X held constant at 0 and 8

	$\hat{Y} = a_{Y \cdot XT}$	$+$	$(b_{YX \cdot T} \cdot X_i)$	$+$	$(b_{YT \cdot X} \cdot T_i)$	$=$	$\hat{Y}$
X held constant at 0:							
if $T_i = 80.00$,	$\hat{Y} = -14.74$	$+$	$(0.08 \cdot 0)$	$+$	$(0.19 \cdot 80.00)$	$=$	0.46
if $T_i = 82.50$,	$\hat{Y} = -14.74$	$+$	$(0.08 \cdot 0)$	$+$	$(0.19 \cdot 82.50)$	$=$	0.94
if $T_i = 85.00$,	$\hat{Y} = -14.74$	$+$	$(0.08 \cdot 0)$	$+$	$(0.19 \cdot 85.00)$	$=$	1.41
if $T_i = 87.50$,	$\hat{Y} = -14.74$	$+$	$(0.08 \cdot 0)$	$+$	$(0.19 \cdot 87.50)$	$=$	1.89
if $T_i = 90.00$,	$\hat{Y} = -14.74$	$+$	$(0.08 \cdot 0)$	$+$	$(0.19 \cdot 90.00)$	$=$	2.36
if $T_i = 92.50$,	$\hat{Y} = -14.74$	$+$	$(0.08 \cdot 0)$	$+$	$(0.19 \cdot 92.50)$	$=$	2.84
if $T_i = 95.00$,	$\hat{Y} = -14.74$	$+$	$(0.08 \cdot 0)$	$+$	$(0.19 \cdot 95.00)$	$=$	3.31
X held constant at 8:							
if $T_i = 80.00$,	$\hat{Y} = -14.74$	$+$	$(0.08 \cdot 8)$	$+$	$(0.19 \cdot 80.00)$	$=$	1.10
if $T_i = 82.50$,	$\hat{Y} = -14.74$	$+$	$(0.08 \cdot 8)$	$+$	$(0.19 \cdot 82.50)$	$=$	1.58
if $T_i = 85.00$,	$\hat{Y} = -14.74$	$+$	$(0.08 \cdot 8)$	$+$	$(0.19 \cdot 85.00)$	$=$	2.05
if $T_i = 87.50$,	$\hat{Y} = -14.74$	$+$	$(0.08 \cdot 8)$	$+$	$(0.19 \cdot 87.50)$	$=$	2.53
if $T_i = 90.00$,	$\hat{Y} = -14.74$	$+$	$(0.08 \cdot 8)$	$+$	$(0.19 \cdot 90.00)$	$=$	3.00
if $T_i = 92.50$,	$\hat{Y} = -14.74$	$+$	$(0.08 \cdot 8)$	$+$	$(0.19 \cdot 92.50)$	$=$	3.48
if $T_i = 95.00$,	$\hat{Y} = -14.74$	$+$	$(0.08 \cdot 8)$	$+$	$(0.19 \cdot 95.00)$	$=$	3.95

$X_i = 0$, we would predict the value of Y to be $\hat{Y} = 0.46$, and given that $T_i = 82.50$ and $X_i = 0$, we would predict the value of Y to be $\hat{Y} = 0.94$. Table 10.4.3 also shows the estimated Y values for X held constant at 8. We can use these values to plot two lines that indicate the partial regression slopes of Y on T holding X constant at 0 and 8, as shown in Figure 10.4.4.

The regression lines we have drawn on the four outside surfaces of the cube define the boundaries of the best-fitting surface that describes the respondents in the three-dimensional scattercube. Now that we have the regression slopes around the four surfaces of the box, we can connect them, thereby forming a flat two-dimensional surface or a plane in three-dimensional space. The sides of the plane follow the lines around the cube along the X and T directions. Thus the plane connects the partial-linear-regression prediction values of Y given X with T held constant and Y given T with X held constant, as is shown in Figure 10.4.5.

Each respondent has a score on Y, a score on X, and a score on T, and the surface of this plane describes and summarizes the respondent values on Y with the least-squares surface. To predict Y values according to the scores on the independent and control variables simultaneously is possible by use of a multiple-linear-regression plane.

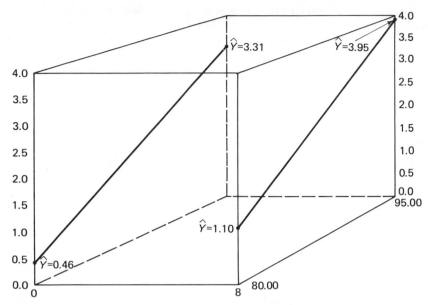

Figure 10.4.4. The partial regression slopes of Y on T holding X constant at 0 and 8.

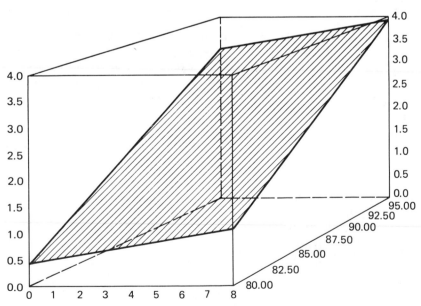

Figure 10.4.5. The linear regression plane of Y given X and T.

228

10.5 USING MULTIPLE LINEAR REGRESSION

Going beyond the three-variable case merely involves extending the principles we have been using. For example, instead of having a set of two-dimensional estimates in three-dimensional space, we would have a set of three-dimensional estimates in four-dimensional space, or a set of four-dimensional estimates in five-dimensional space, and so forth up to a set of $(l - 1)$-dimensional estimates in l-dimensional space. However, we cannot demonstrate such estimates graphically. It is therefore common to deal with only the *values* of the one partial a coefficient and the $(l - 1)$ partial b coefficients. Thus when we have six variables with one dependent variable and five independent variables, we describe them in terms of the multiple-linear-regression formula by calculating one fifth-order partial a coefficient and five fourth-order partial b coefficients.

The use of multiple linear regression has been facilitated to a great extent by the development of high-speed computers. The calculations by hand are arduous, time-consuming, and prone to calculating or transposing errors. Rather than go into the calculations of higher order partial regression coefficients, let us consider some of the ways the techniques can be applied to practical research problems.

Assume that we have been asked to establish a program that is designed to provide professional vocational skills to college seniors. Because of the experimental nature of this project, funds are limited and only a small number of students can be trained initially. One of the goals of this program is to select students who will benefit most from the course. In order to determine the extent to which students may benefit, a pilot study was conducted in which 40 students were given the special program. Our task is to analyze how these 40 pilot subjects did by use of multiple-linear-regression analysis.

The system of variables that we decide to use relates performance on a skill test (Y), to age of respondent (X), the number of science courses taken two years prior to the training (T), and the respondent's aptitude score for the professional skill (V). The multiple-linear-regression formula equation for these four variables can be expressed symbolically:

$$\hat{Y}_i = a_{Y \cdot XTV} + b_{YX \cdot TV} \cdot X_i + b_{YT \cdot XV} \cdot T_i + b_{YV \cdot XT} \cdot V_i$$

Fortunately, a computer calculates the appropriate regression coefficients, and the computer prints out the following values:

$$\bar{Y} = \text{mean performance score} \quad = 28.7$$
$$\bar{X} = \text{mean age} \quad = 20.0$$

$$\bar{T} = \text{mean number of science courses} = 3.0$$
$$\bar{V} = \text{mean aptitude score} = 40.0$$
$$a_{Y \cdot XTV} = 2.0$$
$$b_{YX \cdot TV} = 0.6$$
$$b_{YT \cdot XV} = 0.9$$
$$b_{YV \cdot XT} = 0.3$$

It is useful to know that if we utilize the mean values of the independent and control variables with the regression coefficients, we can determine the mean value of the dependent variable from the regression equation. Here is the multiple-linear-regression equation and the mean values determined by the pilot study.

$$\bar{Y} = \hat{Y} = a_{Y \cdot XTV} + b_{YX \cdot TV} \cdot \bar{X} + b_{YT \cdot XV} \cdot \bar{T} + b_{YV \cdot XT} \cdot \bar{V}$$
$$28.7 = 2.0 + (0.6) \cdot (20.0) + (0.9) \cdot (3.0) + (0.3) \cdot (40.0)$$

This tells us that if we take the mean values for the three independent variables and substitute them into the multiple-linear-regression equation, we would "predict" the mean of Y, skill performance, to be 28.7. As you can see, this is exactly the same as the mean calculated directly from the respondents' scores, and it cannot be otherwise given the nature of the calculations.

Now that we have the essential information for the multiple-linear-regression equation, let us examine how we might put it to use. Suppose that four students wish to enroll in the program, but that we can select only one of them. Given the following applicants and their characteristics, our problem is to decide which one to select.

Student	Age	Science Background (T)	Aptitude Score (V)
Bob	18	3.0	28
Gail	21	3.5	43
George	27	1.0	30
Martha	19	2.0	50

Our task is a simple one, for we merely carry out the multiple-linear-regression equation for each of the four students:

$$\text{Bob} \quad \hat{Y} = 2 + (0.6)18 + (0.9)3.0 + (0.3)28$$
$$= 2 + 10.8 + 2.7 + 8.4 = \underline{\underline{23.9}}$$

$$\text{Gail} \quad \hat{Y} = 2 + (0.6)21 + (0.9)3.5 + (0.3)43$$
$$= 2 + 12.6 + 3.15 + 12.9 = \underline{\underline{30.65}}$$
$$\text{George} \ \hat{Y} = 2 + (0.6)27 + (0.9)1.0 + (0.3)30$$
$$= 2 + 16.2 + 0.9 + 9.0 = \underline{\underline{28.1}}$$
$$\text{Martha} \ \hat{Y} = 2 + (0.6)19 + (0.9)2.0 + (0.3)50$$
$$= 2 + 11.4 + 1.8 + 15.0 = \underline{\underline{30.2}}$$

Since we are interested in selecting the student whose background characteristics lead us to predict that he or she will benefit most from the program, we want to select the student whose predicted performance is the highest. Gail is the student we would select, since her predicted performance score is 30.65.

Let us consider another problem within the same study. This problem focuses on a different kind of prediction. Suppose that 50 students are selected to begin the program and the mean values for each of their background characteristics are as follows:

$$\text{Age } \bar{X} \qquad\qquad\qquad = 22.0$$
$$\text{Science Background Experience } \bar{T} = \ \ 3.0$$
$$\text{Aptitude Score } \bar{V} \qquad\qquad = 39.0$$

Since the program is just beginning, we are interested in predicting how successful these students will be given their background characteristics. Thus our task now is to predict a mean performance score of these 50 students to compare with the mean performance score of the pilot study. Again the task is a simple one, for we merely substitute the values into the multiple-linear-regression equation, and we find that we would predict a mean performance score for this group of 29.6.

$$\hat{Y} = 2 + (0.6)22 + (0.9)3 + (0.3)39 = 2 + 13.2 + 2.7 + 11.7 = 29.6$$

Now suppose that 25 of the students drop out of the program before completion. Since we know the background characteristics of all the students, we decide to analyze those who have dropped out and compare them against those who have remained in the program. We can then make predictions about the performance of those 25 students who remain. The mean values for those students who dropped out are:

$$\text{Age } \bar{X} \qquad\qquad\qquad = 20.0$$
$$\text{Science Background Experience } \bar{T} = \ \ 2.0$$
$$\text{Aptitude Score } \bar{V} \qquad\qquad = 32.0$$

This problem requires that we do some calculating in addition to that for the linear regression. In the first place, we know that the mean score on each of the variables should be multiplied by 50 to give the sum of the respondents' scores on that variable. The sum of the scores on age is 1100; the sum of the scores on the number of science courses taken is 150; and the sum of the aptitude scores is 1950. Next we find the sums for the three variables of the students who dropped out of the program. The sum of scores on age is 500; the sum of scores on the number of science courses is 50; and the sum of the aptitude scores is 800. When we subtract the sums of those who have dropped out of the program from the totals, we have the sums of the scores of those remaining in it. We divide these by the remaining number of students, 25, and find the mean values for each of the three variables for the remaining 25 students.

Age $\bar{X}$ $= \dfrac{1100 - 500}{25} = \dfrac{600}{25} = 24.0$

Science Background Experience $\bar{T} = \dfrac{150 - 50}{25} = \dfrac{100}{25} = 4.0$

Aptitude Score $\bar{V}$ $= \dfrac{1950 - 800}{25} = \dfrac{1150}{25} = 46.0$

Our task becomes a familiar one, and we substitute these values into the multiple-linear-regression equation. The equation shows that we would predict the remaining 25 students to have a mean performance of 33.8. This indicates that we predict that the mean performance score would be 4.2 performance points higher for the remaining students than we would have predicted for the original 50.

Multiple-linear-regression analysis has numerous and varied applications. However, the utility of regression is tied to the related matter of multiple correlation, which we will learn about in Chapter 11.

EXERCISES

1. In your own words, define, describe, or discuss the following terms and give an hypothetical example of each.

$a_{Y \cdot XT}$
partial b coefficient
two variable linear regression equation
partial regression coefficient
multiple-linear-regression formula
test variable
first-order partial
linear prediction
slope

2. If $\bar{Q}$, $\bar{S}$, $\bar{Z}$, and $\bar{M}$ are the means of four variables in a linear relationship, write the multiple linear regression equation for $\bar{Q}$.

3. Using the data presented in problem 4 in Chapter 7, calculate $a_{Y \cdot XT}$, $b_{YX \cdot T}$, and $b_{YT \cdot X}$.

4. In problem 5, Chapter 7, a third variable grade point average (T) is introduced and results in the following values:

$$\bar{T} = 3.0 \qquad\qquad s_{XT} = 6.2$$
$$s_T^2 = 2.5 \qquad\qquad s_{YT} = 10.9$$

Using these data and the results of problem 5 in Chapter 7, calculate:

$b_{YX \cdot T}$
$b_{TY \cdot X}$
$a_{Y \cdot XT}$

5. Suppose you are carrying out a research project in which you are studying the factors that influence student academic performance. You have taken a sample of 100 college students and measured them on interval scales for the following variables.

Y = Grade point average
X = Number of hours spent studying each day
T = Number of hours spent watching T.V. each day
V = Number of college organizations in which one is an active member

The data you have gathered have been collated, tabulated, and cross-tabulated. A computer program produces the following zero-order b coefficient matrix.

Using this matrix, answer the questions below:

Matrix of zero order b coefficients

	Y	X	T	V
Y	1.00	0.90	−0.46	0.33
X	1.00	1.00	−0.62	−0.08
T	−0.66	−0.80	1.00	−0.42
V	0.44	−0.10	−0.38	1.00

What is the first-order partial b coefficient between grade point average and the number of hours spent studying each day:

A. Controlling for the number of hours spent watching T.V. each day?

B. Controlling for the number of college organizations in which one is an active member?

C. What is the first-order partial b coefficient between the number of hours spent studying each day and the number of hours spent watching T.V. each day controlling for the number of college organizations in which one is an active member?

6. Using the multiple linear regression equation and the values in Section 10.5 answer the questions below:

A. Four students wish to enroll in the program, but only one can be selected.

Given each applicant's background characteristics listed below, which one should we select?

Student	Age (X)	Science Background (T)	Aptitude Score (V)
1	19	3	25
2	20	3.2	45
3	24	2.5	50
4	18	1	35

B. Suppose that 50 students are selected for the program. The mean values for each of their background characteristics are given below. What mean performance would we predict for the 50 people?

Age ($\bar{X}$) = 21
Science Background ($\bar{T}$) = 3.3
Aptitude Score ($\bar{V}$) = 40

C. Suppose 25 of the students dropped out before completion. The mean scores for the 25 drop outs are given below. What mean performance would we predict for the 25 *remaining* students?

Age ($\bar{X}$) = 19
Science Background ($\bar{T}$) = 2.0
Aptitude Score ($\bar{V}$) = 30

7. A friend who attends another college is writing a term paper on social attitudes on voting behavior and is using simple linear regression to analyze the data. He has identified 25 attitudes which he thinks affect voting behavior, and he has regressed voting behavior on these 25 attitudes one-at-a-time, i.e., in 25 simple regressions. He states that 6 variables have a marked influence on voting behavior. What flaws do you see in this research? How would you explain them to him?

8. Using words and phrases that your *least* mathematically inclined friend can understand, verbally explain multiple linear regression.

II

Partial and
Multiple
Correlation

Thus far we have learned statistical techniques to help us classify individuals or groups; to measure, predict, estimate, or describe variables; and to measure such things as the linear relationships and correlations between variables. We have also learned that these seemingly diverse methods are interwoven and that aspects of each mesh together in multivariate regression techniques. In this chapter, we use this background to expand the concepts of zero-order total correlation and explanation in the framework of multiple linear regression.

Part of our interest is in developing measures to assess the correlation between variables in light of the influence of a control or test variable. In this regard we are concerned, first, with the zero-order total correlation between two variables, second, with the higher order partial correlations between two variables controlling for one or more other variables, and third, with the overall (multiple) correlation of the independent and control variables combined. The theories and equations to get at these major concerns form the basis of this chapter.

11.1 CONTROL VARIABLES
AND THE CONCEPT OF EXPLANATION

Earlier we discussed the ideas that the general statistical usage for the word "explanation" is descriptive, and that in general "explanation" means "accounts for." More recently we learned that a "control" variable is used to test existing relationships between two others and is "held constant." With these two thoughts in mind, we can conceive of explanation in a more rigorous sense. Namely, we *control* variables in an attempt to *explain* things about their influence.

Not only are there many ways in which we can explain things, but there are also many ways we can look at the concept of explanation. Three of these are important for us: We seek to explain (1) substantive issues, (2) statistical relationships between variables, and (3) total variance in one variable. We know that it is possible to explain a substantive event, occurrence, or phenomenon. Such an explanation may be verbal, graphic, or symbolic; it may take a philosophical or empirical track depending on the nature of the substantive issue in question. For example, at the common-sense level we may "explain" a wet driveway in the morning by citing the rain that fell during the night. Or we may "explain" symbolically how much we care about someone by extensive eye contact and sincere smiling.

Second, moving beyond the level of common sense, sociologists customarily "explain" substantive phenomena (such as trends) graphically. For example, in the study *Voting* the relationship between the dependent variable "vote" and socioeconomic status is "explained" as being partly a function of the political conditions under which each generation comes of age.[1] Figure 11.1.1 graphically indicates that there is a greater tendency among younger voters to vote along social status lines than there is among older voters. The authors "explain" it like this:

> The younger generation raised in the New Deal era showed a high tendency to vote along the socioeconomic class lines associated with the Roosevelt elections . . . Their elders, introduced to politics in the Republican 1920's, are not so far apart on class lines. Acceptance of the political norms current at the time of political initiation does not stop there; it tends to perpetuate itself through succeeding elections.[2]

This example points to an important tie between substance and method: The substantive issue is generally couched in terms of the relationship

[1]Bernard Berelson, Paul Lazarsfeld, and William McPhee, *Voting* (Chicago: The University of Chicago Press, 1954), p. 60.
[2]*Ibid.*, pp. 59 and 61.

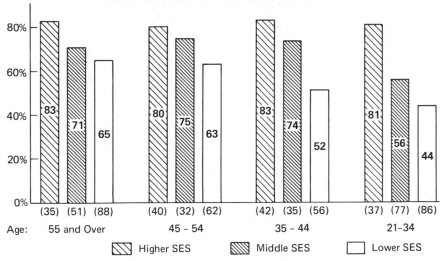

PERCENTAGE REPUBLICAN OF TWO-PARTY VOTE

Age: | 55 and Over | 45 - 54 | 35 - 44 | 21-34

Higher SES   Middle SES   Lower SES

Figure 11.1.1. Age differences indicate a trend toward greater class voting in the New Deal generation.

between two operationally defined variables. This brings us to the third way we can look at explanation.

Earlier we learned that we can decompose the total variance in a dependent variable by comparing the value predicted from its linear relationship with one independent variable, its mean value, and the individual's actual value on it. We saw this in the equations

$$s_Y^2 = \frac{\sum (Y - \bar{Y})^2}{N}$$

$$\hat{Y} = a_{Y \cdot X} + b_{YX} X_i$$

and

$$\frac{\sum (Y_i - \bar{Y})^2}{N} = \frac{\sum (\hat{Y} - \bar{Y})^2}{N} + \frac{\sum (Y_i - \hat{Y})^2}{N}$$

total variance = explained variance + unexplained variance

By placing the explained variance in the numerator and the total variance in the denominator, we can create a ratio that measures the proportion of the variance in Y that is explained by its linear relationship with X.

By expanding the linear regression equation, we can take into account the partial and multiple influence of an independent variable plus any number of test variables. In this framework, let us consider the concept of controlling

237

for test variables and the idea of explained variance with more than two variables.

The multivariate version of how we determine the proportion of variance explained is merely an extension of the two-variable case, and it follows the logic of the link between the zero-order b coefficient and the zero-order correlation coefficient. The first-order partial coefficient $b_{YX \cdot T}$ is calculated by holding constant the control variable T and then examining the regression of Y on X. The three-variable example is a crucial bridge between two-variable statistics and multivariate ones. Once again the three-variable case helps us understand partial correlation.

Remember that the multiple-linear-regression equation

$$\hat{Y} = a_{Y \cdot XT} + b_{YX \cdot T} \cdot X_i + b_{YT \cdot X} \cdot T_i$$

cannot be represented by a single line. Instead, it may be shown as a two-dimensional plane in the three-dimensional space of variables Y, X, and T. On a scattercube we can show variable X across the bottom, variable Y up the side, and variable T along the depth of the cube. The formula for Y describes the predicted values for Y and the plane that best fits the respondents according to the least-squares principle. To portray the respondents in such a cube we used $b_{YT \cdot X}$ to represent the slope of Y on T controlling for X, that is, the slope from front to rear on the sides of the cube. Likewise, $b_{YX \cdot T}$ represents the slope of Y on X controlling for T, that is, the slopes across the front and rear of the cube.

Try to visualize this slope as having two distinct directions; the slope from front to rear is one, and the slope from side to side is another. The two *together* create the plane we discussed earlier, and our task now is to measure the partial correlation for the slope between Y and one of the other variables controlling for either X or T. What we learned about partial b coefficients helps us understand the idea of partial correlation coefficients. We now are interested in measuring the *degree of relationship* between the dependent variable and one of the independent variables controlling for the other independent variables. Put differently, we are interested in that part of the correlation that remains after we have held constant the control variables.

The zero-order b coefficient tells us the regression of one variable on another. Similarly, the zero-order correlation coefficient between Y and X, r_{YX}, tells us *how well* the two variables go together. Furthermore it tells us how closely the respondents are clustered along the regression estimates. If r_{YX} is near ± 1.00, the points (respondents' joint scores) are very close to the linear estimates and the regression line. If, on the other hand, they fall in the same general direction but are scattered further from the line, r_{YX} would be lower. If they are just random points, then r_{YX} would be 0.00.

11.2 THE PARTIAL PRODUCT-MOMENT CORRELATION COEFFICIENT

Let us go directly to the first-order partial correlation coefficient between Y and X controlling for T, $r_{YX \cdot T}$. The *partial correlation coefficient* can be thought of as the average degree of spread of the respondents around the regression line of Y on X at various levels of T. More specifically, for any level of T a unique slope of Y on X could be plotted. The partial $r_{YX \cdot T}$ is a weighted average that measures how well the individual respondents fit the line of Y on X at each level of T.

The first-order partial correlation coefficient $r_{YX \cdot T}$ has the same characteristics as the zero-order product-moment correlation coefficient r_{YX}. Specifically, it ranges from ± 1.00 (perfect positive or negative correlations) to 0.00 (no correlation), and it measures *how well* the single straight line of Y on X fits the points of a scattercube of the two variables controlling for a third variable T.

In order to calculate the first-order partial correlation coefficient, we operate similarly to calculating partial regression coefficients. Thus we use the three zero-order total correlation coefficients of the three variables being examined, r_{YX}, r_{XT}, and r_{YT}. The first-order partial correlation coefficient of Y given X controlling for T can be expressed symbolically:

$$r_{YX \cdot T} = \frac{r_{YX} - (r_{YT}) \cdot (r_{TX})}{\sqrt{1.00 - r_{YT}^2} \cdot \sqrt{1.00 - r_{XT}^2}}$$

Recall that the formula for the first-order partial regression coefficient $b_{YX \cdot T}$ was expressed symbolically as:

$$b_{YX \cdot T} = \frac{b_{YX} - (b_{YT}) \cdot (b_{TX})}{1.00 - (b_{XT}) \cdot (b_{TX})}$$

Examining the formula for $r_{YX \cdot T}$ closely helps us understand what it is measuring. The numerator for $r_{YX \cdot T}$ is very similar to the one for $b_{YX \cdot T}$. It calls for the zero-order correlation coefficients instead of the zero-order b coefficients, but its basic format is identical. Both start with the zero-order total coefficient of Y given X and then subtract the product of the zero-order total coefficients of Y given T and T given X. The denominators of the two are where we find important differences. The first-order partial correlation coefficient uses both Y and X in combination with T. With the first-order partial b coefficients, we had to be cognizant of the order of the subscripts because $b_{YX \cdot T} \neq b_{XY \cdot T}$. However, since $r_{XY \cdot T} = r_{YX \cdot T}$, the order of the variables shown by the subscripts of partial correlations is unimportant.

The denominator for $b_{YX \cdot T}$ is 1.00 minus the product of the zero-order b's of variables X and T, that is, $(1.00 - b_{XT} \cdot b_{TX})$. Thus it contains information about the independent variable X and the control variable T. Clearly it does not contain any information about the *dependent* variable Y. Put differently, the denominator for the first-order partial b coefficient tells us about the independent variable X and about the control variable T. This means that we use the first-order partial $b_{YX \cdot T}$ to predict Y from knowledge of X controlling for T.

With partial correlation we are no longer interested in predicting. Instead we want to know how well Y and X go *together* controlling for T. To measure this we use a different denominator. Since we are interested in measuring how well Y goes with X controlling for T, the denominator for the first-order partial correlation coefficient $r_{YX \cdot T}$ involves the independent variable X, the control variable T, and the dependent variable Y.

In Chapter 8 we learned the steps to calculate zero-order correlation coefficients and met the idea of a matrix of zero-order correlation coefficients. To assist our present calculation, Table 11.2.1 repeats the zero-order correlations between grade point average (Y), number of hours spent studying (X), and high school average (T).

We can use the values in the matrix of zero-order correlation coefficients to calculate the first-order partials. We are interested in (1) the correlation between Y and X controlling for T and (2) the correlation between Y and T controlling for X. Incidentally, these are the correlation coefficients for the sets of predictions we made with the first-order b coefficients; that is, $r_{YX \cdot T}$ is the correlation for $b_{YX \cdot T}$. The first-order partial correlation coefficient of grade point average (Y) on number of hours spent studying (X) controlling for high school average (T) can be calculated with the formula

$$r_{YX \cdot T} = \frac{r_{YX} - (r_{YT}) \cdot (r_{TX})}{\sqrt{1.00 - r_{YT}^2} \cdot \sqrt{1.00 - r_{XT}^2}}$$

$$r_{YX \cdot T} = \frac{0.69 - (0.92) \cdot (0.62)}{\sqrt{1.00 - 0.85} \cdot \sqrt{1.00 - 0.38}} = \frac{0.69 - 0.57}{\sqrt{0.15} \cdot \sqrt{0.62}}$$

$$r_{YX \cdot T} = \frac{0.12}{0.39 \cdot 0.79} = \frac{0.12}{0.31} = \underline{\underline{0.39}}$$

Table 11.2.1. Matrix of zero-order product-moment correlation coefficients for variables Y, X, and T
($N = 8$)

	Y	X	T
Y	1.00	0.69	0.92
X	0.69	1.00	0.62
T	0.92	0.62	1.00

The partial correlation coefficient between Y and X controlling for T can be interpreted as indicating that there is a correlation of $r_{YX \cdot T} = 0.39$ between grade point average and the number of hours spent studying *after* high school average has explained as much as it can of each of the two variables.

We also can carry out the calculations for the first-order partial correlation coefficient of Y and T controlling for X:

$$r_{YT \cdot X} = \frac{r_{YT} - (r_{YX}) \cdot (r_{XT})}{\sqrt{1.00 - r_{YX}^2} \cdot \sqrt{1.00 - r_{TX}^2}}$$

$$r_{YT \cdot X} = \frac{0.92 - (0.69)(0.62)}{\sqrt{1.00 - 0.48} \cdot \sqrt{1.00 - 0.38}} = \frac{0.92 - 0.43}{\sqrt{0.52} \cdot \sqrt{0.62}}$$

$$r_{YT \cdot X} = \frac{0.49}{0.72 \cdot 0.79} = \frac{0.49}{0.57} = \underline{\underline{0.86}}$$

The partial correlation coefficient $r_{YT \cdot X}$ between Y and T after X has explained all it can of each is 0.86. Even though we do not use it for analytical purposes, there is also the partial correlation between X and T controlling for Y. All three first-order partial correlations are shown in the matrix of first-order partial coefficients in Table 11.2.2.

Table 11.2.2.   Matrix of first-order* partial
correlation coefficients for *Y, X,* and *T*

	Y	X	T
Y	1.00	0.39	0.86
X	0.39	1.00	-0.04
T	0.86	-0.04	1.00

*Beyond first-order partials, it is not possible to display the values in a matrix such as this one. This is because there are many *combinations* of independent variables.

The interpretation of the partial correlation becomes clearest when we square the first-order partial as we did the zero-order coefficient. Recall that the zero-order correlation coefficient squared, r_{YX}^2, equals the proportion of variance in Y that is explained by its linear relationship with X. The first-order partial correlation coefficient squared, $r_{YT \cdot X}^2$, equals the proportion of the unexplained variance in Y that is explained by T in addition to that which is explained by X alone. Thus since $r_{YX}^2 = 0.69^2 = 0.48$, and since $r_{YT \cdot X}^2 = 0.86^2 = 0.74$, we can say that X alone explains 48 percent of the variance in Y and that T explains an additional 74 percent of that which is unexplained by X, that is, 74 percent of 52 percent.

Importantly, $r_{YX \cdot T}$ and $r_{YT \cdot X}$ measure the average "spreadoutiveness" of the respondents on Y given an independent variable at constant levels of the control variable. Before we close the curtain on explanation and correlation, let us put things into perspective.

The first-order partial correlation coefficient $r_{YX \cdot T}$ measures the *goodness of fit* of the linear regression of Y on X controlling for T. Thus it measures the degree of spread around the least-squares *lines* of Y on X controlling for T. However, it does not measure how well the respondents fit the *overall plane* created by the variables and the multiple-linear-regression predictions. To get at this, we soon will learn about the multiple correlation coefficient. First, however, we should examine the concept of multivariate explanation.

11.3 EXPLAINED VARIANCE WITH TWO OR MORE VARIABLES

Let us go back to some familiar concepts as we learn about multivariate explanation. In the first place, the variance of variable Y, s_Y^2, represents the *total* variance in Y. Remember we arrive at s_Y^2 by subtracting a single value estimate, that is, the arithmetic mean of variable Y, from each individual score, squaring the difference, summing the squares, and dividing by N. Thus,

$$s_Y^2 = \frac{\sum_{i=1}^{N}(Y_i - \bar{Y})^2}{N} = 100 \text{ percent of the variance in } Y$$

We also have the total variance of X, s_X^2:

$$s_X^2 = \frac{\sum_{i=1}^{N}(X_i - \bar{X})^2}{N} = 100 \text{ percent of the variance in } X$$

We also have the total variance of T, s_T^2:

$$s_T^2 = \frac{\sum_{i=1}^{N}(T_i - \bar{T})^2}{N} = 100 \text{ percent of the variance in } T$$

We also have the covariances of the three variables:

$$s_{XY} = \frac{\sum_{i=1}^{N}(X_i - \bar{X}) \cdot (Y_i - \bar{Y})}{N} = 100 \text{ percent of the covariance in } X \text{ and } Y$$

$$s_{TY} = \frac{\sum_{i=1}^{N}(T_i - \bar{T}) \cdot (Y_i - \bar{Y})}{N} = 100 \text{ percent of the covariance in } T \text{ and } Y$$

$$s_{XT} = \frac{\sum_{i=1}^{N}(X_i - \bar{X}) \cdot (T_i - \bar{T})}{N} = 100 \text{ percent of the covariance in } X \text{ and } T$$

Then we similarly calculate the zero-order correlation coefficient r_{YX}. When squared, r_{YX}^2 represents the proportion of the variance in Y that is explained by its linear relationship with X. This means, of course, that r_{YX}^2 represents some proportion of the total variance of Y. Thus,

$$r_{YX}^2 = \frac{s_{XY}^2}{s_X^2 s_Y^2} = \begin{array}{l} \text{the proportion of } s_Y^2 \text{ that is explained} \\ \text{by its linear relationship with } X \end{array}$$

Since s_Y^2 is 100 percent of 1.00, and since r_{YX}^2 is some proportion of it, the rest of the variance must be equal to $(1.00 - r_{YX}^2)$. This remainder is the proportion of variance in Y that is *not* explained by X. Put differently, it is the error that remains after we predict $\hat{Y}$ knowing values of X. Technically, the expression $(1.00 - r_{YX}^2)$ is called the *coefficient of alienation* because it represents the proportion of the variance in Y that is "alien" to the relationship between it and X. Thus it represents that part of the variance still left after X has explained all it can. In other words, if $r_{YX}^2 =$ the proportion of variance in Y *explained* by its linear relationship with X, then $(1.00 - r_{YX}^2) =$ the proportion of the variance in Y that is *unexplained* by its linear relationship with X.

We can also consider the third variable T and calculate the correlation between it and Y, r_{YT}. When we square it, r_{YT}^2 tells us the proportion of variance in Y that is explained by its linear relationship with T. Naturally, the proportion of variance that is left unexplained, that is, the error in prediction, is equal to $(1.00 - r_{YT}^2)$.

Our analytical efforts aim at explaining as much of the total variance in the dependent variable as possible by use of all the independent variables together. Thus we ought to be able to add together the explanatory powers of several variables, thereby getting a multiple measure of explained variance. For example, suppose that we start with the proportion of the variance in Y that is explained by its linear relationship with X, that is, r_{YX}^2. Then suppose that we add to it the proportion of variance in Y that is explained by its linear relationship with T, that is, r_{YT}^2. These two proportions added together, $r_{YX}^2 + r_{YT}^2$, ought to explain a greater proportion of the total variance in Y than either does alone.

This situation is easy to understand when seen diagrammatically. Diagram A in Figure 11.3.1 on page 244 graphically portrays the total variance of Y, s_Y^2, as an area, which represents 100 percent of its variance. In Diagram B, 100 percent of the variance in X and Y is represented by two areas, one for each variable, and the proportion of variance explained in each overlaps and is

$$\frac{s_{XY}^2}{s_X^2 s_Y^2} = r_{YX}^2$$

This indicates the proportion of variance in Y that is explained by its linear relationship with X, and since $r_{YX}^2 = r_{XY}^2$, r_{XY}^2 indicates the proportion of variance in X explained by its linear relationship with Y.

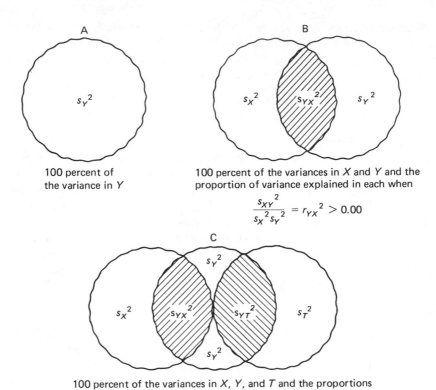

Figure 11.3.1. Diagrammatic representations of the variances in X, Y, and T and the proportions of variance explained when X and T are not correlated.

X and T both may be used to explain proportions of the variance in Y, and if they are *not correlated* with each other, they do not overlap. Diagram C illustrates the case in which the proportion of variance in Y explained by X and the proportion explained by T are separate and distinct proportions; that is, s^2_{YX} and s^2_{YT} do not overlap. This situation occurs (1) when X and Y are correlated and r^2_{YX} is greater than zero, $r^2_{YX} > 0.00$, (2) when Y and T are correlated and r^2_{YT} is greater than zero, $r^2_{YT} > 0.00$, and (3) when X and T are uncorrelated and r^2_{XT} is equal to zero, $r^2_{XT} = 0.00$. In such cases, we can say that the proportion of variance explained is the sum of the two zero-order correlation coefficients squared, and we can use a capital R-squared to indicate the multiple proportion of variance explained. It can be expressed symbolically for this *special* case:

$$R^2_{Y \cdot XT} = r^2_{YX} + r^2_{YT}$$

244

$R^2_{Y \cdot XT}$ is a second-order multiple coefficient that indicates the proportion of variance in the variable to the left of the dot that is explained by its multiple linear relationship with the variables to the right of the dot. This coefficient often is called *multiple R-squared* or *big R-squared* to distinguish it from zero-order total or higher order partial-product-moment-correlation coefficients, which often are called *r-squared* or *small r-squared.*

Naturally, it is possible that X and T are correlated and that one helps explain a proportion of the variance in the other. Figure 11.3.2 illustrates the

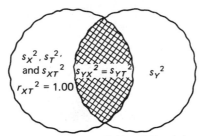

100 percent of the variances in X, Y, and T and the proportion of variance explained in each when

$$\frac{s_{XT}^2}{s_X^2 s_T^2} = r_{XT}^2 = 1.00 \text{ and } \frac{s_{YX}^2}{s_Y^2 s_X^2} = \frac{s_{YT}^2}{s_Y^2 s_T^2} = r_{YX}^2 = r_{YT}^2 > 0.00$$

Figure 11.3.2. Diagrammatic representation of the variances in X, Y, and T and the proportions of variance explained when X and T are perfectly correlated.

extreme and unusual situation when X and T are perfectly correlated. Even though this situation is rare, our discussion of it will help you understand the more common ones, which will be described shortly. In this extreme case of perfect correlation between X and T, we can represent the variance in X and the variance in T with one circle. In other words, the two variances and their covariance completely overlap, and they completely explain each other; therefore, they may be represented by one circle. Thus, in Figure 11.3.2, the circle on the left represents s_X^2, s_T^2, and s_{XT}^2.

Since X and T are perfectly correlated ($r_{XT}^2 = 1.00$), 100 percent of the variance in X is explained by its relationship with T and vice versa. When this is the case, the proportion of variance explained in Y will be the same regardless of whether we use X or T; that is, $r_{YX}^2 = r_{YT}^2$. In such a case, we can express the multiple proportion of variance explained as being equal to either of the zero-order correlation coefficients squared. It can be expressed symbolically for this *special* case:

$$R^2_{Y \cdot XT} = r_{YX}^2 = r_{YT}^2$$

Let us turn now to the more common situation in which X and T are somewhat correlated with each other, and one helps explain a proportion of the variance in the other, that is, $r^2_{XT} > 0.00$. In such a case, the covariances s_{YX}, s_{YT}, and s_{XT} overlap, as can be seen by the cross-hatched shaded area in the center of the three areas in Figure 11.3.3. This means that there is an overlapping proportion of explained variance, that the three variables are intercorrelated, and that each explains part of the others, that is, $r^2_{YX} > 0.00$, $r^2_{YT} > 0.00$, and $r^2_{XT} > 0.00$.

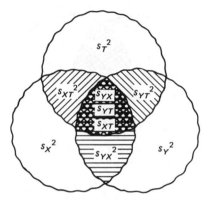

100 percent of the variances in Y, X, and T and the proportion of variance explained in each when s_{YX}, s_{YT}, and s_{XT} overlap, and when $r_{YX}{}^2 > 0.00$, $r_{YT}{}^2 > 0.00$, and $r_{XT}{}^2 > 0.00$

Figure 11.3.3. Diagrammatic representation of the variances in Y, X, and T and the proportions of variance explained when all three are intercorrelated.

The overlapping portion will increase the total proportion of variance in Y explained by the others only once. Put differently, an intercorrelation between the three variables reflects that r_{YX} and r_{YT} overlap because of r_{XT}. This overlap means that the proportion of the total variance in Y that is explained by T alone *is additional to* that explained by X alone and by X and T together. To take this into account, we multiply the product of the three zero-order correlations by 2.00. The reason we use the multiple 2.00 is because we count the proportion of s^2_Y that is explained by the others three times, once each with s^2_{YX}, s^2_{YT}, and s^2_{XT}. Since we want to count this proportion only once, we must subtract two of them in our calculations. The overlapping proportion can be expressed symbolically:

$$2.00 \cdot [(r_{YX}) \cdot (r_{YT}) \cdot (r_{XT})]$$

We can put all of this explanatory information into one statement:

$$r^2_{YX} + r^2_{YT} - 2.00(r_{YX} \cdot r_{YT} \cdot r_{XT})$$

In words the statement says: Add together the two proportions of the variance in Y that is explained by first, X, and second, T. Then subtract two of the three parts of the proportion of the variance in Y that is explained by X and T together. Although as far as it goes this makes good sense, it omits an essential aspect of multiple explanation. Namely, since we are using both X and T to explain the variance in Y, we must consider not only the extent to which X and T go together, but also the extent to which they do not go together—that is, the proportion of the variance in one that is unexplained by its linear relationship with the other. To do this we use the coefficient of alienation, $(1.00 - r_{XT}^2)$, which represents the proportion of variance that is *unexplained* by the linear relationship between X and T. We use this measure of the unexplained variance in X and T as a divisor to derive the proportion of variance in Y that is explained by its multiple linear relationships with the other two variables. In other words, we are interested in the ratio of the proportion of variance in Y that is explained by X and T to the proportion of variance in X that is unexplained by T. This can be expressed conceptually in words:

$$
\begin{array}{l}
\text{proportion of} \\
\text{variance in } Y \\
\text{explained by} \\
X \text{ and } T
\end{array}
=
\dfrac{
\begin{array}{l}
\text{the proportion} \\
\text{of variance in } Y + \\
\text{explained by } X
\end{array}
\begin{array}{l}
\text{the proportion} \\
\text{of variance in } Y - \\
\text{explained by } T
\end{array}
\begin{array}{l}
\text{the proportion} \\
\text{of variance in } Y \\
\text{explained by } X \\
\text{and } T \text{ together}
\end{array}
}{
\begin{array}{l}
\text{the proportion of variance} \\
\text{in } X \text{ unexplained by } T
\end{array}
}
$$

11.4 THE MULTIPLE CORRELATION COEFFICIENT

So far we have learned that it is possible to add together proportions of the explained variance and, by correcting for multiple relationships, we can come up with a measure of the multiple explained variance. As yet, we have only called this measure "multiple R-squared" or "big R-squared," and in Section 11.3 we saw its abbreviation in two very specialized situations. Even so, the idea of this measure makes sense, especially if we think of it in light of the working parts just described. Using the summary of information shown in Table 11.4.1, we can state a general formula to measure the multiple explained variance.

Multiple R-squared can be expressed symbolically for *all* cases:

$$
R_{Y \cdot XT}^2 = \frac{r_{YX}^2 + r_{YT}^2 - 2.00 \cdot [(r_{YX}) \cdot (r_{YT}) \cdot (r_{XT})]}{(1.00 - r_{XT}^2)}
$$

Since we already have the zero-order coefficients called for by this equation, we may substitute them in and calculate multiple R-squared, $R_{Y \cdot XT}^2$. For

Table 11.4.1.   The working parts of the measure of multiple explanation

Symbol	
s_Y^2	= 100 percent of the variance in Y
s_X^2	= 100 percent of the variance in X
s_T^2	= 100 percent of the variance in T
s_{YX}^2	= 100 percent of the covariance between X and Y
s_{YT}^2	= 100 percent of the covariance between T and Y
s_{XT}^2	= 100 percent of the covariance between X and T
r_{YX}^2	= the proportion of s_Y^2 explained by X
	or
	the proportion of s_X^2 explained by Y
r_{YT}^2	= the proportion of s_Y^2 explained by T
	or
	the proportion of s_T^2 explained by Y
r_{XT}^2	= the proportion of s_X^2 explained by T
	or
	the proportion of s_T^2 explained by X
$1.00 - r_{YX}^2$	= the proportion of s_Y^2 not explained by X
	or
	the proportion of s_X^2 not explained by Y
$1.00 - r_{YT}^2$	= the proportion of s_Y^2 not explained by T
	or
	the proportion of s_T^2 not explained by Y
$1.00 - r_{XT}{}^2$	= the proportion of s_X^2 not explained by T
	or
	the proportion of s_T^2 not explained by X
$R_{Y \cdot XT}^2$	= the proportion of s_Y^2 explained by X and T
$1.00 - R_{Y \cdot XT}^2$	= the proportion of s_Y^2 not explained by X and T

example, in the case of the students being examined on present academic performance, we have

$$R_{Y \cdot XT}^2 = \frac{r_{YX}^2 + r_{YT}^2 - 2.00 \cdot [(r_{YX}) \cdot (r_{YT}) \cdot (r_{XT})]}{(1.00 - r_{XT}^2)}$$

$$= \frac{0.69^2 + 0.92^2 - 2.00 \cdot [(0.69) \cdot (0.92) \cdot (0.62)]}{(1.00 - 0.62^2)}$$

$$= \frac{0.476 + 0.846 - 2.00(0.394)}{1.00 - 0.384}$$

$$R_{Y \cdot XT}^2 = \frac{1.322 - 0.788}{0.616} = \frac{0.534}{0.616} = \underline{\underline{0.866}}$$

This can be interpreted as indicating that 88 percent of the variance in Y is explained by its multiple linear relationships with X and T. This means that among the students in our study, high school average and number of hours spent studying account for nearly 90 percent of the variance in grade point average.

There is another way to think of all this, and it will help us in understanding just what multiple explanation is all about. Multiple $R^2_{Y.XT}$ is comprised of the proportion of variance in Y that is explained by its linear relationship with X, that is, r^2_{YX}, plus the proportion in Y that is explained by its linear relationship with T controlling for X, $r^2_{YT.X}$. In order to keep the overlapping proportion of variance explained from being counted twice, thereby inflating how much we think we have explained, we control all three simultaneously. This means that we can use the square of the first-order *partial* correlation between Y and T controlling for X, $r^2_{YT.X}$, to help compute $R^2_{Y.XT}$. The partial correlation coefficient automatically takes into account the possibility that variables X and T might overlap. As we mentioned earlier, the partial measures the *additional* proportion of the unexplained variance in Y that is explained by T. This additional explanation "comes out of" what is *un*explained by X alone. Thus we multiply the first-order partial squared by the proportion of the variance in Y that is unexplained by its linear relationship with X, that is, the coefficient of alienation $(1.00 - r^2_{YX})$.

Because of these characteristics, we can express $R^2_{Y.XT}$ in terms of an alternate formula, which uses the square of the zero-order correlation coefficient between Y and X, r^2_{YX}, plus the square of the first-order partial correlation of Y and T controlling for X, $r^2_{YT.X}$, multiplied by the proportion of variance in Y that is unexplained by X, the coefficient of alienation $(1.00 - r^2_{YX})$. First we can express it verbally:

$$\begin{pmatrix} \text{the proportion of} \\ \text{variance in } Y \\ \text{explained by } X \\ \text{and } T \end{pmatrix} = \begin{pmatrix} \text{the proportion} \\ \text{of variance in } Y \\ \text{explained by } X \end{pmatrix} +$$

$$\begin{pmatrix} \text{the proportion of} \\ \text{variance in } Y \\ \text{explained by } T \text{ in} \\ \text{addition to that} \\ \text{already explained by } X \end{pmatrix} \cdot \begin{pmatrix} \text{the proportion} \\ \text{of variance in } Y \\ \text{unexplained by } X \end{pmatrix}$$

It also can be expressed symbolically:

$$R^2_{Y.XT} = r^2_{YX} + r^2_{YT.X} \cdot (1.00 - r^2_{YX})$$

Once again we have the values necessary to calculate multiple R-squared with this formula:

$$R_{Y \cdot XT}^2 = r_{YX}^2 + r_{YT \cdot X}^2 (1.00 - r_{YX}^2)$$
$$= 0.69^2 + [0.86^2 \cdot (1.00 - 0.69^2)]$$
$$= 0.476 + [0.740 \cdot (1.00 - 0.476)]$$
$$= 0.476 + [0.740 \cdot (0.524)]$$
$$= 0.476 + 0.390$$
$$R_{Y \cdot XT}^2 = \underline{\underline{0.866}}$$

Up until now, our interest has been on the multiple proportion of variance explained rather than on the multiple relationships between the variables. As it stands multiple R-squared is a very useful measure. However, we can go one easy step further and use multiple R-squared to help measure the goodness of fit of a set of least-squares predictions to the data. All that we have to do is take the positive square root of $R_{Y \cdot XT}^2$, and we derive the *multiple correlation coefficient*:

$$\text{multiple correlation coefficient} = R_{Y \cdot XT} = \sqrt{R_{Y \cdot XT}^2}$$

The multiple correlation coefficient measures the degree to which the respondents' scores fit the *overall plane* created by a set of predictions, rather than the degree to which they fit *one regression line* along one surface of the plane. Put differently, the multiple correlation coefficient measures the correlation between the actual values of the dependent variable Y_i and those values that we predict by using the least-squares equation

$$\hat{Y} = a_{Y \cdot XT} + b_{YX \cdot T} \cdot X_i + b_{YT \cdot X} \cdot T_i$$

Once again, the idea is to compare Y_i and $\hat{Y}$, but now we want to measure the *multiple regression* predicted scores compared to all the respondents' scores. Thus the multiple correlation coefficient is the zero-order correlation coefficient between the actual values (Y_i) for each respondent on the dependent variable and the predicted values ($\hat{Y}$) resulting from the least-squares equation.

In the example about students and academic performance, we use the value of $R_{Y \cdot XT}^2$, thus,

$$R_{Y \cdot XT} = \sqrt{R_{Y \cdot XT}^2} = \sqrt{0.88} = 0.94$$

This means that the goodness-of-fit measure of our regression predictions indicates a very strong positive multiple relationship of $R_{Y \cdot XT} = 0.94$.

The extension of multiple explained variance and correlation to numerous independent and control variables is straightforward, even though the calculations become quite complex. In general, the coefficient $R^2_{Y \cdot XTS \ldots K}$ represents the proportion of variance in the first subscript that is explained by its multiple linear *relationship* with all the independent variables shown to the right of the dot. For example, the four and five variable formulas can be expressed symbolically as follows:

$$R^2_{Y \cdot XTS} = R^2_{Y \cdot XT} + r^2_{YS \cdot XT} (1.00 - R^2_{Y \cdot XT})$$

and

$$R^2_{Y \cdot XTSV} = R^2_{Y \cdot XTS} + r^2_{YV \cdot XTS} (1.00 - R^2_{Y \cdot XTS})$$

Before going on you should go back and reread the chapter, paying special attention to the diagrams and tables. One way to understand this material better is to explain it in plain English to your *least* mathematically inclined friend. One try at giving a *verbal explanation* of this multivariate process usually makes it clearer. Explaining it in words will also help you understand the appealing simplicity of this analytical framework.

EXERCISES

1. In your own words, define, describe, or discuss the following terms and give an hypothetical example of each.

    control variable
    variance
    degree of relationship
    perfect positive correlation
    $r_{XY \cdot T}$
    coefficient of alienation
    $R^2_{Y \cdot XT}$
    multiple correlation coefficient
    partial product-moment correlation coefficient

2. Using the data presented in problem 4 in Chapter 7, calculate $r_{YX \cdot T}$, $r_{YT \cdot X}$, and $r_{Y \cdot XT}$.

3. The following zero-order correlation coefficient matrix is generated by a computer in response to your further study of problem 5 in Chapter 10. Using the matrix below answer the questions which follow.

    Y = Grade point average
    X = Number of hours spent studying each day
    T = Number of hours spent watching T.V. each day
    V = Number of college organizations in which one is an active member

    A. What is the first-order partial product-moment correlation between grade point average and the number of hours spent studying each day:
        1. Controlling for the number of hours spent watching T.V. each day; and

      2. Controlling for the number of college organizations in which one is an active member.

B. What is the first-order partial product-moment correlation coefficient between the number of hours spent studying each day and the number of hours spent watching T.V. each day controlling for the number of college organizations in which one is an active member?

Matrix of zero-order correlation coefficients

	Y	X	T	V
Y	1.00	.95	− .55	.38
X	.95	1.00	− .71	− .09
T	− .55	− .71	1.00	− .40
V	.38	− .09	− .40	1.00

4. A sociology research team recently reported that they found a multiple relationship between student *age*, *A*, the length of *time* spent preparing for class, *T*, the amount of previous *experience* in the subject field, *E*, and academic *performance*, abbreviated *P*, all interval scales. They reported the following zero-order product moment correlation coefficients:

$r_{AGE, TIME}$ = .70
$r_{AGE, EXPERIENCE}$ = .60
$r_{AGE, PERFORMANCE}$ = .40
$r_{TIME, EXPERIENCE}$ = .30
$r_{TIME, PERFORMANCE}$ = .50
$r_{EXPERIENCE, PERFORMANCE}$ = .80

A. What is the first-order partial product-moment correlation coefficient between student age and academic performance:
    1. Controlling for the length of time spent preparing for class;
    2. Controlling for the amount of previous experience in the subject field?

B. What is the first-order partial product-moment correlation coefficient between the length of time spent preparing for class and academic performance:
    1. Controlling for age;
    2. Controlling for the amount of previous experience in the subject field?

C. What is the first-order partial product-moment correlation coefficient between the amount of previous experience in the subject field and academic performance:
    1. Controlling for age;
    2. Controlling for the length of time spent preparing for class?

5. Using the data in problem 4 answer the following questions.

A. What proportion of the variance in academic performance is explained by its linear relationship with:
    1. Student age;
    2. The length of time spent preparing for class;
    3. The amount of previous experience in the subject field?

  B. What proportion of the variance in academic performance is explained by its multiple linear relationship with:
  1. Student age and the length of time spent preparing for class;
  2. Student age and the amount of previous experience in the subject field;
  3. The length of time spent preparing for class and the amount of previous experience in the subject field?

6. Explain the value of calculating first-order partials in analysis. What additional information is gained by this effort?

7. $r_{YX \cdot T}$ and $r_{YT \cdot X}$ measure the average "spreadoutiveness" of the respondents on Y given an independent variable at constant levels of the control variable. Explain this concept to a non-mathematical friend using a hypothetical example.

8. Discuss the meaning and utility of second, third, fourth, and fifth order multiple correlation coefficients.

part V

Inferential

Statistics

12

An Introduction to Statistical Inference

Thus far we have learned about descriptive statistics that enable us to describe respondents numerically according to operationally defined variables. The statistics we have learned cover a wide range of descriptive measures. These can be used to investigate the characteristics of large bodies of data gathered through the research process. We have studied classificatory, univariate, bivariate, and multivariate, descriptive statistics. For this section of the book, we will devote our attention to *inferential statistics*. Inferential statistics are those we apply to samples in order to make inferences about the populations from which they come.

In this chapter we will learn some of the essential elements underlying inferential statistics. Some of the material in this chapter stems from some fairly complex mathematical theories and procedures. However, it is more important to understand the *concepts* than the mathematics. Therefore, computational demands will be minimal.

Since probability is an important underlying aspect of the material covered in later chapters, we first will learn some of the basic laws of probability. Probability concepts are essential in relating the characteristics of a sample to those of the population from which the sample has been drawn. Thus we also will learn about sampling and the testing of statistical hypotheses.

12.1 STATISTICS VERSUS PARAMETERS

Throughout this book we have been concerned with the research process and the descriptive role statistics play in it. The general pattern of the research process proceeds from the formulation of some general problem, to the development of specific questions, to the gathering of evidence, to a statistical description of the data we have gathered. In addition to the descriptive values of these measures, we also are concerned about whether the statistics we have computed are truly representative of the population from which our sample was selected within certain probabilistic limits. In other words, we are shifting our concern from descriptive statistics, and we are now interested in learning how we can use the characteristics of a probability "sample" to make inferences about the characteristics of a "population." This is called *inferential statistics*.

The term *population* is used to describe any finite or infinite collection of individual cases, that is, the aggregate or totality of values. The population may be comprised of college students, a set of occupations, the work force in the United States, the books in a library, the cities within the United States, the nations of the world, or a set of scores on a Graduate Record Exam. Theoretically, therefore, the population reflects a fixed set of units or elements that are the basis of a research study. Populations may be either *finite* or *infinite*. Those in which we will be interested are finite, although many of the remarks in this chapter and those following will utilize statistical procedures that are based on the assumption that the population is infinite. Furthermore, a population can be either (1) *defined* or (2) *enumerated*. An enumeration should follow the selection rules of the definition; however, infinite populations can only be defined.

The term *sample* is used to describe a subset of units that is part of a population. From a nonstatistical point of view, a sample may be any subset of a population. For our purposes, however, a sample is a *deliberately selected* subset that is the object of investigation. That is, we use a sample to study characteristics or properties of the larger population.

It is important to recognize that a population is not necessarily an aggregation of individual people, as is commonly thought. Long ago, we learned that individual cases possess traits or characteristics that we call variables. All the traits or characteristics that are measurable represent the *population* of that characteristic. Thus, although we may think of the individuals as comprising the population, it is actually the possible set of measures possessed by each individual that makes up many populations of these measures. Importantly, the concept of population can be extended to ever-decreasing subsets of some larger population and still be treated as a population from which we can draw a sample.

In order to draw a sample, we must have a population to begin with. This means that we should have (1) a total enumeration (or universe) of that which we are going to investigate and (2) a subset of that totality that adequately represents it. The concepts of a population and a sample are common to everyday life as well as to formal statistics. For example, we all are familiar with public opinion polls that report that a "representative sample" was questioned on a given subject and that the sample responded in a certain fashion. Experience has shown that many such samples well may be representative because such polls are often quite accurate in their predictions. These polls draw a sample from the population and then make projections about the population from information obtained from the sample. It is this process with which we will be concerned throughout the remainder of this chapter.

Before we learn some of the basic elements of sampling, let us first differentiate between what we measure in the population contrasted to what we measure in the sample. Specifically, we want to distinguish between the characteristics of each by the use of the different symbols that designate various measures. A characteristic of the *population* is called a *parameter*, and a characteristic of the *sample* is called a *statistic*. We use parameters to describe the variables in the population, and we use statistics to describe the variables in the sample. A parameter is to the population what a statistic is to the sample.

Let us compare the symbols for the measures with which we are already familiar. Table 12.1.1 on page 260 presents the measure and the appropriate symbols for it for both the population and the sample. Generally, we use Roman letters (a, b, etc.) as abbreviations for statistics and Greek letters (α, β) as abbreviations for parameters.

Thus far the values we have learned to compute on samples have often been referred to as "predicted" or "estimated" values. Importantly, our version of prediction has to do with predicting or estimating a parameter in the population. Thus, we generally treat parameters as if they were fixed values, even though in reality they may change over time. For example, when we examine the variable age, we are interested in the parameter as it exists *at that time* rather than in the fact that age distributions change over time. Even though parameters are fixed values, most often they are unknown values. Statistics, on the other hand, are not fixed values (they may change from sample to sample), and they are generally known or easy to compute.

Importantly, we are never actually certain just how representative a sample is compared to the population or how well the statistic estimates the appropriate parameter. Even though the sample statistic may or may not be close to the population parameter, we try to insure that it is "representative" by making certain probabilistic assumptions about the population parameters. Then we draw a probability sample, the details of which will be explained soon. From the population values we may test the randomness of our sample.

Table 12.1.1. Names and symbols for statistical inference abbreviations and symbols

Verbal Description

Characteristics	Sample Statistics	Population Parameters
Number of cases	N	M
Mean of Y (Y bar) (mu)	$\bar{Y}$	μ_Y
Estimated value of Y (Y predicted) (psi)	$\hat{Y}$	ψ_Y
Standard deviation of Y (s) (sigma)	s_Y	σ_Y
Variance of Y (s squared) (sigma squared)	s_Y^2	σ_Y^2
Covariance of Y and X	s_{YX}	σ_{YX}
α of Y predicted from $X, T, \ldots, K$	$a_{Y \cdot XT \ldots K}$	$\alpha_{Y \cdot XT \ldots K}$
b of Y on X, controlling for $T, \ldots, K$ (*lower*-case b) (beta)	$b_{YX \cdot T \ldots K}$	$\hat{\beta}_{YX \cdot T \ldots K}$
Beta weight of Y on X, controlling for $T, \ldots, K$ (*upper*-case B) (beta)	$B_{YX \cdot T \ldots K}$	$\beta_{YX \cdot T \ldots K}$
Correlation of Y and X, controlling for $T, \ldots, K$ (*lower*-case r) (rho)	$r_{YX \cdot T \ldots K}$	$\rho_{YX \cdot T \ldots K}$
Correlation squared of Y and X, controlling for $T, \ldots, K$ (*lower*-case r) (rho)	$r_{YX \cdot T \ldots K}^2$	$\rho_{YX \cdot T \ldots K}^2$
Multiple correlation of Y, controlling for $X, T, \ldots, K$ (*upper*-case R) (rho)	$R_{Y \cdot XT \ldots K}$	$P_{Y \cdot XT \ldots K}$
Multiple correlation-squared of Y, controlling for $X, T, \ldots, K$ (*upper*-case R) (rho)	$R_{Y \cdot XT \ldots K}^2$	$P_{Y \cdot XT \ldots K}^2$

More commonly, however, we use sample statistics to estimate the population parameters, *knowing* that we have taken a probability sample. In other words, in most cases we rest our faith on the probability sampling procedure rather than upon the "representativeness" of a particular sample.

12.2 PROBABILITY

Long ago, we learned that to calculate "expected" frequencies, we multiply the relevant marginal frequencies of one variable times the relevant marginal frequencies of the other variable and divide the product by N. The extent to which expected and observed frequencies differ is the essence of the matter of statistical independence versus dependence. Our concern with the association between variables is related to the concept of statistical independence and observed versus expected frequencies and the likelihood of certain findings. Importantly, we deal with sets of variables assuming that the values of one variable depend upon the values of the other variables. No matter what we *assume* about the variables, it is not until we examine the actual marginal frequency distributions that we can arrive at expected cell values. Furthermore, no matter what values we *expect*, it is only after we cross-tabulate the variables that we know what the observed values are. However, we can calculate the likelihood or "probability" of certain events. Statisticians use probability theory as the underlying framework for inferential statistics, so an understanding of a few basic principles will help you to understand better the ideas behind statistical inference.

Probability involves a body of mathematical theory that has to do with chance, likelihood, or risk. When someone says "The odds are that I'll have an enjoyable weekend," it is a probability statement. Probability theory is something that most people have an intuitive grasp of without using mathematics. For instance, suppose that out of ten people in a room, we know that four of them are males and six of them are females. If we randomly select one of them from the room, how likely is it that the student will be a female? It does not require a graduate math background to figure that the chances are 6 out of 10. Even though the concepts and computations will become more complex, this is the essence of probability.

It is important to distinguish between two approaches to probability. First, there is the *empirical* approach to probability in which we assign relative frequencies on the basis of empirical evidence, that is, evidence based on observation or experimentation. Second, there is the *classical* or *a priori* approach to probability in which we assign relative frequencies on the basis of a priori knowledge, that is, knowledge based on deduction or reasoning. In a priori probability our concern is with the mathematical properties of possible outcomes given the event in question. For example, tossing a coin results in either a head or a tail, thus we can deduce that the probability of tossing, say, a head is 1 out of 2. In real life situations and in social statistics we generally assign expected relative frequencies based on evidence we gather. After we have such evidence or an estimate of it, we then use classical a priori probability to test our statements. In other words, we combine the approaches of empirical and classical probability for inferential statistics.

One of the difficulties with probabilistic properties and the rules of probability is that by themselves, they do not mean very much to many people. They are much clearer and more useful if incorporated in an example that emphasizes a number of the rules at once. Consider the information in Table 12.2.1, which is a two-way frequency distribution showing the cross-tabulated frequencies for the variables age and class in college for a random sample of 100 college students.

Let us start by putting things in explicit terms and by defining *probability*. Since an infinite number of selections provide an infinite number of events, the probability of event A occurring *over the long run* is the proportion of times that the specific event A is observed to the total number of possible events (N). Put differently, the probability of event A is the *ratio* of A to N.

Table 12.2.1. Class in college and age

Class in College	Age		
	19 and under	over 19	Total
Freshman	30	10	40
Sophomore	17	13	30
Junior	10	10	20
Senior	3	7	10
Total	60	40	100

We express the probability of A symbolically:

$$\Pr(A) = \frac{\text{the number of observed } A \text{ events}}{\text{the total number of possible events}} = \frac{A}{N}$$

For example, if we designate the probability of selecting a senior as event A, then,

$$\Pr(A) = \frac{A}{N} = \frac{10}{100} = 0.10$$

There are several mathematical properties that form the basic rules of probability. If you can get a good grasp of these rules, not only will you understand better the material in this book but you may also be more proficient at the race track, when playing cards, rolling dice, or pitching pennies.

If event A cannot occur, its probability is zero, i.e., 0.00. If event A can and must occur, its probability is equivalent to 100 percent of the events, i.e., 1.00. Thus the probability of event A occurring may be equal to or greater than zero and equal to or less than one. Since the probability of event A is the ratio of A to N and is a proportion between 0.00 and 1.00, the probability that event A does not occur is the difference between 100 percent and the probability of event A. Thus the sum of the probability of event A plus the probability of its "complement" equals 100 percent of the probability, or 1.00. As this discussion indicates, probabilities are always expressed as fractions or proportions.

All the above features of probability can be summarized and expressed symbolically:

Statements of Probability

1. The probability of event A [$\Pr(A)$] $= \dfrac{A}{N}$.

2. The probability that event A does not occur, A' [called "A prime" or the complement of A] $\Pr(A') = \dfrac{A'}{N} = 1.00 - \dfrac{A}{N}$.

Properties of Probability

1. Since events A plus events A' are all possible events, $\Pr(A) + \Pr(A') = 1.00$.
2. If event A cannot occur, $\Pr(A) = 0.00$.
3. If event A can and must occur, $\Pr(A) = 1.00$.
4. Thus, $0.00 \leq \Pr(A) \leq 1.00$.
5. Thus, $\Pr(A') = 1.00 - \Pr(A)$.

12.3 RULES OF PROBABILITY

Single-Draw Probability Rules

If events A and B *cannot* possibly occur simultaneously in the same test, they are mutually exclusive, as shown in Figure 12.3.1. For instance, if we select *one* person from our sample, it is possible to get a freshman or a sophomore or a junior or a senior, but it is impossible to get both in one *draw* (selection) because the categories of class in school are mutually exclusive. The probability of getting A or B is equal to the probability of A plus the probability of B.

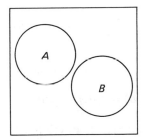

Figure 12.3.1.   Events A and B are mutually exclusive.

Addition Rule 1: If events A and B are mutually exclusive,

$$\Pr(A \text{ or } B) = \Pr(A) + \Pr(B)$$

For example, the probability of selecting a junior (event A) or a senior (event B) in *one* draw may be stated,

$$\Pr(A \text{ or } B) = \Pr(A) + \Pr(B)$$
$$= 20/100 + 10/100$$
$$= 30/100 = \underline{0.30}$$

Now, suppose that events A and B *can* occur simultaneously in the same test, and that they may be partially and/or totally inclusive, as shown in Figure 12.3.2. For instance, if we select *one* person from the sample, it is possible for us to draw a 19-year-old freshman or a 19-year-old sophomore, because the combined categories of age and class may be partially and/or mutually inclusive. In other words, all freshmen are not necessarily 19 years old, and all 19-year-olds are not necessarily freshmen. Thus we need a correction factor for the probability of A and B occurring simultaneously and counting the overlapping probability of the two more than once. The probability of getting A and/or B is equal to the probability of A plus the probability of B minus the probability of A and B together.

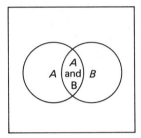

Figure 12.3.2.   Events A and B are partially inclusive.

Addition Rule 2: If events A and B are partially or totally inclusive,

$$\Pr(A \text{ and/or } B) = \Pr(A) + \Pr(B) - \Pr(A \text{ and } B)$$

For example, the probability of selecting a 19-year-old (event A) and/or a freshman (event B) in *one* draw may be stated,

$$\Pr(A \text{ and/or } B) = \Pr(A) + \Pr(B) - \Pr(A \text{ and } B)$$
$$= 60/100 + 40/100 - 30/100$$
$$= 70/100 = \underline{0.70}$$

Multiple-Draw Probability Rules

Let us now consider the probability of properties for two or more events. If the occurrence of event A in no way affects the occurrence of event B and vice versa, then we call events A and B *independent*. For instance, if event A is the selection of one student who is returned to the sample before event B, which is the selection of a second student, the two events are independent because the two selections do not affect each other. If we make two independent selections and either or both events A and B can occur, the probability of A and B equals the probability of A times the probability of B.

Multiplication Rule 1:

If events A and B are independent, and if they can occur in two draws,

$$\text{Pr}(A \text{ and } B) = \text{Pr}(A)\,\text{Pr}(B)$$

For example, the probability of selecting a sophomore (event A) and a junior (event B) in *two* draws with replacement may be stated,

$$\text{Pr}(A \text{ and } B) = \text{Pr}(A)\,\text{Pr}(B)$$
$$= (30/100)\,(20/100)$$
$$= 600/10,000 = \underline{0.06}$$

Whenever two events *do* affect each other, we call them *dependent*. Dependent events require us to use the *conditional* probability rule, which is needed when the first selection changes the probability for the second selection, the second changes the probability of the third, and so forth. Conditional probability is represented by a vertical line, $|$, between the two events. The conditional probability of event B *given* that event A has already occurred is expressed symbolically $\text{Pr}(B|A)$. If events A and B are dependent and either or both can occur in two draws, the probability of A and B equals the probability of A times the probability of B given A, or the probability of B times the probability of A given B.

Multiplication Rule 2:

If events A and B are dependent, and if they can occur in two draws,

$$\text{Pr}(A \text{ and } B) = \text{Pr}(A)\,\text{Pr}(B|A)$$

For example, the probability of selecting a 19-year-old senior in draw number 1 (event A) and another 19-year-old senior in draw 2 (event B) without replacement may be stated,

$$\text{Pr}(A \text{ and } B) = \text{Pr}(A)\,\text{Pr}(B|A)$$
$$= (3/100)\,(2/99)$$
$$= 6/9,900 = \underline{0.00061}$$

Give some thought to what happens when we have two draws. There are several possible ways two draws can occur. First we have the outcome of draw number 1. Then we have the outcome of draw number 2. Clearly, there are several possible combinations. For example, on the first draw, we can get a freshman and on the second draw another freshman; or, on the first draw we can get a freshman and on the second draw a sophomore; or, on the first draw we can get a freshman and on the second draw a junior; or, on the first draw we can get a freshman and on the second draw a senior; and so forth.

If events A and B are independent and if either or both can occur in two draws, and if we are interested in the probability of getting either an A and/or a B, we must consider the possible ways in which this could occur, i.e., AA, AB, BA, and BB.

Multiplicative Addition Rule 1:

If events A and B are independent and if either or both can occur in two draws, and if we want the probability of either an A and/or a B,

$$\Pr(A \text{ and/or } B) = \Pr(A \text{ and } A) + \Pr(A \text{ and } B) + \Pr(B \text{ and } A)$$
$$+ \Pr(B \text{ and } B)$$

$$\Pr(A \text{ and } A) = \Pr(A) \ \Pr(A) +$$
$$+ \Pr(A \text{ and } B) = \Pr(A) \ \Pr(B) +$$
$$+ \Pr(B \text{ and } A) = \Pr(B) \ \Pr(A) +$$
$$+ \Pr(B \text{ and } B) = \Pr(B) \ \Pr(B)$$

For example, the probability of selecting either a 20-year-old freshman (event A) and/or an 18-year-old sophomore (event B) in two draws with replacement may be stated,

$$\Pr(A \text{ and/or } B) = \Pr(A \text{ and } A) + \Pr(A \text{ and } B) + \Pr(B \text{ and } A)$$
$$+ \Pr(B \text{ and } B)$$

$$\Pr(A \text{ and } A) = \Pr(A) \ \Pr(A)$$
$$= (10/100)(10/100) = 100/10,000 +$$
$$+ \Pr(A \text{ and } B) = \Pr(A) \ \Pr(B)$$
$$= (10/100)(17/100) = 170/10,000 +$$
$$+ \Pr(B \text{ and } A) = \Pr(B) \ \Pr(A)$$
$$= (17/100)(10/100) = 170/10,000 +$$
$$+ \Pr(B \text{ and } B) = \Pr(B) \ \Pr(B)$$
$$= (17/100)(17/100) = \underline{289/10,000}$$

$$\Pr(\text{freshman age 20 and/or a} = \overline{729/10,000}$$
$$\text{sophomore age 19 in two}$$
$$\text{draws)} \qquad = \underline{0.0729}$$

A further complication is that we can either return each person to the sample or leave the person out after drawing him or her. If we return the person we have *sampling with replacement*, and if we leave him or her out we have *sampling without replacement*.

Multiplicative Addition Rule 2:

If events A and B are dependent and if either or both can occur in two draws, and if we want the probability of either an A and/or a B,

$$\text{Pr}(A \text{ and/or } B) = \text{Pr}(A \text{ and } A) + \text{Pr}(A \text{ and } B) + \text{Pr}(B \text{ and } A)$$
$$+ \text{Pr}(B \text{ and } B)$$

$$\text{Pr}(A \text{ and } A) = \text{Pr}(A) \ \text{Pr}(A|A) +$$
$$+ \text{Pr}(A \text{ and } B) = \text{Pr}(A) \ \text{Pr}(B|A) +$$
$$+ \text{Pr}(B \text{ and } A) = \text{Pr}(B) \ \text{Pr}(A|B) +$$
$$+ \text{Pr}(B \text{ and } B) = \text{Pr}(B) \ \text{Pr}(B|B)$$

For example, the probability of selecting either a 20-year-old freshman (event A) and/or an 18-year-old sophomore (event B) in two draws without replacement and in any order may be stated,

$$\text{Pr}(A \text{ and } A) = \text{Pr}(A) \ \text{Pr}(A|A)$$
$$= (10/100)(\ 9/99) = \ \ 90/9{,}900 +$$
$$+ \text{Pr}(A \text{ and } B) = \text{Pr}(A) \ \text{Pr}(B|A)$$
$$= (10/100)(17/99) = 170/9{,}900 +$$
$$+ \text{Pr}(B \text{ and } A) = \text{Pr}(B) \ \text{Pr}(A|B)$$
$$= (17/100)(10/99) = 170/9{,}900 +$$
$$+ \text{Pr}(B \text{ and } B) = \text{Pr}(B) \ \text{Pr}(B|B)$$
$$= (17/100)(16/99) = \underline{272/9{,}900}$$

Pr(freshman age 20 and/or $\qquad = \overline{702/9{,}900}$
sophomore age 19 in two
draws) $\qquad\qquad\qquad = \underline{0.0709}$

There are many extensions of probability theory, and some of them involve very complex mathematical procedures. For the purposes of this book, however, the basic ideas we have just learned are sufficient to help you understand the logic of statistical inference because these ideas can be applied to the statistician's need and desire to use sampling methods.

12.4   SAMPLING

So far we have talked a good deal about sampling without specifying what we mean by a probability sample or a representative sample.[1] Sampling may

[1]There are many excellent treatments of sampling which range from a popular version by Slonim to a technical version by Hansen, Hurwitz, and Madow. Students interested in more detailed discussions can read Morris James Slonim, *Sampling* (New York: Simon and Schuster, 1960); Hubert M. Blalock, Jr., *Social Statistics*, 2nd edition (New York: McGraw-Hill Book Company, 1972), pp. 509–530; M. H. Hansen, W. N. Hurwitz and W. G. Madow, *Sample Survey Methods and Theory*, Vol. 1 (New York: John Wiley and Sons, Inc., 1953); and William G. Cochran, *Sampling Techniques*, 2nd edition (New York: John Wiley and Sons, Inc., 1953).

be defined as the process of selecting units of observations from a population for describing, analyzing, or testing the measures and relationships within a research project. We mentioned earlier the common claim by public opinion polls that a representative sample, which is very small in proportion to the population, can be used to make accurate predictions. The use of sampling is not restricted to data gathered by pollsters. One crucial use of sampling with which we are all familiar involves the common blood test. Physicians are able to analyze the characteristics and patterns of an individual's entire blood system from just a few drops of blood. Quite clearly, it would be unhandy and impractical to test every drop of blood contained in the human body in order to learn the blood's characteristics.

Through well-developed theories of sampling and appropriate steps for selecting a sample, we can study the traits and characteristics of a population far more economically than would be the case if we had to study every element individually. Most social statistics call for precisely executed *probability* samples. In a probability sample every element in the population has a *known probability* of being selected. It is only when the probabilities of selection are known that it is appropriate to use statistical inference. Even though a non-probability sample may turn out to be a very *representative* sample, without knowledge of the probabilities of selection it is not possible to calculate the risks of error we have taken. This is a major aspect of inferential statistics.

Four basic types of probability sampling are of importance to sociologists. They are: (1) simple random sampling; (2) stratified random sampling; (3) cluster sampling; and (4) systematic sampling.

A *simple random sample* is a sample in which every element and every combination of elements in the population has a *known* and *equal* probability of being selected. A *biased sample* is one in which every element in the population does *not* have an equal chance of being selected. In biased samples, instead of random error there is systematic error. Clearly, we want a sample that is representative of the population, so that our statistics will be representative estimates of the parameters. It is through random sampling of a population that samples over the long run will be representative of that population. Some of the random samples will not be representative, while others will be very representative.

In order to select a simple random sample, we must have a total enumeration or definition of the population we are going to investigate. We must be certain that from this list, on any given draw, the probability of all the remaining elements being drawn are equal, regardless of which elements have already been selected. We start with a list of M elements in the population and then randomly select N of these elements for our sample. Remember, selection *must* insure that each individual element in the population has an equal probability of selection, and that every possible combination of N elements has an equal probability of comprising our sample.

As we mentioned earlier, there are two forms of random sampling, one in which the elements are replaced after selection (called "sampling with replacement"), and the other in which the elements are not replaced (called "sampling without replacement"). Naturally, sampling without replacement continuously increases the probability of selection, because one less element remains each time an individual element is drawn. At the same time, sampling without replacement does not allow for every *possible combination* of elements an equal probability of selection. Thus only a sample drawn with replacement truly meets the criteria of a simple random sample. The basic model for inferential statistics is based on simple independent random sampling *with* replacement. Most sociologists violate this basic assumption; however, the model is still useful for inferential purposes. The reason sociologists can violate the replacement assumption is because the populations from which they draw samples generally are quite large relative to the size of the sample.

Most often sociologists select random samples without replacement, and this may present serious consequences if the sample is large relative to the size of a finite population. In a limited size population, the sampling variances of the statistics depend upon the number in the population (M) as well as the number in the sample (N). In such cases, it is common to use the finite population multiplier $(1 - (N/M))$ to correct for the high ratio between the sample size and the population size.

A *stratified random sample* is a sample in which the elements are divided into subpopulations (called *strata*), and then a random sample from within each strata is selected. Stratified random sampling is used when the analyst wishes (1) to make comparisons among subsets of the population, (2) to reduce the cost of a study, or (3) to reduce the sampling error. Clearly, in order to conduct a stratified random sample, we need more than a mere list. We need information on relevant variables so we can differentiate between subpopulations. There are three types of stratified random samples. The first type allows us to select samples that are proportionate to the size of each stratum in the population. The second type allows us to select samples that are of equal size, no matter what the size of each stratum. The third type allows us to select samples that are optimum for a fixed overall, or total, sample of a certain size; this is called *optimum-allocation-stratified-random sampling*. Our aim is to select a sample in such a way that the size for each stratum will give us the most precise estimates given the total size of the sample. The key to stratified random sampling is in selecting strata that are internally homogeneous, but that are externally heterogeneous, that is, as different as possible from stratum to stratum.

A *cluster sample* is a sample in which the elements are divided into a number of natural units (called *clusters*), and then a single cluster is selected at random and all the individuals within it become the sample. For example, school districts, apartment complexes, city blocks, and census divisions comprise types of natural units. Our hope is to obtain clusters that are as nearly as possible externally homogeneous, but in which each cluster is internally heterogeneous. The purpose of cluster sampling is to reduce expenses such as traveling and interviewing by dealing with a single cluster.

A *systematic sample* is a sample in which the elements of a population are enumerated in a specific order from 1 to M and are sampled by selecting the initial unit or element at random from the first K units and then selecting every Kth unit or element from then on. For example, if we wish to draw a sample of 100 from a list of 5000, we would select the first element by a random procedure and then take every 50th individual on the list. Systematic sampling is not nearly so complex or difficult as simple random sampling, and in many instances it produces equivalent samples to those that would be selected by the simple random method. For example, even though surnames may be clustered to some extent according to ethnic background, alphabetized lists generally can be systematically sampled, producing an approximately simple random sample. Systematic random sampling is not appropriate when the elements have been arranged according to some rank order because the level at which the first selection occurs will influence all subsequent selections. For example, if we were selecting 100 elements from a population of 5000 that was arranged in numerical order on age from the youngest to the oldest, we would have systematically younger values if we were to select the first person listed and then every 50th person, compared to the values if we were to select the 50th person listed and then every 50th person. However, when appropriate lists are available, because of simplicity and ease of selection systematic sampling is often used as a substitute for simple random sampling.

It is important to recognize that suitable sampling techniques enable a researcher to reduce the overall cost of a project, cut down on the number of researchers, and obtain comprehensive data quickly. At the same time, proper sampling may actually increase the degree to which some estimates approximate the true figure of the population. In all sampling there is the risk that the sample estimate is not representative of the population being studied. The connection between probability sampling procedures and probability theory is the basis of inferential statistics. Not only are sampling procedures important to inferential statistics, but also sampling characteristics are important.

12.5 THE USE OF THE NULL HYPOTHESIS

A *statistical hypothesis* is a statement about an event whose outcome is unknown or a prediction of some future event. We set up such a statement and then "test" its accuracy with statistical techniques. Usually the hypothesis we test is called the *null hypothesis*, abbreviated H_0.

Technically, the null hypothesis is the "population hypothesis" with the understanding that any difference between the corresponding statistic ($\bar{X}$) and the hypothesized value of the parameter (μ_X) is attributable to random sampling variation. The null hypothesis is generally used to test that there is *no* difference between sample values. Most of the time we call alternatives to the null hypothesis the *research hypotheses*, abbreviated H_1, H_2, H_3, and so forth. Generally, these alternatives are tested as being "the opposite of" the null hypothesis H_0.

We use the null hypothesis as a "setup," that is, as a hypothesis we expect to be able to reject. First we state H_0 *as if* it were true. Then we use a *statistical test* to see how likely our findings would be if H_0 actually were true. A *statistical test* is one we use only for hypothesis testing and not as a descriptive measure of our sample data. Put simply, we calculate a statistic from our sample, and then we "test" it by comparing the sample value with the appropriate probabilistic value in the sampling distribution of the statistic.

Once we state a null hypothesis, there are four essential steps used to test it. The *first* step in testing a null hypothesis is to define the variables, their limits, and all possible outcomes *prior* to testing. The *second* step is to determine the operational definitions of the variables so as to be able to measure the values of the outcomes *prior* to testing. The *third* step in testing a null hypothesis involves the rejection of the hypothesis. If we hypothesize something and it turns out that the findings fall in the critical region, then we *reject* the hypothesis. For example, suppose we hypothesize that the population mean on a test measure of the tolerance of deviant behavior is 75, and suppose that the mean grade of a sample actually turns out to be 90. Depending on our prior decisions, we might reject the hypothesis that the mean grade is 75. Formally stated, the *third* step in testing a statistical hypothesis is to decide *prior* to testing which outcomes will lead to the rejection of a given hypothesis. The *fourth* step in testing a null hypothesis is to observe the event, to record the outcomes, and then to decide whether or not to reject the null hypothesis.

Step one calls for us to state the definition clearly. For example, suppose that the test of tolerance of deviant behavior is intended to measure "one's willingness to tolerate people who are labeled deviant," and that we are interested in how tolerant the population is. First, all the outcomes must be anticipated prior to testing. Thus we might decide that the range is from complete tolerance to complete intolerance.

Second, we decide to operationally define "tolerance" by the score each person gets on the test, and a person's score operationally measures tolerance from 0 to 100. We know prior to testing that we will not give anyone a minus grade or a grade above 100. This means that, among others, it is possible to get means of, say, 0, 10, 30, $32\frac{1}{4}$, 60, $74\frac{1}{2}$, 80, 100, and so forth. Thus we can anticipate every possible outcome prior to testing.

The third step is to decide which outcomes will or will not be rejected, and we must do so prior to testing. Let us now state a null hypothesis: $H_0 =$ "The mean grade of all persons (the population) for this test will be 75." We could use such a statement to indicate which outcomes will be rejected and which accepted prior to testing. Suppose that the mean of the sample turns out to be 84. Strictly speaking, we would reject the null hypothesis.

However, instead of hypothesizing an exact value, suppose that we give ourselves some leeway and hypothesize that the mean will be somewhere between 65 and 85. Now, a sample mean of 84 falls in the range. Clearly the hypothesis is true, and we do not reject it. In such a case we have set several outcomes that comprise an interval score, and we will reject none of these outcomes.

Our example is very simplified, but, in general, hypothesis testing goes essentially like the procedure we have just stated. The idea behind testing a hypothesis is that it is a test about a population using partial information from a sample. We state a null hypothesis, know all the possible outcomes, state a clear-cut operational definition, draw a sample, observe the event, and make a decision about rejection.

When we estimate or predict from a sample there is always the chance of making an error because of an unrepresentative sample. Such a risk may be taken into account by using certain fundamentals of probability theory and judging our chances of error when deciding about the rejection of the hypotheses. For example, suppose that we have a distribution of respondents on some variable, and that we have 5 percent in each *tail* of the distribution as shown in Figure 12.5.1. The one-tailed probability is 5 percent that we will

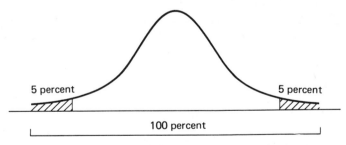

Figure 12.5.1.

select someone with a value in one end or the other, that is, in one of the shaded sections. The chance that we will select someone in either end is 10 percent. Thus, there is a two-tailed 10 percent chance that a given respondent will have a value in either one tail or the other.

Let us now imagine that a set of grades from an exam are symmetrically distributed and range from 50 to 100 with a mean of 75, as shown in Figure 12.5.2. Quite clearly, this distribution indicates that more people scored in the vicinity of 75 than in the vicinities of 100 or 50. Extreme cases, such as those who get 100, are the ones who "break the curve." Other extreme cases, such as those who get 50, are the ones on the bottom who "automatically fail." In social statistics we are often interested in the probability of obtaining a value that is to a specific degree unusual or rare. *Symmetrical distributions* enable us to consider both tails of the distribution by doubling the one-tail probability.

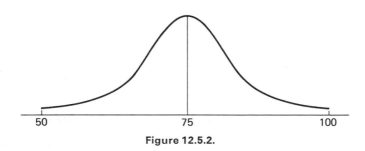

Figure 12.5.2.

We are in a dilemma when we must decide whether to use hypotheses about one or two tails of the population. Thus far we have discussed setting up a null hypothesis H_0 and a research or alternative hypothesis H_1. In general, the H_0 hypothesizes certain values for the population parameter in question, and the H_1 hypothesizes that the population parameter in question is *different from* the value hypothesized.

For example, in a two-tailed test, the hypothesis is that the mean value of a sample is rejected if $\bar{Y}$ falls into one or the other tail of its sampling distribution. In a one-tailed test, we reject the hypothesis that the mean value is "at least" the hypothesized amount only if $\bar{Y}$ is too small. We do this because larger sample values of $\bar{Y}$ would substantiate rather than prove false the "at least" hypothesis. In other words, we may state an alternative hypothesis that is *directional*. In such cases, first, we hypothesize that the population parameter is different from the hypothesized value, and second, we hypothesize the direction of the difference.

In other words, with a one-tailed test we have only *one* portion of the curve into which the sample value must fall for the rejection of the null hypothesis. By selecting a one-tailed test, we concentrate the proportion of "rare" events at one end of the distribution, thereby increasing the directional area for rejecting the null hypothesis. For example, if we decide that the most extreme

5 percent of the distribution will be considered "rare," we can locate either 2.5 percent in each tail or 5 percent in one tail only.

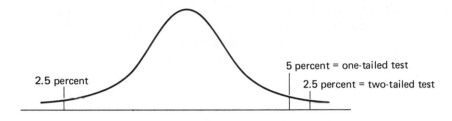

Suppose that we want to determine if a sample mean for weekly earnings of, say, three randomly selected nonagricultural industries presented in Table 12.5.1 actually is $156.25. We would reject the null hypothesis that it is *at*

Table 12.5.1. Average weekly earnings for selected nonagricultural industries

Specific Individual Industry	Average Weekly Earnings
Apparel and Accessory Stores	89
Banking	117
Communication	175
Credit Agencies	117
Eating and Drinking Places	61
Electric, Gas, and Sanitary Services	179
Food Stores	124
General Building Construction	210
Heavy Construction	195
Insurance Agents, Brokers, Services	139
Railroad and Trucking Transportation	215
Special Trade Construction	254

$$\sum_{i=1}^{N} Y_i = 1{,}875$$

$$\bar{Y} = 156.25$$

Source: Adapted from U.S. Bureau of the Census, *Statistical Abstracts of the United States: 1971* (92nd edition). Washington, D.C., 1971, No. 345, p. 219.

least $156.25 only if the sample mean is too small. If, instead, the sample mean is too high, we test the null hypothesis $H_0: \mu_Y = \$156.25$ against the alternative hypothesis $H_1: \mu_Y < \$156.25$. Thus we either accept H_0, or we accept H_1 as the alternative.

EXERCISES

1. In your own words, define, describe, or discuss the following terms and give an hypothetical example of each:

   statistical inference
   parameter
   statistic
   probability sample
   null hypothesis
   estimated value
   μ_Y
   σ_Y
   β

2. Discuss the differences between random sampling, cluster sampling, and systematic sampling. Give examples of research projects that might utilize each type.

3. Verbally explain the difference between a population and a sample. Why do you think it is important for this differentiation to be made? Why employ a different notation system?

4. Assume that you have been duped into a card game in your best friend's dorm, and that you have the feeling that you may lose because of the smooth dealing of someone who looks overly innocent. Since you are nervous about things, you insist that he shuffle the deck very well and hope for the best. Figure your chances for winning by calculating the following probabilities.[2]

   A. What is the probability of getting four aces in four draws without replacement from that well-shuffled deck of cards?

   B. What is the probability of getting at least one king in three draws with replacement from that well-shuffled deck of cards?

   C. What is the probability of getting at least one jack and at least one queen in three draws with replacement from that well-shuffled deck of cards?

5. Assume that you select a random sample with replacement of 500 students from your college and question them about their tastes in music. You find that 200 of them are wild about steel drum music and 300 are not. Among those who are wild about steel drum music, only 50 of them also like American folk music. You also find that there are 200 students who do not care for either steel drum music or American folk music. Finally the rest of the students sampled care only for American folk music.

   If you now select one of these students at random, and if event A is that he or she is wild about steel drum music, and if event B is that he or she cares for American folk music, calculate the following probabilities:

   A. I. $\Pr(A) =$ _____   III. $\Pr(A|B) =$ _____
       II. $\Pr(B) =$ _____   IV. $\Pr(B|A) =$ _____

   [2]This and some of the following problems bear close correspondences to some presented by Blalock, *op. cit.*, pp. 146–147.

B. What is the probability of getting a student who likes one of the two but not both?

C. Numerically verify that
$Pr(A \text{ and/or } B) = Pr(A) Pr(B|A) = Pr(B) Pr(A|B)$.

6. Assume that you are involved in a research project investigating academic performance and educational aspirations, depending upon the sex of the respondent. You select a simple independent random sample of 1,000 and classify the respondents according to the three variables as follows:

Sex	Educational Aspirations	Academic Performance (G.P.A.)			Total
		Low	Medium	High	
Male	High	15	50	60	125
	Medium	60	80	35	175
	Low	100	70	30	200
Female	High	15	80	80	175
	Medium	50	90	60	200
	Low	60	30	35	125
Total		300	400	300	1000

Suppose you were to draw individual students (one draw) at random from this population of 1,000 and then replace them. What is the probability of getting a student

A. With low academic performance?
B. With medium academic performance?
C. With high academic performance?
D. With low educational aspirations?
E. With medium educational aspirations?
F. With high educational aspirations?
G. Who is male?
H. Who is female?

7. Suppose you were to draw individual students, as we did in Problem 6, from the same population, what is the probability of getting

A. A male with low academic performance?
B. A female with high academic performance?
C. A male with medium academic performance?

8. Suppose you were to draw individual students, as we did in Problem 6, from the same population, what is the probability of getting a student

A. Who has low educational aspirations and high academic performance?
B. Who has medium educational aspirations and medium academic performance?
C. Who has high educational aspirations and low academic performance?
D. Who is female with medium educational aspirations and low academic performance?
E. Who is male with high educational aspirations and high academic performance?

F. Who is a female with neither low educational aspirations nor low academic performance?

9. Suppose you were to draw individual students, as we did in Problem 6, from the same population, what is the probability of getting a student

A. Who is either a male with high educational aspirations and high academic performance or a female with low educational aspirations and low academic performance?

B. Who is either a female with medium educational aspirations or a male with low educational aspirations and high academic performance?

C. Who is neither a female nor a male with medium or high educational aspirations?

D. Who is a male with medium educational aspirations and medium academic performance?

E. Who is a female with medium educational aspirations and medium academic performance?

F. Who is either male or female with medium educational aspirations and medium academic performance?

G. Who has high educational aspirations and low academic performance given that he is a male?

H. Who has low academic performance given that she is a female?

I. Who is a female with medium academic performance given that she has low educational aspirations?

10. Suppose you were to draw a random sample of five students (five draws) from this population of 1,000, replacing them after each draw. What is the probability of getting five students

A. With low academic performance?

B. With medium academic performance?

C. With high academic performance?

D. With low educational performance?

E. With medium educational performance?

F. With high educational performance?

G. Who are male?

H. Who are female?

I. Who are males with low academic performance?

J. Who are females with high academic performance?

K. Who are males with medium academic performance?

L. Who have low educational aspirations and high academic performance?

M. Who have medium educational aspirations and medium academic performance?

N. Who have high educational aspirations and low academic performance?

O. At least two of whom are female?

P. At least three of whom are males with medium educational aspirations and medium academic performance?

Q. One of whom is a female with high educational aspirations and two of whom are either males or females with high educational aspirations and high academic performance?

13

Hypothesis Testing

There are many ways to carry out statistical tests. For our purposes, the important steps involve drawing a sample, analyzing the data produced by it, and making inferences about the population based on information from the sample. Then we can test the statistical inferences we have made in two basic ways. First, we can test specific hypotheses. This occurs when a hypothesis is stated about some parameter and then is tested by examining sample data. Second, we can test statistical inference by making an estimate about the value of some parameter.

In this chapter we learn about hypothesis testing in light of the general concept of a sampling distribution. Then we give special attention to the binominal sampling distribution. At times we have simple dichotomies such as "yes or no," "success or failure," "heads or tails," and we can hypothesize the probability of one state or the other. When this is the case, and when the number of trials is relatively small and the trials are independent of each other, then it is appropriate to use the binomial distribution.

After we have become familiar with the concept of sampling distributions, we will learn how to carry out statistical tests in which we accept or reject the hypothesis. This is done with a certain risk of being wrong, so we will learn, in turn, about the types of error that can exist in testing a statistical hypothesis.

13.1 THE CONCEPT OF A
SAMPLING DISTRIBUTION OF A STATISTIC

Earlier we learned about the special sampling distribution called the normal distribution. Now let us turn our attention to the general concept of the sampling distribution of a statistic. This concept involves the combination of some elementary probability theory with some elementary principles of sampling. A *sampling distribution* is different from a sample because it is theoretically, not empirically, derived, and it is based on the *probability* of events happening over the *long run*.

Let us begin by stating the definition formally. The sampling distribution of a statistic is a theoretically expected distribution of *any statistic* for a large number of random samples of a specific size. A sensible question to ask is "What is meant by a theoretically expected distribution of *any statistic*?" Much of the rest of this section will provide an answer to this question.

A series of steps comprise and demonstrate two important mathematical theorems. Consider what we would get if we actually did draw numerous independent random samples from a given population and each time calculated summary statistics. In the first place, we would have unique population parameters. Second, we would have numerous sample statistics, that is, different values for different samples. Finally, we would have *distributions* of sample statistic values. These distributions would be the sampling distributions of the statistics.

Let us examine a distribution of sample means, keeping in mind that it is possible to use any statistic as the basis for a sampling distribution. Suppose that we decide to carry out a research project investigating family size in our state using the average number of children as the measure of family size. From the data provided by the registrar of vital statistics, we find that the population of families is not normally distributed along the interval scale of the number of children, and that the mean number for the entire state is $\mu_Y = 2.75$, the variance is $\sigma_Y^2 = 0.16$, and the standard deviation is $\sigma_Y = 0.40$. It is not normally distributed, in that there are many families clustered closer to the mean below it and not too many far above it.

Now, suppose that we decide to draw random samples of $N = 50$. Sample 1 turns out to have a mean of $\bar{Y}_1 = 3.00$, a variance of $s_{Y_1}^2 = 0.25$, and a standard deviation of $s_{Y_1} = 0.50$. After we return those samples to the population, we select another random sample of 50. Sample 2 turns out to have a mean of $\bar{Y}_1 = 2.40$, a variance of $s_{Y_2}^2 = 0.09$, and a standard deviation of $s_{Y_2} = 0.30$. Since we drew *simple random samples*, such a difference between the statistic values is not too surprising—we already know that different samples from the same population may produce different values for each statistic. The two sample means can be entered on a frequency distribution of mean values.

Back to drawing samples. Suppose that we go on selecting more samples until we have, say, 100 samples of size 50. We would have 100 sample means, at least some of which are not likely to be equal. We could plot the 100 mean values on a frequency distribution of mean cumulative average number of children, as shown in Figure 13.1.1.

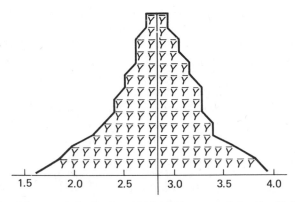

Figure 13.1.1. A frequency distribution of 100 sample means of size $N = 50$ for average number of children.

By connecting the peak of each frequency in Figure 13.1.1, we can construct a frequency histogram. You should remember that even though the population of families is not normally distributed on the variable family size, the more samples we draw, the more the distribution approaches a normal distribution of sample statistic values.

Instead of using the frequencies, we could convert the distribution of sample means into a distribution of probabilities. This would give us a *probability distribution* of the 100 sample means. Such a probability distribution resembles the frequency distribution that we just constructed, except that now we plot the *probabilities* of the statistic values. To calculate each probability, we would divide the frequency of a particular mean value $\bar{Y}$ by the number of samples, 100. For example, there are 2 sample means of 2.0, 8 sample means of 2.5, and 13 sample means of 2.7. Thus the probability of drawing samples that produce means of 2.0 is $2/100 = 0.02$, 2.5 is $8/100 = 0.08$, 2.7 is $13/100 = 0.13$.

If we were to select an *infinite* number of samples of 50, we could construct a distribution of the mean (or any other statistic for that matter) for all the possible values of the population. A probability distribution for an infinite number of samples for any statistic is called the *sampling distribution of the statistic*. We can portray the sampling distribution of the mean as a normal curve as shown in Figure 13.1.2.

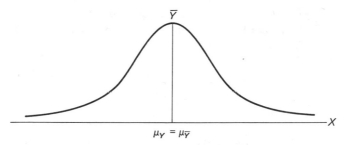

Figure 13.1.2. A probability distribution curve of the mean.

13.2 THE SAMPLING DISTRIBUTION OF THE MEAN

The sampling distribution of the mean will be precisely normal, and the mean of the sampling distribution will be equal to the population mean, even though many of the individual sample means may not be.

So that we can understand this concept more fully, let us simplify matters a bit and use a tiny population of five people, alphabetically abbreviated. Each of the five watches television a different amount of time, from 1 to 5 hours a week. This means that on variable X we have

Person	Number of hours spent watching TV (X)
A	1
B	2
C	3
D	4
E	5
	$\sum_{i=1}^{N} X_i = 15 \qquad \mu_X = 3.00$

The population mean $\mu_X = 3.00$. Let us draw every possible simple random sample of size $N = 3$ from this population and compute the mean for each sample.

Sample	$\sum_{i=1}^{N} X_i$	$\bar{X}$
ABC	$1 + 2 + 3 = 6$	2.00
ABD	$1 + 2 + 4 = 7$	2.33
ABE	$1 + 2 + 5 = 8$	2.67
ACD	$1 + 3 + 4 = 8$	2.67
ACE	$1 + 3 + 5 = 9$	3.00
BCD	$2 + 3 + 4 = 9$	3.00
ADE	$1 + 4 + 5 = 10$	3.33
BCE	$2 + 3 + 5 = 10$	3.33
BDE	$2 + 4 + 5 = 11$	3.67
CDE	$3 + 4 + 5 = 12$	4.00
	$\sum_{i=1}^{N} X_i = 30.00 \qquad \mu_{\bar{X}} = 3.00$	

Since only two of the sample means are equal to the population mean, we have *random distortion* in eight of the ten samples. However, if we sum all the sample means and divide by the number of samples, we get the mean of the sampling distribution, and it *is* equal to the mean of the population. Since we generally calculate a mean from only one sample, we base our estimate on the characteristics of the sampling distribution from the sample. Specifically, because the overall mean of all possible simple random sample means is equal to the population mean, we call the mean of a simple random *sample* an *unbiased* estimate of the *population* mean.

A sample is systematically distorted if we select it according to certain specifications rather than by random selection. In random sampling the mean of the sample might be wrong, but it will *not be biased*, because it will not have *systematic* distortion. If we systematically choose by other-than-random methods, the sampling distribution may be distorted, and the samples would produce biased estimates.

We can now put this concept into general terms. Suppose that we draw a simple independent random sample of any size from a normal population. For operationally defined interval or ratio variables, we can calculate arithmetic means for such a sample. If we were to draw an infinite number of simple random samples of the same size from the same population, each sample would have its own mean $\bar{Y}$, variance s_Y^2, and standard deviation s_Y for all interval or ratio variables. By using the value of *any* summary statistic for each sample, we could plot a frequency distribution of the sample statistic for the infinite samples. Then we could convert it to a probability distribution of a sample statistic.

It is important to emphasize again that the samples we are talking about now are not actually gathered. Instead, we are setting up a procedure that will enable us to examine empirical findings more easily. Our research interests are in *one* sample obtained from an honest-to-goodness population, and on the computation of a specific statistic such as the mean, the variance, or the correlation coefficient. Thus, for all practical purposes, we seldom (if ever) "construct" a sampling distribution of a statistic, especially since there are tables that have been worked out to show them.

We assume that we begin with a normal population with a mean μ_Y, a variance σ_Y^2, and a standard deviation of σ_Y. Remember, normality is an abstraction, and a "normal" population does not necessarily exist in empirical reality. Instead, normality is a statistical device that we use to help examine *probabilistic* relationships. The distribution of an *infinite* number of sample means along the levels of an interval or ratio variable will be clustered around the population mean μ_Y. Thus, the sampling distribution of sample means produces a *normal curve* whose mean is equal to the population mean, that is, $\mu_{\bar{Y}} = \mu_Y$.

We can also measure the variability of the sampling distribution. Some fairly complex math is required to calculate the variance of the sampling distribution, and for our purposes we need not go through it. Instead, let us merely state that for an infinite number of samples of size N, the variance of the sampling distribution is σ_Y^2/N, and the standard deviation of the sampling distribution is $\sigma_Y/\sqrt{N}$. This last term, the standard deviation of the sampling distribution, is called the *standard error* and is expressed symbolically as $\sigma_Y/\sqrt{N} = \sigma_{\bar{Y}}$. These measures are summarized in Table 13.2.1.

Table 13.2.1. Symbols for the measures of the sampling distribution of sample means

Population $\left\{ \begin{array}{l} \\ \\ \\ \end{array} \right.$	$\mu_Y = $ the mean of the population $\sigma_Y^2 = $ the variance of the population $\sigma_Y = $ the standard deviation of the population
Sample $\left\{ \begin{array}{l} \\ \\ \\ \end{array} \right.$	$\bar{Y} = $ the mean of each sample $s_Y^2 = $ the variance of each sample $s_Y = $ the standard deviation of each sample
Sampling Distribution $\left\{ \begin{array}{l} \\ \\ \\ \end{array} \right.$	$\mu_{\bar{Y}} = $ the mean of the sampling distribution $\sigma_Y^2/N = \sigma_{\bar{Y}}^2 = $ the variance of the sampling distribution $\sigma_Y/\sqrt{N} = \sigma_{\bar{Y}} = $ the standard deviation of the sampling distribution or the standard error

It is essential to realize that we are dealing with three separate but related distributions, only one of which is derived empirically from the information we gather in our sample. The first two are familiar to us and have been the subject of past discussion. The first distribution is empirically derived and is called the *sample distribution* of scores within a single sample. Its *statistics* are the familiar mean $\bar{Y}$, the variance s_Y^2, and the standard deviation s_Y. The second distribution is an assumed abstraction and is called the *population distribution*. Its *parameters* are the mean μ_Y, the variance σ_Y^2, and the standard deviation σ_Y.

The third distribution is new to us in this chapter, but it has many familiar characteristics. It is an abstraction that is theoretically derived, and it is called the *sampling distribution* of a statistic. The mean of the sampling distribution $\mu_{\bar{Y}}$ is identical with the population mean, μ_Y. However, its variance depends upon the size of the sample, since $\sigma_{\bar{Y}}^2 = \sigma_Y^2/N$. Similarly, its standard deviation is $\sigma_Y/\sqrt{N}$ and is called the *standard error* of the sampling distribution $\sigma_{\bar{Y}}$. The parameters and statistics of the three distributions are summarized in Table 13.2.2 on page 284.

An important feature in the formulas for the variance and standard deviation of the sampling distribution involves the denominator, which contains the sample size N. Since N equals the number of respondents in the sample, it is always a whole number. Thus any quotient automatically decreases as the denominator becomes larger. This means that as the sample size becomes larger and the denominator of the formula for the standard error also becomes larger, the variance and the standard error will become smaller. Put differently, the larger the sample, the smaller the standard error of its sampling

Table 13.2.2. The symbols indicating the characteristics of samples, populations, and sampling distributions

	Sample (Frequencies)	Population (Universe)	Sampling Distribution of a Statistic
Mean	$\bar{Y}$	μ_Y	$\mu_{\bar{Y}}$
Variance	s_Y^2	σ_Y^2	$\sigma_Y^2/N = \sigma_{\bar{Y}}^2$
Standard Deviation	s_Y	σ_Y	$\sigma_Y/\sqrt{N} = \sigma_{\bar{Y}}$
Derivation	Empirical	Theoretical	Theoretical

distribution. Thus, as N goes up, $\sigma_{\bar{Y}}$ goes down. For example, suppose that we have a population of job applicants, the mean value for which on an aptitude test is $\mu_Y = 80$, with a variance of $\sigma_Y^2 = 324$, and standard deviation $\sigma_Y = 18$. With two sampling distributions, one for samples of $N = 20$ and one for samples of $N = 100$, we can calculate the standard error of the sampling distribution, as shown in Table 13.2.3.

Table 13.2.3. Computations for the standard error of the sampling distributions

Sample Size	$N = 20$	$N = 100$
Mean of the Sampling Distribution and the Population	$\mu_{\bar{Y}} = \mu_Y = \quad 80$	80
Standard Deviation of the Population	$\sigma_Y = \quad 18$	18
Standard Error of the Sampling Distribution	$\sigma_{\bar{Y}} = \sigma_Y/\sqrt{N} = 18/\sqrt{20}$ $\sigma_Y/\sqrt{N} = 18/4.5$ $\sigma_{\bar{Y}} = \quad 4.0$	$18/\sqrt{100}$ $18/10$ 1.8

The standard error when $N = 20$ is 4.0, and when $N = 100$, it is 1.8. Obviously, if everything else is equal, the larger sample provides us with a lower standard error. This may be easier to see in diagrammatic form. Figure 13.2.1 shows that as the sample gets larger, the standard error gets smaller.

The series of steps that we have just been through illustrates two mathematical theorems, the Central Limit Theorem and the Law of Large Numbers.[1]

[1]For a mathematical treatment of the two theorems see John G. Kemeny, J. Laurie Snell, and Gerald L. Thompson, *Introduction to Finite Mathematics*, 2nd ed. (Englewood Cliffs, New Jersey: Prentice-Hall, Inc., 1966), pp. 175–182 and 201–207; and for a more detailed discussion of the statistical application see Hubert M. Blalock, Jr., *Social Statistics*, 2nd ed. (New York: McGraw-Hill Book Company, Inc., 1972), pp. 181–184 and 220–222.

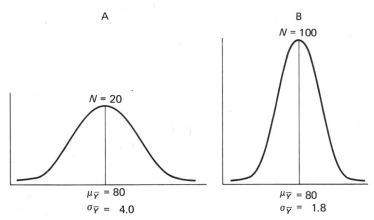

Figure 13.2.1. Normal sampling distributions of sample means when $\mu_Y = 80$ and $\sigma_Y = 18$

These two mathematical theorems are based on the laws of probability. They serve as the basic features of quantification that enable us to justify the use of social statistics for making inferences.

Before stating the theorems, we will give a brief synopsis of the steps described thus far. Because the *dispersion* of sample means *decreases* as we *increase* the sample size, no matter what the distribution, it makes common sense that as the sample size increases, the mean for one sample has a better chance of approximating the mean of the population. Common sense is not the only reason this the case, and the two theorems tell us why.

The Central Limit Theorem is: If a large number of simple random samples of size N are drawn from a normal population which has a mean μ_Y, a variance σ_Y^2, and a standard deviation σ_Y, the mean of the theoretical sampling distribution of sample means will be equal to the mean of the population of original scores from which the sample was drawn. The sampling distribution will be normal with a mean of $\mu_{\bar{Y}}$, a variance $\sigma_{\bar{Y}}^2$ of σ_Y^2/N, and a standard error of $\sigma_{\bar{Y}}$ of $\sigma_Y/\sqrt{N}$.

The Law of Large Numbers is: As the size N of a sample drawn from *any* population that has a mean μ_Y, a variance σ_Y^2, and a standard deviation σ_Y increases toward infinity, the sampling distribution of sample means approaches normality with mean μ_Y and variance σ_Y^2/N.

When we combine the Law of Large Numbers with the Central Limit Theorem, we can state a general sampling theorem. For any population of scores whether or not they are normally distributed, the sampling distribution of the mean will approach a normal distribution as N becomes large. This means that we can deal with populations that are *not* normally distributed. The ability to do this is essential for social science research, since most of the populations with which we deal are *not* normal.

This general sampling theorem has an additional value for inferential statistics: If N is large enough, we can relax the assumption of normality and use the normal curve to carry out our statistical tests. This feature of large samples is especially important because we usually base our estimates of a given population mean on a *single sample* drawn from that population.

Together, the two theorems tell us that we are likely to approximate more closely the parameters in question as the sample size is increased. Furthermore, it does not matter whether or not the population is normal. Thus, since in most social research samples we have a large N, we can generally treat sampling distributions of sample statistics as if they were normal.

The concept of sampling distributions is extremely important because we almost always *estimate* population parameters from sample statistics. Generally, therefore, we are concerned with *how much* error there is between a given sample statistic and the corresponding hypothesized population parameter. Because of this concern, we compare our sample results with the *theoretically expected* results given by the appropriate sampling distribution. To make such a comparison we must first specify the sampling distributions for various statistics. When we do, we can use statistical models whose mathematical properties are known, such as the normal distribution. Another commonly used model is the binomial sampling distribution, and a clear understanding of it facilitates understanding of other more complex models.

13.3 THE BINOMIAL SAMPLING DISTRIBUTION

Technically, the binomial distribution is a theoretically expected probability distribution of a certain number of random samples that are selected from a population. For example, suppose that we are interested in selecting a sample of size 10 from a total population of 5000 of which the dichotomous variable sex is split 50/50 with 2500 females and 2500 males. We know the probability of getting a female is $\text{Pr} = 0.50$ in the population. Suppose that we also want to know the probabilities of getting any specific number of females (or males) in an *infinite number of samples* of size 10. In other words, what are our chances of getting exactly three females or of getting exactly six males over the long run?

Let us begin by listing all the possibilities. If we draw a random sample of 10 students, we could get anywhere from 0 to 10 females (or males). Diagrammatically, this fact can be located on a continuum, as shown in Figure 13.3.1. The problem is to determine the probabilities for each possible number.

The basic theorem of the binomial distribution is a probability statement: The probability of getting exactly x successes (one category or the other) is equal to the *number* of ways of getting x successes times the probability of any

Figure 13.3.1. The number of females (or males).

given sequence of x successes and $N - x$ failures. It can be expressed symbolically:

$$\text{Pr}(x) = \binom{N}{x} \cdot p^x \cdot q^{(N-x)}$$

Probability of getting exactly x successes	=	Number of ways of getting x successes	·	Probability of any given sequence
$\text{Pr}(x)$	=	$\binom{N}{x}$	·	$p^x \cdot q^{(N-x)}$

Let us begin by looking at "the *number* of ways of getting x successes." This part of the formula is included because in simple random sampling we draw one person at a time from the population. Thus on draw 1, we can get a female or a male, on draw 2, we can get a female or a male, etc., up through draw 10. With this in mind, we can speculate about the total number of *possible ways* we can get any number of females or males in 10 draws. For example, it is possible to draw 0 females and 10 males in only *one* way:

draw number: 1 2 3 4 5 6 7 8 9 10

respondent's sex: M M M M M M M M M M

The number of ways it is possible to draw 1 female and 9 males is a bit more intricate to enumerate since there are 10 ways, namely:

draw number: 1 2 3 4 5 6 7 8 9 10

respondent's sex: F M M M M M M M M M
                   M F M M M M M M M M
                   M M F M M M M M M M
                   M M M F M M M M M M
                   M M M M F M M M M M
                   M M M M M F M M M M
                   M M M M M M F M M M
                   M M M M M M M F M M
                   M M M M M M M M F M
                   M M M M M M M M M F

It becomes arduous beyond this point, because for each increase in the number of females up until there is an equal number of males and females,

287

that is, 5 and 5, there is a multiplicative increase in the number of possible ways. Specifically, we can get exactly 2 females and 8 males in 10 draws in 45 ways. For instance, the first and last 3 ways would be:

draw number:	1	2	3	4	5	6	7	8	9	10
respondent's sex:	F	F	M	M	M	M	M	M	M	M
	F	M	F	M	M	M	M	M	M	M
	F	M	M	F	M	M	M	M	M	M

.

.

.

	M	M	M	M	M	M	F	M	M	F
	M	M	M	M	M	M	M	F	M	F
	M	M	M	M	M	M	M	M	F	F

We could go on like this enumerating for 3, 4, 5, 6, 7, 8, 9, and 10 females, each with the appropriate number of males. Rather than enumerating each set of possible ways and counting them, we use a straightforward formula. Technically, to find the number of ways of getting x successes, we calculate the ratio of the factorial for the total number of draws to the product of the factorial for the exact number in one category times the factorial for the exact number in the other category. This statement is very mathematical, but it is really quite simple when expressed symbolically. Here are the definitions of the terms in the formula:

N = the total number of draws, e.g., the sample size.

x = the number of successes, e.g., the number of cases in one category.

$N - x$ = the number of failures, e.g., the number of cases in the other category.

! = the symbol for factorial multiplication. It tells us to multiply the value preceding the ! by all the whole numbers less than it but greater than zero, e.g., if $N = 8$, then $N! = 8 \cdot 7 \cdot 6 \cdot 5 \cdot 4 \cdot 3 \cdot 2 \cdot 1 = 40,320$.

We can now express this ratio in terms of the following formula:

$$\text{The number of ways of getting exactly } x \text{ successes in } N \text{ draws} = \frac{N!}{x!(N - X)!} = \binom{N}{x}$$

Let us carry out five combinations for our sample of 10. Because we are dealing with a population probability of 0.50, the number of possible combinations increases and decreases symmetrically. This means that once the number in one category and the number in the other category pass the midpoint, the possible number of ways of getting x successes decreases exactly as it increased. Notice that we can cancel terms in the numerator and denominator *before* multiplying them, making our work much easier.

The number of ways to get exactly 0 successes
$$= \binom{10}{0} = \frac{\cancel{10} \cdot \cancel{9} \cdot \cancel{8} \cdot \cancel{7} \cdot \cancel{6} \cdot \cancel{5} \cdot \cancel{4} \cdot \cancel{3} \cdot \cancel{2} \cdot \cancel{1}}{(10 \cdot 9 \cdot 8 \cdot 7 \cdot 6 \cdot 5 \cdot 4 \cdot 3 \cdot 2 \cdot 1)} = 1$$

The number of ways to get exactly 1 success
$$= \binom{10}{1} = \frac{10 \cdot 9 \cdot 8 \cdot 7 \cdot 6 \cdot 5 \cdot 4 \cdot 3 \cdot 2 \cdot 1}{1 \cdot (9 \cdot 8 \cdot 7 \cdot 6 \cdot 5 \cdot 4 \cdot 3 \cdot 2 \cdot 1)} = 10$$

The number of ways to get exactly 2 successes
$$= \binom{10}{2} = \frac{\overset{5}{\cancel{10}} \cdot 9 \cdot 8 \cdot 7 \cdot 6 \cdot 5 \cdot 4 \cdot 3 \cdot 2 \cdot 1}{\cancel{2} \cdot 1 \cdot (8 \cdot 7 \cdot 6 \cdot 5 \cdot 4 \cdot 3 \cdot 2 \cdot 1)} = \frac{45}{1} = 45$$

The number of ways to get exactly 3 successes
$$= \binom{10}{3} = \frac{\overset{3}{10} \cdot 9 \cdot \overset{4}{\cancel{8}} \cdot 7 \cdot 6 \cdot 5 \cdot 4 \cdot 3 \cdot 2 \cdot 1}{\cancel{3} \cdot \cancel{2} \cdot 1 \cdot (7 \cdot 6 \cdot 5 \cdot 4 \cdot 3 \cdot 2 \cdot 1)} = \frac{120}{1} = 120$$

The number of ways to get exactly 4 successes
$$= \binom{10}{4} = \frac{10 \cdot \overset{3}{\cancel{9}} \cdot 8 \cdot 7 \cdot 6 \cdot 5 \cdot 4 \cdot 3 \cdot 2 \cdot 1}{\cancel{4} \cdot \cancel{3} \cdot \cancel{2} \cdot 1 \cdot (6 \cdot 5 \cdot 4 \cdot 3 \cdot 2 \cdot 1)} = \frac{210}{1} = 210$$

The number of ways to get exactly 5 successes
$$= \binom{10}{5} = \frac{\overset{2}{\cancel{10}} \cdot 9 \cdot \overset{3}{\cancel{8}} \cdot 7 \cdot \overset{2}{\cancel{6}} \cdot 5 \cdot 4 \cdot 3 \cdot \overset{3}{\cancel{2}} \cdot 1}{\cancel{5} \cdot \cancel{4} \cdot \cancel{3} \cdot \cancel{2} \cdot 1 \cdot (5 \cdot 4 \cdot 3 \cdot 2 \cdot 1)} = \frac{252}{1} = 252$$

Binomial values can be presented in tabular form as shown in Appendix Table E. The binomial expansion is symmetrical, so when N goes above 10 the x values up to N are truncated from the table. For example, when $N = 14$, the midpoint is when $x = 7$ and the number of ways = 3432. Notice that in the last column the binomial values end at $\binom{14}{10} = 1001$. The remaining values may be found by using those at the other end of the continuum, that is,

$$\binom{14}{11} = \binom{14}{3} = 364, \binom{14}{12} = \binom{14}{2} = 91, \binom{14}{13} = \binom{14}{1} = 14,$$

and

$$\binom{14}{14} = \binom{14}{0} = 1$$

Now that we can calculate "the number of ways of getting exactly x successes," let us turn to the matter of finding "the *probability* of any given sequence of x successes and $N - x$ failures." Remember, we are making random draws, which, by definition, are independent. Furthermore, we are assuming that we carry out sampling *with* replacement; thus we have a stable probability of success over many draws. Therefore, we can designate the probability of an x draw as being equal to the proportion of cases in category x, that is, $p =$ the probability of getting a success in one draw from the population. Then we can set the proportion of cases in the other category $(1.00 - p)$ equal to q, that is, $q =$ the probability of getting a nonsuccess in one draw from the population.

The probability of getting any sequence of successes and failures in a specified number of draws is now a matter of multiple probabilities. Specifically, to calculate the probability of any given sequence of x successes and $(N - x)$ failures, we multiply the population probability of x by itself x number of times, that is,

$$(p \cdot p \cdot p \cdot \ldots \cdot p) = x \text{ number of } p\text{'s times each other} = p^x$$

Now we multiply the population probability of $(N - x)$ by itself $(N - x)$ number of times, that is,

$$(q \cdot q \cdot q \cdot \ldots \cdot q) = (N - x) \text{ number of } q\text{'s times each other} = q^{(N-x)}$$

The product of these two products is the probability of any sequence of x successes and $N - x$ failures, and may be expressed symbolically,

$$\text{Probability of a given sequence} = p^x \cdot q^{(N-x)}$$

What we now have is a straightforward probability problem. Since we are interested in the probability of exactly x successes, we multiply the number of ways of getting exactly x successes by the probability of any given sequence of x and $N - x$ events. Now we can set all parts of the binomial theorem into defined terms.

$\Pr(x) =$ the probability of exactly x successes.
   $N =$ the total number of draws, e.g., the sample size.
   $x =$ the number of successes, e.g., the cases in one category.
$N - x =$ the number of failures, e.g., the cases in the other category.
   $p =$ the proportion of cases in one category.
   $q =$ the proportion of cases in the other category $(1.00 - p)$.

The theorem we stated a few pages ago can again be expressed symbolically:

$$\Pr(x) = \binom{N}{x} \cdot p^x \cdot q^{(N-x)}$$

Probability of getting exactly = x successes	Number of ways of getting x successes	Probability of · any given sequence.

$$\Pr(x) \quad = \quad \binom{N}{x} \quad \cdot \quad p^x \cdot q^{(N-x)}$$

Let us carry out the calculations for our sample of 10. We already know the number of ways we can get various combinations of respondents from the binomial sampling distribution; the rest of the steps are as follows:

$$\Pr(0) = \binom{10}{0} \cdot (1/2)^0 \cdot (1/2)^{(10-0)}$$
$$= 1 \cdot (1) \cdot (1/1024) = 1/1024 = \underline{0.001}$$

$$\Pr(1) = \binom{10}{1} \cdot (1/2)^1 \cdot (1/2)^{(10-1)}$$
$$= 10 \cdot (1/2) \cdot (1/512) = 10/1024 = \underline{0.010}$$

$$\Pr(2) = \binom{10}{2} \cdot (1/2)^2 \cdot (1/2)^{(10-2)}$$
$$= 45 \cdot (1/4) \cdot (1/256) = 45/1024 = \underline{0.044}$$

$$\Pr(3) = \binom{10}{3} \cdot (1/2)^3 \cdot (1/2)^{(10-3)}$$
$$= 120 \cdot (1/8) \cdot (1/128) = 120/1024 = \underline{0.117}$$

$$\Pr(4) = \binom{10}{4} \cdot (1/2)^4 \cdot (1/2)^{(10-4)}$$
$$= 210 \cdot (1/16) \cdot (1/64) = 210/1024 = \underline{0.205}$$

$$\Pr(5) = \binom{10}{5} \cdot (1/2)^5 \cdot (1/2)^{(10-5)}$$
$$= 252 \cdot (1/32) \cdot (1/32) = 252/1024 = \underline{0.246}$$

We can summarize the binomial probabilities of getting a specific number of one category by chance for the dichotomous variable sex in a probability histogram. This histogram displays the *theoretically expected probability* distribution (the sampling distribution) of the exact number of females drawn in a random sample of ten as shown in Figure 13.3.2.

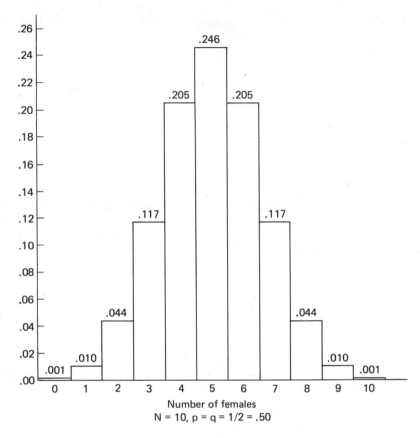

Figure 13.3.2.

Our theoretically expected probability distribution indicates that just less than one-quarter (0.246) of the time we can expect to get a perfect 50-50 split of males and females. Similarly, the histogram shows just how often we can expect to get an outcome different from the "expected" $p = q = 1/2 = 0.50$ dichotomy. If we total all the probabilities other than the five females-five males event, we find that we can expect to be different from that 50-50 split 75.4 percent of the time in an infinite number of random samples of 10.

When we get outcomes such as eight females and two males or one female and nine males, and if we are certain that the population is really 50 percent male and 50 percent female, we may suspect that our method of random selection may be faulty. Alternatively, if we know the sampling has been random but have just "guessed" that the population is split 50-50, an unlikely outcome throws doubt on that guess (hypothesis). As a matter of fact, the more unusual or improbable an event is, the more likely we are to believe that the event is not caused by chance, but rather that it is caused by the characteristics of the group sampled. This raises the important issue of *probabilistic inference*, which is the very basis for inferential statistics. We are concerned with just how likely or probable any given event is *if* our hypotheses and other assumptions are correct.

In our sample of 10, for example, if we wish to determine how rare an event it would be to get nine or more females or males in a sample of 10, we must take into account that there are other events, just as rare or rarer, in both tails of the distribution, and we sum the probabilities of all outcomes that are "improbable." This sum gives the probability of the specific events as well as the events that would be even less likely if the hypothesis were true. Thus,

$$\Pr(0) + \Pr(1) + \Pr(9) + \Pr(10) = (1/1024) + (10/1024) + (10/1024)$$
$$+ (1/1024) = 22/1024 = \underline{0.0215}$$

13.4 STATISTICAL TESTING

There are five practical steps in statistical testing[2]:

1. Listing our assumptions.
2. Determining the appropriate sampling distribution.
3. Choosing the level of significance and specifying the critical region.
4. Computing the test statistic.
5. Deciding about rejection of the null hypothesis.

(1) The first step in statistical testing involves listing our assumptions about the population with which we are concerned and about the sampling procedures we use. Assumptions can be of two general types, those that are relatively certain and those that are relatively questionable. Those assumptions of which the researcher is relatively certain are referred to as "the model"; and those assumptions that are relatively questionable are referrred to as "the hypotheses." The hypotheses are the assumptions we want to test.

For example, suppose that we are interested in the population of students at our college and that we have drawn a large simple random sample. Also suppose that we are interested in the proportion of students in each year of college that wants to major in sociology. We might assume (1) that the proportion in the entire college is about equal in all four classes and (2) that the sample is truly a random one. If our data show that a considerably higher

---

[2]These steps follow the logic, order, and sequence of those presented by Hubert M. Blalock, Jr., *Social Statistics*, 2nd edition (New York: McGraw-Hill Book Company, 1972), pp. 155–66.

percentage of freshmen and sophomores than juniors and seniors wish to major in sociology, we might conclude that at least one of our two assumptions ought to be rejected because it is false.

It is possible that our sampling method was biased and that we do not have a truly random sample. However, since we have full knowledge about our sampling procedure, and since we carried out the sampling with great care to insure that the sample was random, we are likely to be willing to accept the assumption of *random selection* as our *model*. If so, we focus on the different proportion of sociology majors. Since we have no control over the population distribution, we are likely to believe that the assumption of equal distribution is relatively questionable. Thus we decide to test the questionable assumption or the *null hypothesis* of *equal distribution*. As we have pointed out, we rest our faith on the *procedure* of probability sampling rather than on the test itself.

(2) The second step in statistical testing is determining the appropriate sampling distribution. In order to derive a sampling distribution of a statistic, we use theoretically distributed probabilities to specify how likely all *possible* sample statistic values are if our assumptions are correct. Thus we can associate certain probabilities with each possible value, and then we compare our actual empirical findings with the probabilities of the sampling distribution. Since we know the chance probability of specific values if our assumptions are true, we can make a decision about our willingness to risk rejecting the assumptions.

A little thought makes it clear that there are two possibilities if we get an *improbable* statistic value. First, the assumptions may be true and we have encountered a rare sample statistic from one or the other tail of the probability sampling distribution. Second, one of the assumptions may be false and we have a sample statistic from a different distribution. Sadly, even though the laws of probability specify the proportion of times *in the long run* that we are likely to find such a value, we are never exactly certain which of these two possibilities is the case with any given sample statistic.

(3) A concern with sample statistics leads us to the third step in statistical testing which is choosing the level of significance and specifying the critical region. We can separate statistic values into two possible types. On the one hand, there are those values that would make us reject the null hypothesis; on the other hand, there are those values that do not enable us to reject the null hypothesis. The *critical region* is made up of those improbable statistic values that seem to be so unlikely that finding them would make us reject the null hypothesis.

We must decide in advance which improbable events would cause us to doubt our hypotheses and make it impossible for us to make inferences about the population from the sample. Later we will learn that we specify which unlikely outcomes will cause us to fail to accept certain kinds of hypotheses.

For now, however, we should know that *prior to testing* we decide upon certain unlikely outcomes, and, if they are obtained, then we do not accept certain hypotheses about the population.

Such unlikely outcomes comprise what is the critical region. When we set a specific level or score as the *cutting point* of improbable events, we determine the critical region. Thus, the cutting point establishes what is called a *level of significance*. Technically, a level of significance is the probability that the sample outcome falls in the critical region if certain hypotheses are true. Once the level of significance is specified, the cutting point is determined by the sum of the probabilities of all the outcomes within the critical region.

(4) The fourth step in statistical testing is computing the *test statistic*, the sampling distribution of which will be used in the test. We examine the data from our sample and use them to calculate a statistic, the possible values of which vary according to the laws of probability. Then we compare the value of the statistic with the sampling distribution and determine the probability of our sample statistic occurring if the null hypothesis were true.

(5) The fifth step in statistical testing is deciding whether to reject or fail to reject the null hypothesis. If our findings fall in the critical region, we reject the null hypothesis; if our findings do not fall in the critical region, we back into "accepting" the null hypothesis; however, technically, we do not really "accept" it. If one of the possible alternative hypotheses is true, the null hypothesis will be false, but we may fail to recognize it. Thus we say that we "fail to reject" the *null hypothesis* rather than saying that we accept it. Keep in mind that we never know for certain whether or not we have made a sound decision. Instead, we rest our faith on the *procedures* of probabilistic inference.

Rejection of the null hypothesis leads us to test several alternative hypotheses as offering possible acceptable conclusions about our findings. As we mentioned earlier, we call the test hypothesis the "null" hypothesis, H_0, and we call the alternatives the "research" hypotheses, H_1, H_2, H_3, etc. What we do is use the null hypothesis as a straw man, which we set up for the purpose of knocking down, that is, rejecting. A logical question that now arises involves our faith in probability procedures and the extent to which we may be in error.

13.5 TYPES OF ERROR

Our concern with the risks we take by rejecting or not rejecting a given outcome is embodied in the question "How likely are we to get a specific outcome if our assumptions are correct?" Any seemingly unlikely sample can be a rare event, or it can result because of a false hypothesis. When we establish a level of significance we have a chance of making an error in our decision about rejecting a hypothesis because of a "rare event" in the critical region.

If we select a sample and its statistics turn out to be "rare" events, we may err and reject a true null hypothesis because the sample values fall into the critical region. This improper rejection of a true H_0 is called a *type I error*. Put differently, a type I error occurs when we reject a set of true assumptions because the statistic values fall into the critical region.

There is a second way we can be in error, and it comes about when the sample gives a false picture. If the sample is not representative of the population, but only *appears* to be, we may fail to reject a false null hypothesis. This is referred to as "affirming the consequent" or, more commonly, "failing to reject" a null hypothesis that is actually false. This improper failure to reject a false H_0 is called a *type II error*. In other words, a type II error occurs when we fail to reject a set of false assumptions because the statistic values do not fall into the critical region.

The two types of errors can be shown diagrammatically. Suppose that we are studying population A and that we can represent the sampling distribution of a statistic as a curve with a level of significance and a critical region as shown in Figure 13.5.1. A null hypothesis will be rejected if sample outcomes

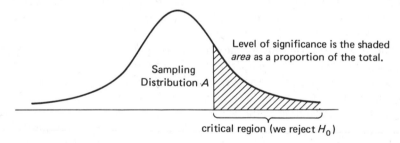

Figure 13.5.1.

fall into the critical region. This rejection will occur even though the assumptions may actually be "true" ones; that is, that the sample values truly come from sampling distribution A. Such rejection constitutes a type I error, the risk of which is determined by a specified level of significance.

Now, suppose that there is a second sampling distribution that is based on the same population but with different assumptions. This is what is meant by alternative hypotheses, and in this case we mean H_0 versus H_1. The second sampling distribution, B, can overlap with A, and one tail of its probability curve can fall into the same range that population A occupies. When this is the case, we have a certain risk of erroneously accepting a sample value as being a "true" one from sampling distribution A when it actually falls in one of the tails of sampling distribution B's curve, as is shown in Figure 13.5.2.

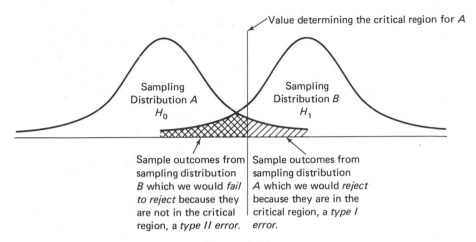

Value determining the critical region for A

Sampling Distribution A H_0

Sampling Distribution B H_1

Sample outcomes from sampling distribution B which we would *fail to reject* because they are not in the critical region, a *type II error.*

Sample outcomes from sampling distribution A which we would *reject* because they are in the critical region, a *type I error.*

Figure 13.5.2.

Seeing how these two types of errors can occur helps in understanding them. However, it also helps to think about what they mean in terms of what we *decide to say* based on our sample, compared to what is actually the case.

Table 13.5.1 indicates that a type I error is saying that the null hypothesis is false when it actually is true; that is, we would reject a true null hypothesis. A type II error is saying that the null hypothesis is true when it is actually false; that is, we would fail to reject a false null hypothesis.

Table 13.5.1. Types of error in hypothesis testing

Based on our findings if we say H_0 is	If H_0 actually is	
	True	False
False (we reject it)	Type I error	Correct decision
True (we fail to reject it)	Correct decision	Type II error

The level of significance is defined as the probability of a type I error. The probability of a type II error depends upon which alternative hypothesis is actually true, and this is a rather complex matter. The investigator sets and controls the probability of a type I error; the probability of a type II error is a consequence of this decision. Thus, for a given alternative, the greater the probability of a type I error, the smaller the probability of a type II error.

Since this is the case, we generally set a specified level of significance. Then we choose the most *powerful* test available given the assumptions about the data. Hopefully this will minimize the probability of a type II error.

We set the level of significance because we want to be conservative in our rejection of the null hypothesis. We therefore prefer to select a rejection level that enables us to make the probability of a type I error as small as possible without greatly inflating the probability of a type II error. We prefer making the decision to accept a false null hypothesis to making the decision to reject a true one.

In order to measure the probability of making a type II error, we must determine the *power* of our statistic. The notion of power makes sense: The smaller the probability of making a type II error, the more powerful our test. Estimating the probability of making a type II error, and thereby determining the power of a statistic, involves specifying alternative hypotheses in terms of the research problem at hand. Importantly, alternatives are not "tested" to get the probability of a type II error, but the alternatives need to be *specified*.

In general, when using the same statistic with the same level of significance, a one-tailed test is more powerful than a two-tailed test against an alternative in the "right" direction. This is because in a one-tailed test the area is concentrated in one tail of the normal curve. It is also generally true that a statistic will become more powerful as the sample size increases because, as the sample becomes larger, the standard error becomes smaller. Since the smaller the standard error, the more powerful the test, we are more likely to reject false hypotheses with a larger sample.

Let us return briefly to the problem of a type I error. Technically, the probability of a type I error is obtained by summing the probabilities of all the statistic values that fall within the critical region. Commonly the significance level is set arbitrarily and this level is used to determine the cutting point for the critical region. Significance can be set at any level specified by the researcher, and there are no hard and fast rules about selecting one. It is common, however, for sociologists to use the 0.05, 0.01, or 0.001 levels of significance.[3] Most research reports specify the level of significance selected by the researcher.

In summary, after proceeding through the first four steps of statistical testing, we must take the fifth one and either reject or fail to reject the assumptions that form our hypothesis. If the outcome is in the critical region, we reject the null hypothesis with a known probability of a type I error. If, on the other hand, the outcome does not fall into the critical region, we do not reject the null hypothesis, thereby taking a chance on making a type II error.

---

[3] For a fuller discussion of this question see *American Sociologist*, 4 (May 1969), pp. 131–40; especially the list of references on pp. 139 and 140.

EXERCISES

1. In your own words define, describe, or discuss the following terms and give an hypothetical example of each:
   probability distribution
   Central Limit Theorem
   binomial sampling distribution
   simple random sample
   random distortion
   systematic distortion
   normal curve
   binomial expansion
   level of significance

2. Describe verbally the Law of Large Numbers, and explain its value to sampling for a research project.

3. Explain the procedure involved in the rejection of a null hypothesis.

4. True and False Questions
   A. We use the word "hypothesis" to mean, among other things, guesses, hunches, notions, and ideas.
   B. A statistical hypothesis is a statement about an event whose outcome is unknown or about a prediction of some future event.
   C. The binomial distribution is a theoretically expected probability distribution of a certain number of random samples which are selected from a population.
   D. A sampling distribution is empirically derived and is based on events that happened in one sample.
   E. If our findings fall in the critical region, we do not reject the null hypothesis.
   F. Bias refers to random distortion of a given statistic resulting from a sampling procedure.

5. Assume that there are half as many males as females in your college, and that you decide to select a random sample with replacement of five students for a sociology research project. Using the binomial distribution answer the following questions.
   A. What is the probability of getting five males and no females?
   B. What is the probability of getting three males and two females?
   C. What is the probability of getting one male and four females?

6. Assume that you are carrying out a research project in the sociology of sport, and that you are interested in the extent to which an athlete's color is related to his performance and value to the team as measured by the most valuable player awards. In your college ten percent of the student body is black, and the intercollegiate varsity sports include football, soccer, track, basketball, tennis, golf, and baseball. Your findings indicate that the most valuable player award for each sport was given to the following color players:

Black M.V.P	White M.V.P.
football	soccer
track	tennis
basketball	golf
baseball	

A. How many blacks would you expect to find if color is *not* related to team value as determined by the M.V.P. awards?

B. What is the probability of getting exactly two blacks in two draws with replacement from:
   1. The student body?
   2. The M.V.P. list?

C. What is the probability that your findings could happen by chance?

7. In your own words define and describe type 1 and type 2 errors and explain the differences between them.

8. Enumerate the steps in statistical testing, and in your own words explain what they accomplish and how they do so.

14

Comparisons, Tests, and Estimates

In this chapter we learn some of the basic ideas of "testing" as it relates to inferential statistics. Our major interest will be in the ways we make inductive statements about a population based on findings from an actual sample. It is important to realize that both comparing samples and estimating parameters provide alternatives to testing hypotheses about sample values.

First we will learn about the use of the Z statistic in single sample tests in which we compare values from a single sample with parameters from the population, specifying how we decide whether to accept or reject the null hypothesis. Second, we will learn about the use of a special sampling distribution called Student's t distribution and about the connection between t and the concept of degrees of freedom. Third, we will learn about estimating population parameters and how to establish confidence intervals.

14.1 COMPARISONS WITHIN SAMPLES AND THE Z DISTRIBUTION

When we compare the sample means and proportions within a single sample with hypothesized means and proportions in the population, this comparison is called a *single-sample test*. Single-sample tests have somewhat limited use-

301

fulness in sociological research; however, the ideas and general principles underlying them help clarify many aspects of comparisons *between* samples, which is a more important analytical procedure for sociologists.

We can compare statistics from a sample with knowledge of their sampling distribution. Then we can test the null hypothesis with the Z statistic, testing whether or not the difference between the null hypothesis and the findings is statistically significant. Put differently, we can draw inferences about the population from the sample findings by use of the sampling distribution of the relevant statistic.

Let us start with an example in which we know the population parameters μ_Y, σ_Y^2, and σ_Y. Recalling the Law of Large Numbers and the Central Limit Theorem, if N is large enough, we can treat the sampling distribution of sample means as a normal distribution with a mean $\mu_{\bar{Y}}$, a variance $\sigma_{\bar{Y}}^2$, and a standard error $\sigma_{\bar{Y}}$. Because the sampling distribution is normal, we can figure probabilities in terms of the normal curve. We can also determine the probability of a given set or range of sample means for a given N drawn at random from the population in question. However, in practice one never attempts to determine the probability of a given mean. Earlier we learned that the values of any variable can be converted to the normally distributed Z scale, and that we can determine the probabilistic values according to the corresponding Z scores.

Let us test the sampling procedure for a research project in which our interest is in academic performance as measured by grade point average. Suppose that we have a population of students that has a mean grade point average of 2.75, a variance of 0.16, and a standard deviation of 0.40. Suppose that we have information about a sample of 100 students for another research project and that we can use data from this sample. Also suppose that the study focused on grade point average and found the sample mean to be 2.89. Being skeptical about the sampling procedure, we decide to test the null hypothesis of simple, independent random sampling. Making use of the steps of statistical testing we will carry out this problem.

1. Listing Our Assumptions

Level of Measurement: Grade point average is an interval scale.
Model: We assume a normal population, with

$$\mu_Y = 2.75$$
$$\sigma_Y^2 = 0.16$$
$$\sigma_Y = 0.40$$

Null Hypothesis (H_0): We will test the hypothesis that the sample is a simple random one.

Alternative Hypothesis (H_1): If we reject H_0 we then assume that the sample was not a simple independent random one.

2. Determining the Appropriate Sampling Distribution

Because N is large enough, we can relax the assumption of normality of the population distribution and assume that the sampling distribution will be approximately normal; therefore, we can use the normal table and the Z statistic.

3. Choosing the Level of Significance and Specifying the Critical Region

Remember, we are testing the nondirectional hypothesis that the sample is random. Thus we decide to use a two-tailed test. In order to reduce the probability of a type I error, we decide to use the 0.01 significance level. This means that we want 1 percent of the sampling distribution to be in the critical region. Thus the probability of a type I error is 0.01. The two shaded tails are the critical region of 1 percent of the area of the sampling distribution, as shown in Figure 14.1.1.

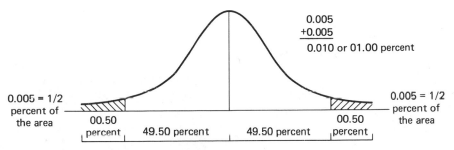

Figure 14.1.1. One percent of the area divided equally between the area above the mean and the area below the mean.

To specify this 1 percent critical region, we turn to the normal table in the Appendix. Since we must divide the 1 percent between the area below the mean and the area above the mean, we locate 49.50 in each tail. Since we want the critical region to contain not more than 1 percent of the area in the normal table, we find the proportion 4951. Moving from the proportion of 4951 in the body of the table to the left margin, we find a Z score of 2.5 as the base unit, and by moving across the top we find the column value to be 0.08. Thus, for $4951 = 2.58$, therefore, a Z score of ± 2.58 indicates that about 99 percent of the area is under the curve between the two scores. This means that if we get a Z value from the sample that is equal to or greater than 2.58, that is, if

$Z \geq 2.58$, we will reject the null hypothesis that the sample is a random one. Figure 14.1.2 shows these steps diagrammatically.

Remember, our confidence lies in the procedures of a two-tailed test at the 0.01 significance level. In essence, we rest our faith on the odds. We know that if the same procedure were used over and over again, 99 percent of the time it would include the true value, and 1 percent of the time it would not. Thus, even though for any given sample we are either right or wrong, a confidence level of 0.99 means that our procedure is known to work 99 times out of 100.

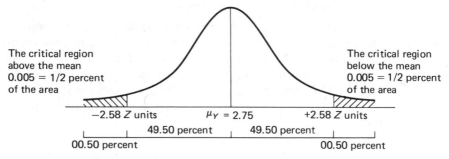

Figure 14.1.2. Two-tailed critical region at the 0.01 significance level.

4. Computing the Test Statistic

Now we must convert the grade point average scores to standard scores so that we can use the Z statistic. For individual scores, the formula for the Z score was expressed symbolically

$$Z = \frac{Y_i - \bar{Y}}{s_Y}$$

However, since we are using a sampling distribution of means, we have to adapt the formula for Z. It is still a given value minus the mean of the distribution divided by the standard deviation of the distribution. But in this instance the given value happens to be a sample mean. Thus, for sampling distributions, the Z formula may be expressed symbolically

$$Z = \frac{\bar{Y} - \mu_Y}{\sigma_{\bar{Y}}} = \frac{\bar{Y} - \mu_Y}{\sigma_Y/\sqrt{N}}$$

In words, we convert the observed value to a Z score by taking the sample value $\bar{Y}$ and comparing it to the parameter value μ_Y and then dividing it by the standard error of the sampling distribution $\sigma_{\bar{Y}} = \sigma_Y/\sqrt{N}$.

In order to calculate Z, let us list the essential information.

$$\mu_Y = 2.75 \qquad \sigma_Y = \quad 0.40$$
$$\bar{Y} = 2.89 \qquad N = 100$$

Thus,

$$Z = \frac{\bar{Y} - \mu_Y}{\sigma_Y/\sqrt{N}} = \frac{2.89 - 2.75}{0.40/\sqrt{100}} = \frac{0.14}{0.40/10} = \frac{0.14}{0.04} = \underline{\underline{3.50}}$$

The Z score $= 3.50$ indicates that our sample mean $\bar{Y} = 2.89$ is 3.50 standard errors greater than the population mean, $\mu_Y = 2.75$.

5. Deciding About Rejection of the Null Hypothesis

Since our sample mean $\bar{Y} = 2.89$ falls in the critical region at the 0.01 significance level, that is, since $Z > 2.58$, we reject the null hypothesis H_0 of simple random sampling, and we accept the alternative hypothesis H_1 that the sample is not truly a random one. Put differently, the Z table tells us that a Z of 3.50 means that 49.98 percent of the area above the mean is included; thus, the probability that we would select a simple independent random sample producing a mean grade point average of 2.89 is only 0.0002. Our decision, then, is that from a population whose parameters are mean $\mu_Y = 2.75$, variance $\sigma_Y^2 = 0.16$, and standard deviation $\sigma_Y = 0.40$, a random sample of size 100 would produce a value of 2.89 only two times out of 10,000. We hardly need to say that it makes sense to reject the null hypothesis.

14.2 THE ESTIMATED STANDARD ERROR AND DEGREES OF FREEDOM

The Z statistic is appropriate if σ_Y is known. A sensible question to ask at this point is what happens if σ_Y is unknown, which, by the way, is usually the case. When σ_Y is unknown, we want to use an *unbiased estimate* of it. To get this estimate, we have to take what may appear to be some elaborate steps.

In the first place, take on faith that the following formula is an unbiased estimate of σ_Y^2:

$$\hat{\sigma}_Y^2 = \frac{\sum_{i=1}^{N}(Y - \bar{Y})^2}{N - 1} = \text{an unbiased estimate of } \sigma_Y^2$$

Since this is the case, statisticians often use the following formula as an unbiased estimate of σ_Y:

$$\hat{\sigma}_Y = \sqrt{\frac{\sum_{i=1}^{N}(Y - \bar{Y})^2}{N - 1}} = \text{an unbiased estimate of } \sigma_Y$$

Now think back to the formula for the Z score. For sampling distributions, we calculate Z for populations with a known σ_Y as follows:

$$Z = \frac{\bar{Y} - \mu_Y}{\sigma_Y / \sqrt{N}} = Z \text{ when } \sigma_Y \text{ is known}$$

In the case where σ_Y is not known, it makes sense to substitute the unbiased estimate of it in the denominator, that is, $\hat{\sigma}_Y / \sqrt{N}$.

Actually, the unbiased estimate is easier to compute than this formula implies, and a few algebraic manipulations help clarify why. Besides, they lead to some important concepts, more about which will come up soon. Let us carefully examine the estimated standard deviation divided by the square root of N, $\hat{\sigma}_Y / \sqrt{N}$.[1] We can display the entire formula in terms of two fractions:

$$\frac{\hat{\sigma}_Y}{\sqrt{N}} = \frac{\sqrt{\dfrac{\sum\limits_{i=1}^{N} (Y - \bar{Y})^2}{N - 1}}}{\sqrt{\dfrac{N}{1}}}$$

Instead of working with two square roots, one in the numerator and one in the denominator, we can put the entire equation under one radical, thus,

$$\frac{\hat{\sigma}_Y}{\sqrt{N}} = \sqrt{\frac{\dfrac{\sum\limits_{i=1}^{N} (Y - \bar{Y})^2}{N - 1}}{\dfrac{N}{1}}}$$

To simplify this expression, we invert and multiply the terms under the radical.

$$\frac{\hat{\sigma}_Y}{\sqrt{N}} = \sqrt{\frac{\sum\limits_{i=1}^{N} (Y - \bar{Y})^2}{N - 1} \cdot \frac{1}{N}} = \sqrt{\frac{\sum\limits_{i=1}^{N} (Y - \bar{Y})^2 (1)}{(N - 1) \cdot (N)}}$$

If we divide both the numerator and denominator by N, we eliminate it from the denominator, and it appears in the numerator as the divisor. We can then separate the equation into one square root in the numerator and one square root in the denominator.

$$\frac{\hat{\sigma}_Y}{\sqrt{N}} = \sqrt{\frac{\dfrac{\sum\limits_{i=1}^{N} (Y - \bar{Y})^2}{N}}{N - 1}} = \frac{\sqrt{\dfrac{\sum\limits_{i=1}^{N} (Y - \bar{Y})^2}{N}}}{\sqrt{N - 1}}$$

---

[1]The logic developed here follows that of Hubert M. Blalock, Jr., *Social Statistics*, 2nd edition (New York: McGraw-Hill Book Company, 1972), pp. 188–93.

Looking closely at the numerator of the equation, we can see that it is the familiar sample standard deviation of Y, which we learned to calculate long ago. So, when σ_Y is unknown, we can use what is known, that is, the *standard deviation of the sample* and divide it by the square root of $N - 1$. Thus,

$$\frac{\hat{\sigma}_Y}{\sqrt{N}} = \frac{\sqrt{\dfrac{\sum\limits_{i=1}^{N} (Y - \bar{Y})^2}{N}}}{\sqrt{N - 1}} = \frac{s_Y}{\sqrt{N - 1}}$$

Even though we know the formula for this estimate, we are not quite ready to put it to use in an example. First we must examine the estimated standard error $\hat{\sigma}_Y/\sqrt{N}$ which is an unbiased estimate of the actual standard error $\sigma_Y/\sqrt{N}$. This estimate is equal to the standard deviation of the sample divided by the square root of $(N - 1)$; that is,

$$\frac{\hat{\sigma}_Y}{\sqrt{N}} = \frac{s_Y}{\sqrt{N - 1}}$$

You may wonder why we did not use $\sqrt{N}$ instead of $\sqrt{N - 1}$ in the denominator of the unbiased estimate. The reason is important, so to see just why, we will bring together some material that we covered long ago. Earlier we learned that the sum of the differences of each individual score from the mean on a given variable always equals zero, that is, $\sum\limits_{i=1}^{N} (Y_i - \bar{Y}) = 0$. Let us assign a symbol to these differences so that $(Y_i - \bar{Y}) = D_i$. We know that the sum of D_i is always zero, that is, $\sum\limits_{i=1}^{N} D_i = 0$. In other words, there is a single possible answer when we subtract from each score Y_i the mean $\bar{Y}$ and sum the differences; we always get 0. If $N = 5$, we are free to assign arbitrarily any value to *all but one* of the D_i's. In other words, the values are independent of each other until we use up all our "degrees of freedom" and link them to the mean. For example, let us arbitrarily assign $D_1 = 12$, $D_2 = -18$, $D_3 = 10$, and $D_4 = 3$, leaving D_5 unspecified. Since $\sum\limits_{i=1}^{N} D_i = 0$, D_5 must be -17. Thus we have only four *degrees of freedom* when $N = 5$, because the number of degrees of freedom for $\sum\limits_{i=1}^{N} (Y_i - \bar{Y})$ is $(N - 1)$.

When examining one variable, we determine the degrees of freedom by subtracting the number 1 from the number of unknown elements. In our example, we have four degrees of freedom to set the differences equal to anything we want. We are free to assign any values arbitrarily to any four of the differences, but when we get to the last one, we do not have any choice. Put differently, if $D_1 + D_2 + D_3 + D_4 + D_5 = 0$, once we assign arbitrary values to any four of them, then the fifth one is determined. Thus, on the fifth element we do not have any degrees of freedom left to assign an arbitrary value.

Technically, the number of degrees of freedom equals the number of unknown elements N minus the number of independent equations that link them together. An independent equation is one that is calculated using the individual values of each element that are *independent* until linked by the equation. For example, we need only one independent equation to tie together the N unknown elements Y_1, Y_2, Y_3, Y_4, . . . , Y_N, to the sample mean $\bar{Y}$. That one equation calculates the mean,

$$\frac{\sum_{i=1}^{N} Y_i}{N} = \bar{Y}$$

In this case, the unknown elements are independent until we link them together with the *one* equation for the mean.

Since the number of degrees of freedom equals the number of elements minus the number of independent equations that tie them together, in our example this equals $(N - 1)$. Put quite simply, a certain *degree of freedom* means that we are *not completely* free to assign values arbitrarily, but that we are free only to a certain *degree*.

14.3 STUDENT'S *t* DISTRIBUTION

If we do not know the value of σ_Y, we must change from the Z distribution, for we can no longer use the Z score and the normal table. Fortunately, we can use the *estimate* of the standard error divided by the square root of $N - 1$ with a distribution similar to Z. This alternative distribution is called Student's t distribution. When σ_Y is unknown or when we have a sample smaller than about 120, we use the t score. The formula can be expressed symbolically:

$$t = \frac{\bar{Y} - \mu_Y}{s_Y/\sqrt{N - 1}}$$

We are now ready to carry out an example that calls for the t statistic. Suppose that in a research project we are interested in the percentage of graduating seniors who go on to college. An association of suburban high schools has decided that the standard mean percentage of college-bound seniors should be 87 percent. Since we do not know the value of the standard deviation, σ is unknown. We are specifically interested in studying a group of twenty schools that was selected at random from the total population. In our sample we find that the mean percentage of college-bound seniors is 85 percent, with a standard deviation of $s_Y = 4$ percentage points. We decide to test the directional hypothesis that the mean percentage is lower than the expected standard.

1. Listing Our Assumptions

Level of measurement: Percent College-Bound Seniors is an interval scale.

Model: We assume a normal population and random sampling.

Null Hypothesis (H_0): We will test the hypothesis that the mean percentage is at least 87.

Alternative Hypothesis (H_1): If we reject H_0, then we assume that the percentage is lower than the expected standard.

2. Determining the Appropriate Sampling Distribution

Because we have a small sample and because we do not know the value of σ, we use the distribution of the t table and the t statistic in the Appendix.

3. Choosing the Level of Significance and Specifying the Critical Region

For a directional hypothesis, we use a one-tailed test, and we decide to use the 0.05 significance level. Thus we want only 5 percent of the distribution to be in the critical region.

To specify the 5 percent critical region, we turn to the t table in the Appendix. The values of the t statistic are listed in the body of the table. We select the appropriate column for a one-tailed test at the 0.05 significance level. The degrees of freedom are calculated by subtracting the number of independent equations from the sample size, thus

$$\text{degrees of freedom} = (N - 1) = (20 - 1) = 19.$$

We move down the left-hand column of the table to the "df 19" row. Then we move across to the 0.05 one-tailed test column, in which cell we find the value for $t = 1.729$. Thus if the score from our sample produces a t score equal to or greater than 1.729, we reject the null hypothesis. Figure 14.3.1 on page 310 shows the critical region diagrammatically.

4. Computing the Test Statistic

Our next task is to convert the mean percentage high school average to the t scale so that we may use the t statistic value for comparison. Recall that the formula for t is

$$t = \frac{\bar{Y} - \mu_Y}{s_Y / \sqrt{N - 1}}$$

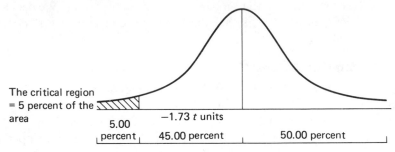

The critical region
= 5 percent of the
area

−1.73 t units

5.00
percent

45.00 percent

50.00 percent

Figure 14.3.1. One-tailed critical region at the 0.05 significance level.

Stated verbally, we convert the observed value to a t score by taking the sample value $\bar{Y}$ and comparing it to the parameter value μ_Y and then dividing it by our estimate of the standard error of the sampling distribution. To carry out the calculation of t, let us list the essential information:

$$\mu_Y = 87.00 \qquad s_Y = 4.00$$
$$\bar{Y} = 85.00 \qquad N = 20$$

Thus,

$$t = \frac{\bar{Y} - \mu_Y}{s_Y/\sqrt{N-1}} = \frac{85-87}{4/\sqrt{20-1}} = \frac{-2}{4/\sqrt{19}}$$

$$t = \frac{-2}{4/4.36} = \frac{-2}{0.92} = \underline{\underline{-2.174}}$$

This indicates that our sample mean $\bar{Y}$ of 85.00 is 2.174 standard errors less than the hypothesized population mean $\mu_Y = 87.00$.

5. Deciding About Rejection of the Null Hypothesis

Since our t score falls into the critical region at the 0.05 significance level for a one-tailed test, that is, since $t > \pm 1.729$, we reject the null hypothesis, H_0. Thus we accept the alternative hypothesis H_1, that the actual percentage of college-bound seniors is lower than the expected standard of the association of suburban high schools.

As pointed out earlier, in practice we use the t score primarily as a small sample statistic. We do so because even if the population variance σ_Y is unknown we use the Z table when N is larger than approximately 120.

Thus far we have learned about hypothesis testing, significance levels, and degrees of freedom. These lead into the next section, in which we will learn about estimating parameters.

310

14.4 ESTIMATING PARAMETERS

There are two ways that we can estimate parameters. First, we can estimate with a specific point by choosing a single number that is an estimate. For example, we might estimate that the mean age of our statistics class is twenty. This one specific age can be used as a *point estimate* of the mean age of students in the class. If we estimate the mean age of the class to be twenty, we either are or are not accurate. Second, we can estimate by choosing a range of numbers, or an interval. Generally, an interval estimate gives a better chance of being accurate. Suppose that instead of using one age, we estimate the mean age to be somewhere between nineteen and twenty-one. This range of ages can be used as an *interval estimate* of the mean age of students. Let us discuss both estimates in turn, starting with point estimation.

Remembering that a parameter is a characteristic of a population and that a statistic is a characteristic of a sample, we want a point estimate to be the *best* single-statistic value to estimate a parameter value. The catch is that "best" must be defined.

In parameter estimation, "best" is determined by two criteria, *bias* and *efficiency*. Literally, bias means a particular inclination a person holds that prevents an unprejudiced judgment of an issue. Statistically, bias implies tendencies *away from* the average and of being "off to one side" in terms of the laws of probability.

Bias refers to *systematic distortion* of a given statistic resulting from a sampling procedure. This may be contrasted with random distortion, which occurs because of fluctuations in particular sample values but not in the sampling distribution. For example, the arithmetic mean of a simple random sample is called an *unbiased* estimate of the population mean. This is not because the mean of any given sample is identical with the population mean. Rather, it is because the *mean of the sampling distribution* of sample means is equal to the mean of the population, provided that the sample was selected by simple random sampling.

Bias exists when the mean of the sampling distribution is *not* equal to the population parameter, that is, when $\mu_{\bar{y}} \neq \mu_Y$. We never actually know if a given sample is or is not biased. Instead, it is the procedure of random sampling and our knowledge of probability that enable us to say that *in the long run* our results are or are not likely to be biased. Figure 14.4.1 shows the true population mean μ_Y, and a biased sampling distribution mean $\mu_{\bar{y}}$.

Earlier we learned that we use $(N - 1)$ in the denominator to calculate an *unbiased* estimate of the standard error of the sampling distribution. This is because we calculate the standard deviation s_Y with a formula that uses the *mean*, which we now know is an unbiased estimate. We correct for bias by

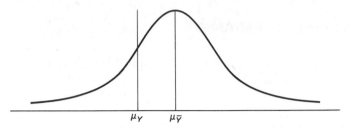

Figure 14.4.1. The sampling distribution of a biased estimate.

use of the sample standard deviation divided by $(N - 1)$, $s_Y/\sqrt{N - 1}$, providing an unbiased estimate.

The second criterion used to determine whether a statistic is "best" is its
efficiency. Efficiency focuses on how closely the sampling distribution is clustered around the true parameter value. In general, the lower the standard
error, the more efficient the statistic, and vice versa. The standard error is
therefore directly related to efficiency. The notions of bias and efficiency
should be thought of together, and we can illustrate them both at once, as is
shown in Figure 14.4.2.

Suppose that we have the population parameter μ_Y and the mean of the
sampling distribution $\mu_{\bar{Y}}$. In order to have an unbiased sampling distribution,
its mean $\mu_{\bar{Y}}$ should be identical to the population mean μ_Y. Figure 14.4.2
shows two sampling distributions. Diagram A represents an estimate that is
unbiased but only moderately efficient. Diagram B represents an estimate that
is biased but relatively quite efficient. The sampling distribution mean in
Figure 14.4.2.A is identical to the population mean; thus it is not biased.

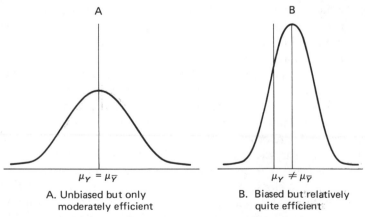

A. Unbiased but only
moderately efficient

B. Biased but relatively
quite efficient

Figure 14.4.2. Sampling distributions of two estimates which occupy an equal area and have
constant sample size.

312

However, because it is somewhat "flat," it is not too efficient. The sampling distribution mean of B is not the same as the population mean; thus it is biased; however, because of the way it is clustered, the distribution is quite efficient. Generally, unless it is heavily biased, we strive for the most efficient estimate rather than the least biased one since *in the long run* we are more likely to be closer to the parameter in question.

So far we have talked about using a single point, such as the mean of a sample, as an estimate of a parameter, and we called this point estimation. We also explained that we can estimate a parameter to be between certain limits, and we called it interval estimation.

14.5 ESTABLISHING CONFIDENCE INTERVALS

Interval estimation makes common sense. If we were estimating the average age of a group of people, we could estimate by choosing one unique number, say, thirty years of age. There is a good chance that we might be wrong. Instead, suppose we estimate the average age to be in the interval between twenty-nine and thirty-one years. There is a much better chance that we have covered the correct population value; therefore, we would have more confidence in our estimate. Suppose we now estimate the average age to be between twenty-five and thirty-five years. We would have an even better chance of being correct; that is, we would have greater confidence in our estimate. Such an interval often is called a *confidence interval* because we have a certain confidence that the interval contains the true population parameter value. However, greater confidence is "bought" at the price of lower precision; that is, we have wider, less precise intervals.

To obtain a confidence interval, we "construct" an interval on either side of a single-point estimate. We go both below and above the point estimate by multiplying the standard error of the sampling distribution times the *Z* or *t* score that corresponds to the confidence level chosen.

The first thing we must decide upon is a confidence level. A confidence level has to do with the likelihood of drawing a sample with specified characteristics if we drew repeated samples. It is similar to a significance level, and it involves the laws of probability. Confidence levels are ordinarily very high, such as 95 or 99 percent, whereas significance levels are ordinarily very low, such as 5 or 1 percent. The 95 percent confidence level specifies the 0.05 significance level. As with significance levels, we rest our faith in the procedure. For example, if we specify a 95 percent confidence level, we have confidence in the random sampling procedure to the extent that the procedure will produce an interval that fails to include the parameter only 5 percent of the time.

Since we know how to specify a point estimate, part of our task is already done. When we are using a sample mean as the point estimate, we must go above the mean a certain distance to specify the upper limit of the confidence interval, and we must go below the mean the same distance to specify the lower limit of the confidence interval. To do this, we use a formula that utilizes the mean $\bar{Y}$, the symbol $\pm$, the appropriate Z or t score to indicate the number of standard error units for the chosen confidence level, and the standard error of the sampling distribution, $\sigma_Y/\sqrt{N}$ or $s_Y/\sqrt{N-1}$. Let us look at these steps in more detail.

1. If σ_Y is known, we use the appropriate Z score to specify our confidence level. Thus,

$$\text{confidence interval for the mean} = \bar{Y} \pm Z \cdot \frac{\sigma_Y}{\sqrt{N}}$$

2. If σ_Y is unknown or if we have a small sample, we use the appropriate t score to specify our confidence level. Thus,

$$\text{confidence interval for the mean} = \bar{Y} \pm t \cdot \frac{s_Y}{\sqrt{N-1}}$$

Earlier, we decided to reject the sample as not being random because the mean GPA of $\bar{Y} = 2.89$ was too much above the population mean of $2.75 = \mu_Y$, with a standard deviation of $\sigma_Y = 0.40$. Suppose that we decide to draw our own random sample of 100 and find that the mean GPA of our sample is $\bar{Y} = 2.81$. Now suppose that we want to construct a 95 percent confidence interval for our sample mean. Since we want to know the interval above *and* below the value of $\bar{Y}$ and are not making a directional hypothesis, we use a two-tailed set of Z scores.

Since the normal curve includes 100 percent of the distribution, we know that for the 95 percent confidence level, we must use the "middle" 95 percent. In other words, we want to have two equal areas at each end of the curve, with the total area of the two tails containing 5 percent. Each tail of the distribution, therefore, must contain 2.5 percent of the area, as shown in Figure 14.5.1.

Turning to the normal table in the appendix, we find the Z value that corresponds to an area between the mean and that score that equals 47.50 percent of the total area. In the body of the table we see the value 4750 and in the left-hand and top margins we see that it equals a Z score of 1.96. Since $\bar{Y} = 2.81$, $\sigma_Y = 0.40$, and $N = 100$, the 95 percent confidence interval would be:

$$\bar{Y} \pm Z \cdot \frac{\sigma_Y}{\sqrt{N}} = 2.81 \pm 1.96 \cdot \frac{0.40}{\sqrt{100}} = 2.81 \pm 1.96 \cdot (0.04)$$

$$\text{Lower limit} = 2.81 - 1.96 \cdot (0.04) = 2.81 - 0.0784$$
$$= 2.7316 \simeq 2.73$$

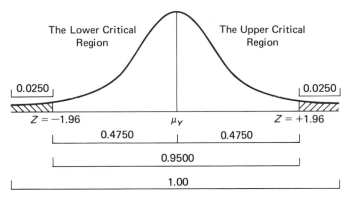

Figure 14.5.1. Two-tailed critical region at the 0.05 significance level.

$$\text{Upper limit} = 2.81 + 1.96 \cdot (0.04) = 2.81 + 0.0784$$
$$= 2.8884 \simeq 2.89$$

We can now state that the confidence interval of our mean $\bar{Y} = 2.81$ runs from 2.73 to 2.89 at the 0.95 confidence level. This means that given a sample mean $\bar{Y} = 2.81$ with the population standard deviation $\sigma_Y = 0.40$, and since the population mean is fixed with the interval varying from sample to sample, and even though the boundaries of the intervals so constructed will vary from sample to sample, there is only a 5 percent probability that an interval constructed in this way will fail to include the population mean. Put differently, our procedure enables us to say that we will be correct 95 percent of the time *in the long run* when we *estimate* that the actual mean lies within the 95 percent confidence interval just established. In this case, we have one of those samples whose confidence interval includes the population parameter $\mu_Y = 2.75$. Usually we are not in a position to know this, and we seldom know whether or not our confidence interval actually contains the population parameter μ_Y.

We can, of course, carry out similar steps and construct confidence intervals for t scores. For the sake of simplicity and for ease of comparison, let us use the same mean high school average example that we discussed earlier in this chapter. Since $\bar{Y} = 85.00$, $s_Y = 4.00$, and $N = 20$, we use the confidence interval formula

$$\bar{Y} \pm t \cdot \frac{s_Y}{\sqrt{N-1}} = 85 \pm t \cdot \frac{4}{\sqrt{20-1}}$$

All that remains to be done is specifying the value of t, and that depends on the level of confidence we choose. Suppose that we decide upon the 0.99 confidence level. The value for t from the t distribution table is $t = 2.86$. We can now carry out the calculations.

$$\bar{Y} \pm t \cdot \frac{s_Y}{\sqrt{N-1}} = 85 \pm 2.86 \cdot \frac{4}{\sqrt{19}} = 85 \pm 2.86 \cdot \frac{4}{4.36}$$

$$\text{Lower limit} = 85 - 2.86 \cdot 0.92 = 85.00 - 2.63$$
$$= 82.37 \simeq 82.4$$
$$\text{Upper limit} = 85 + 2.86 \cdot 0.92 = 85.00 + 2.63$$
$$= 87.63 \simeq 87.6$$

Thus we can say that our estimate of the population mean based on our sample mean is 85.00, with confidence limits of 82.4 and 87.6 at the 0.01 level of significance. Again, we place our faith in the procedure of random sampling and statistical testing. As with the Z score example, we have a sample whose confidence interval includes the population parameter $\mu_Y = 87.00$.

Remember, however, the t test at the 0.05 significance level was such that we rejected the sample mean $\bar{Y} = 85.00$ as being too low. The fact that we find one test including the population mean and one test not including it helps emphasize an important point. Namely, confidence intervals and the level of significance do not show that our procedure has established the probability that our sample means are correct ones. Our statements about a given confidence interval must be cautious ones, and we must realize that our faith lies in the procedure, not in the actual sample value.

To determine confidence intervals for Q, for large samples we apply the general large numbers theorem, which enables us to use the Z distribution. If we use the 95 percent confidence interval for Q_{XY}, we select the Z value of 1.96 and use it in the following formula:

$$\text{Confidence limit} = Q_{XY} \pm (1.96)\sqrt{\frac{(1.00 - Q_{XY}^2)^2\left(\frac{1}{a} + \frac{1}{b} + \frac{1}{c} + \frac{1}{d}\right)}{4}}$$

Consider Table 14.5.1, a fourfold table in which sex and earnings have been cross-tabulated for U.S. Census Data.[2]

$$Q_{XY} = \frac{bc - ad}{bc + ad}$$

$$Q_{XY} = \frac{[(2754)(2128)] - [(2166)(413)]}{[(2754)(2128)] + [(2166)(413)]} = \underline{+0.735}$$

Using the confidence limit formula for Q_{XY}, we can compute the 95 percent confidence interval.

[2]This example is taken directly from James A. Davis, *Elementary Survey Analysis* (Englewood Cliffs, New Jersey: Prentice-Hall, Inc., 1971), p. 34.

Table 14.5.1.   Sex and earnings in 1959 among those reporting some earnings

		Earnings		
		Under $4000	$4000 or more	Total
Sex	Male	2166	2754	4920
	Female	2128	413	2541
		4294	3167	$7461 = N$

Source: Adapted from James A. Davis, *Elementary Survey Analysis* (Englewood Cliffs, New Jersey: Prentice-Hall, Inc., 1971), Table 2.1, p. 34.

Confidence limit

$$= 0.735 \pm (1.96) \sqrt{\frac{(1.00 - 0.735^2)^2 \left(\frac{1}{2166} + \frac{1}{2754} + \frac{1}{2128} + \frac{1}{413}\right)}{4}}$$

Confidence limit

$$= 0.735 \pm (1.96) \sqrt{\frac{(0.211)(0.00046 + 0.00036 + 0.00047 + 0.00242)}{4}}$$

Confidence limit

$$= 0.735 \pm (1.96) \sqrt{\frac{(0.211)(0.00371)}{4}}$$

$$= 0.735 \pm (1.96) \sqrt{0.0001956}$$

$$= 0.735 \pm (1.96)(0.0140)$$

$$= 0.735 \pm 0.0274$$

Thus we obtain an upper confidence limit of $+0.762$ and a lower confidence limit of $+0.708$ around the sample value of $+0.735$. Our inferential judgment, therefore, would be that the Q association between earnings and sex in the U.S. population falls between the Q values of $+0.708$ and $+0.762$ at the 95 percent confidence level.

Confidence intervals may be constructed for other statistics, and although their computations are beyond the scope of this book, the basic principle underlying them is essentially the same. It involves testing the significance of the statistic using the null hypothesis that the population parameter is zero. Then, confidence intervals are constructed on the basis of the decision regarding the null hypothesis.

EXERCISES

1. In your own words define, describe, or discuss the following terms and give an hypothetical example of each:

    Z statistic
    significance level
    t statistic
    estimated standard error
    normal table
    degrees of freedom

2. Using the random sample described in Section 14.3, test the null hypothesis that the sample is a simple random sample at the .05 significance level with the t statistic.

3. Explain verbally how to set confidence intervals.

4. Explain the similarities and differences between the Z statistic and the Student's t statistic. When would you use each?

5. Point out the two ways in which you can estimate parameters.

6. Explain the term "bias" as it refers to systematic distortion.

7. Assume that you are carrying out the research project described in this chapter. Recall, the registrar has provided the following parameters for the population of 5,000 students.

For 5,000 students

Mean GPA	$= \mu_Y =$	2.75
Variance	$= \sigma_Y^2 =$	.16
Standard Deviation	$= \sigma_Y =$	.40

Suppose that we draw three simple random samples of (1) 100, (2) 200, and (3) 500 students from the population of 5,000 as follows:

	Sample		
	1	2	3
Mean GPA	$\bar{Y}_1 = 3.10$	$\bar{Y}_2 = 2.45$	$\bar{Y}_3 = 2.95$
Variance	$s_{Y_1}^2 = .81$	$s_{Y_2}^2 = .25$	$s_{Y_3}^2 = .04$
Standard Deviation	$s_{Y_1} = .90$	$s_{Y_2} = .50$	$s_{Y_3} = .20$
Size	$N_1 = 100$	$N_2 = 200$	$N_3 = 500$

Test the null hypothesis that each sample is a simple random sample at the .05 significance level with

A. the Z statistic.

B. assuming that we do not have the population parameters, the t statistic.

C. Compare the results and discuss their implications.

8. Using the data presented in Problem 7, compute the confidence intervals for the mean GPA of each sample at the .05 significance level using

A. the Z statistic.

B. assuming that we do not have the population parameters, the t statistic.

C. Compare the results and discuss their implications.

9. Suppose you are evaluating the curricula of a random sample of 10 Social Statistic courses selected from the population of all such courses in a state university system. Each course keeps a record of the percentage of students which pass and a standard has been set that the mean passing percentage for all the courses ought to be 75 percent. In the sample, you find the mean percentage to be 67 percent and the standard deviation to be 6 percent. Compute the t statistic.

A. Do you have reason to suspect for the population of statistics courses as a whole that the percentage of students who pass is below the standard expected?

B. Using the previous calculations determine what size sample is needed in order for a percentage difference of 8 to be significant.

10. Using the data presented in Problem 9, compute the confidence interval for the sample mean using

A. the 95 percent level confidence interval.

B. the 99 percent level.

C. Discuss the findings and their implications in light of the decisions based on A and B.

15

Measuring Differences Between Statistics

Throughout much of this book we have been concerned with measuring relationships between variables. At first our major concern was to describe and summarize the relationships once we measured them. Later our concern was to learn some of the ways we can use statistics gathered from a sample to make inferences about the population from which the sample was drawn. Most of the tests discussed up to this point focus on the values of a single sample compared to the values of its population.

Common sense tells us that we do not always know the characteristics and parameters of a population, or we would not need the statistics of a sample. This is largely because we make inferences about population parameters directly *from* sample statistics. Since we are often in the bind of not knowing the parameters, it is common to measure the differences between the statistics of more than one sample, and then to make inferences based on the differences. Thus we use the statistics gathered from two or more samples and compare them according to hypothesized differences from the population(s) they represent.

In this chapter, we first explore the idea of a sampling distribution of differences, which is merely a variation on a theme we already know. We then learn how to measure the differences between means of two samples.

15.1 EXPLORING DIFFERENCES BETWEEN SAMPLES

By extending the concept of a sampling distribution of a statistic, we can understand the concept of a sampling distribution of *differences between statistics*.[1] Let us first *conceptualize* the ideas behind exploring differences between sample statistics in general. We start by assuming that there are two populations that possess unique sets of parameters. If we draw a random sample from each and if they are truly representative, our statistic values would equal the parameter values, that is, $\mu_{Y_1} = \bar{Y}_1$ and $\mu_{Y_2} = \bar{Y}_2$.

In order to measure the differences between sample statistics, we must first draw a simple random sample from one normal population and calculate descriptive summary statistics, such as means, variances, standard deviations, regression coefficients, and correlation coefficients, for the relevant variables. The next step is to draw a second sample from another normal population and compute the same statistics. Then we compare the values from sample 1 with those from sample 2 by subtracting a given statistic of sample 2 from the same statistic of sample 1. Each such difference would represent the difference between *sample statistics*. We can express the differences between the statistics mentioned above for variables X and Y symbolically:

Difference between sample means $= (\bar{Y}_1 - \bar{Y}_2), (\bar{X}_1 - \bar{X}_2)$

Difference between sample variances $= (s_{Y_1}^2 - s_{Y_2}^2), (s_{X_1}^2 - s_{X_2}^2)$

Difference between sample standard deviations $= (s_{Y_1} - s_{Y_2}), (s_{X_1} - s_{X_2})$

Difference between sample correlation coefficients $= (r_{YX_1} - r_{YX_2})$

Difference between sample regression coefficients $= (b_{YX_1} - b_{YX_2})$

Think of the difference between two statistics as a single value, that is, the *score* of the difference. We may represent the difference between any two statistics with the abbreviation D.

1st difference between statistic 1 minus statistic 2 $= D_1$

2nd difference between statistic 1 minus statistic 2 $= D_2$

3rd difference between statistic 1 minus statistic 2 $= D_3$

[1] Helpful discussions on the matter of distributions of differences may be found in John E. Freund, *Modern Elementary Statistics*, 3rd edition (Englewood Cliffs, New Jersey: Prentice-Hall, Inc., 1967), pp. 219–301; Hubert M. Blalock, Jr., *Social Statistics*, 2nd edition (New York: McGraw-Hill Book Company, 1972), pp. 219–41; Richard P. Runyon and Audrey Haber, *Fundamentals of Behavioral Statistics*, 2nd edition (Reading, Massachusetts: Addison Wesley Publishing Company, 1971), pp. 194–203; and John E. Freund, *Mathematical Statistics*, 2nd edition (Englewood Cliffs, New Jersey: Prentice-Hall, Inc., 1971), pp. 314–42.

Thus D_k represents the kth difference between any two statistic values. As we did with single scores, we can plot each D_k value (the difference between the two statistics) along the interval scale of the variable in question.

If we were to draw an infinite number of samples and measure the differences between two statistic values, we would derive a *sampling distribution of differences*, the mean of which would be equal to the difference between the two corresponding population parameters. Put differently, since our interest is on two normal populations, each with a unique set of parameters, we want to examine the differences between sample statistics from each population. It turns out that the sampling distribution of differences will have a mean, μ_D, that is equal to the difference between the two corresponding population parameters, μ_D = parameter 1 − parameter 2. These steps are analogous to those we took when we plotted the single statistic values for an infinite number of samples.

It may not be obvious at first glance, but if the samples are drawn from the same population, the mean of the sampling distribution of differences and the difference between the two corresponding population parameters will be 0, that is, μ_D = parameter 1 − parameter 2 = 0.

Let us turn to the derivation of the standard error and the variance of the sampling distribution. The *mean* of the sampling distribution of differences involves the straightforward subtraction of one statistic from another; however, the *variance* and the *standard deviation* of the sampling distribution of differences are not so easily computed, for they do not represent the *differences* between two values. The sampling distribution of differences between sample statistics is derived from the difference between the two population parameters rather than from either of them individually. When computing the variance of a sampling distribution of differences, we must recognize that both population variances contribute their *own share* to the overall variance of the sampling distribution. So, instead of subtracting one statistic from another, we add the variances together, that is, $\sigma_1^2 + \sigma_2^2$.

Since the samples may be of unequal size, we may correct for the number in each sample by dividing the variance of population 1 with the N of sample 1 and the variance of population 2 with the N of sample 2. Thus we can represent the variance of the sampling distribution of differences symbolically

$$\sigma_D^2 = \frac{\sigma_1^2}{N_1} + \frac{\sigma_2^2}{N_2}$$

The standard error of the sampling distribution of differences is the square root of the variance of the sampling distribution. It can be expressed symbolically

$$\sigma_D = \sqrt{\frac{\sigma_1^2}{N_1} + \frac{\sigma_2^2}{N_2}}$$

Earlier we learned that we can analyze populations of any form because as the sample N becomes large the sampling distribution of a statistic is normal. We can extend that theorem and say: For any two populations, whether or not they are normally distributed, the sampling distribution of differences between statistics will approximate a normal distribution as the sample N becomes large, say, 120 or more.

This means that if a large number of simple random samples of sizes N_1 and N_2 are drawn from two populations that have means μ_1 and μ_2, variances of σ_1^2 and σ_2^2, and standard deviations σ_1 and σ_2, then the mean of the sampling distribution of differences will be equal to the difference between the statistics of the two populations, the variance of the sampling distribution of differences will be equal to the sum of the population variances divided by the sample size, and the standard error of the sampling distribution of differences will be equal to the square root of its variance. Thus the sampling distribution of differences will be normal with a mean μ_D, a variance σ_D^2, and a standard error σ_D.

15.2 STATISTICS FOR MEASURING DIFFERENCES BETWEEN SAMPLES

If we treat differences as if they were individual scores, we can use the normal distribution for analytical purposes. First we must convert the "difference scores" between the statistics to Z scores and then use the normal curve for testing hypotheses. To do this, we adapt the Z statistic to differences between scores. We can express the Z score conversion equation for differences symbolically

$$Z = \frac{(\text{statistic}_1 - \text{statistic}_2) - (\text{parameter}_1 - \text{parameter}_2)}{\text{standard error of the sampling distribution}}$$

$$Z = \frac{D_{1-2} - \Delta_{1-2}}{\sqrt{\dfrac{\sigma_1^2}{N_1} + \dfrac{\sigma_2^2}{N_2}}}$$

Let us now consider statistics used in examining the differences between means. Not only are the tests for the differences between means fairly easy to understand, it also is common for sociologists to be interested in such differences for analytical purposes.

In order to study the differences between the means from two populations, we compare the mean of one sample with the mean of another. To make such comparisons, we use the means from two simple random samples. It is important that the samples are independent from each other; that is, there must be independence *between* them. In other words, the selection of one sample cannot in any way depend upon the selection of the other, and we must have samples that are independent both within themselves and between each other. There are some technical problems in such sampling procedures that are not within the scope of this text. For our purposes, you need only keep this requirement in mind, even though we do not develop it further.[2]

The Law of Large Numbers and the Central Limit Theorem can be adapted specifically to examine sample means. As N becomes large, the sampling distribution of the difference of means $\bar{Y}_1 - \bar{Y}_2$ is approximately normal, with

$$\text{a mean} \qquad \mu_D = \mu_{Y_1} - \mu_{Y_2}$$

$$\text{a variance} \qquad \sigma_D^2 = \frac{\sigma_{Y_1}^2}{N_1} + \frac{\sigma_{Y_2}^2}{N_2}$$

and

$$\text{a standard error} \quad \sigma_D = \sqrt{\frac{\sigma_{Y_1}^2}{N_1} + \frac{\sigma_{Y_2}^2}{N_2}}$$

As in the general case we learned earlier, when we have full information about the population means, variances, and standard errors, we can convert these values to Z scores and use the Z distribution. The Z score for the sampling distribution of the difference of means can be expressed symbolically

$$Z = \frac{(\bar{Y}_1 - \bar{Y}_2) - (\mu_{Y_1} - \mu_{Y_2})}{\sqrt{\frac{\sigma_{Y_1}^2}{N_1} + \frac{\sigma_{Y_2}^2}{N_2}}}$$

In actual practice we almost never know the population parameters. Thus we generally estimate the value of the standard error of the sampling distribution of the difference of means, and we use the t test rather than the Z test. Before we discuss the t test, let us learn how to estimate the standard error of the sampling distribution. The estimate can be stated symbolically

$$\text{Estimate of } \sigma_{\bar{Y}_1 - \bar{Y}_2} = \hat{\sigma}_{\bar{Y}_1 - \bar{Y}_2}$$

As we learned earlier, the estimated variance divided by N is equal to the sample variance divided by $N - 1$. Similarly, we can estimate the standard error of the sampling distribution of the difference of means from the sample values. To do this when we assume that the two populations have different

[2]For a more detailed discussion of sampling and some of its problems see Blalock, *Social Statistics*, Chapter 21; and Morris H. Hansen, William N. Hurwitz, and William G. Madow, *Sample Survey Methods and Theory*, Vol. 1. (New York: John Wiley & Sons, Inc., 1953).

standard deviations, we construct an estimate of the standard error of the sampling distribution of the difference of means by finding the square root of the sum of the variances of the two samples divided by the respective $(N - 1)$ of each sample. That is,

$$\hat{\sigma}_{\bar{Y}_1 - \bar{Y}_2} = \sqrt{\frac{s_{Y_1}^2}{N_1 - 1} + \frac{s_{Y_2}^2}{N_2 - 1}}$$

This formula provides the appropriate estimate for measuring the difference of means for two samples from information that is available from each sample.[3] Let us now state a version of the *t* test by which we can compare the difference between the means of two samples with the difference between the means of two populations. It can be expressed symbolically

$$t = \frac{(\bar{Y}_1 - \bar{Y}_2) - (\mu_{Y_1} - \mu_{Y_2})}{\hat{\sigma}_{\bar{Y}_1 - \bar{Y}_2}}$$

Before we carry out the steps for statistical testing, we will examine one part of the *t* statistic that may be a bit confusing. We are concerned with the hypothesized population means μ_{Y_1} and μ_{Y_2}. The second term in the numerator of the formula for the *t* statistic is the difference between the population mean of sample 1 minus the population mean of sample 2. We actually test the *null* hypothesis that there is *no* difference between μ_{Y_1} and μ_{Y_2}. Put differently, since all the values for the *t* statistic may be gathered from the sample values except the population means, we treat the population means as being equal, that is, no difference between them, or $\mu_{Y_1} - \mu_{Y_2} = 0$. This hypothesis enables us to use only the statistic values from the sample, and they are already known.

15.3 MEASURING DIFFERENCES BETWEEN THE MEANS OF TWO SAMPLES

Suppose we are interested in examining the difference between students' past academic achievement according to their majors. We decide to compare the past academic performance of math and the natural science majors with that of humanities and social science majors. Past academic performance is operationalized by measuring high school average; our interest is in com-

---

[3]There is another version of the *t* test. The present method is called an unpooled estimate of the standard error. In general, it is a more conservative estimate than the pooled formula provides. For a fuller discussion see Blalock, *Social Statistics*, pp. 224–28.

paring the two types of majors according to the mean high school average of each type. Examination of the data provides the following findings.

Table 15.3.1. **High school average and type of college major**

Math and Natural Science Majors	Humanities and Social Science Majors
$N = 50$	$N = 20$
$\bar{Y}_1 = 91$	$\bar{Y}_2 = 85$
$s_{Y_1}^2 = 64$	$s_{Y_2}^2 = 100$
$s_{Y_1} = 8$	$s_{Y_2} = 10$

1. Listing Our Assumptions

Level of measurement: High School Average Y is an interval scale.
Model: We assume that we have two simple random samples.
Null Hypothesis (H_0): We hypothesize that the population means are equal; thus, $H_0: \mu_{Y_1} - \mu_{Y_2} = 0$.
Alternative Hypothesis (H_1): We hypothesize that the population means are not equal. Thus, $H_1: \mu_{Y_1} \neq \mu_{Y_2}$.

2. Determining the Appropriate Sampling Distribution

Because we have two small samples, and because we do not know the values of σ_{Y_1} and σ_{Y_2}, we use the t distribution and the t table. In order to use the t table, we must first compute the degrees of freedom appropriate for a two-sample test. The degrees of freedom for the two samples may be represented

$$df \text{ sample } 1 = N_1 - 1 = 50 - 1 = 49$$
$$df \text{ sample } 2 = N_2 - 1 = 20 - 1 = 19$$

We can approximate the degrees of freedom for the sampling distribution of the difference of means as

$$df \text{ sample } 1 \text{ and } 2 = N_1 + N_2 - 2 = 50 + 20 - 2 = 68$$

3. Choosing the Level of Significance and Specifying the Critical Region

We decide to use the 0.01 level of significance. Since the alternative hypothesis H_1 does not predict that μ_{Y_1} is lesser or greater than μ_{Y_2}, we do not predict what the direction of the difference may be; thus we choose a two-tailed test.

Checking the t table, we find that with approximately 60 degrees of freedom at the 0.01 level of significance, the t score must equal 2.660. This means that the critical region is made up of all values in the two tails of the t distribution below -2.660 or above $+2.660$. Thus we decide in advance to reject H_0 if our t score is equal to or greater than 2.660.

4. Computing the Test Statistic

Now we must convert the differences between the means for the two groups of students on the variable high school average to the t scale so that we may use the t distribution to analyze the difference. The formula for t is

$$t = \frac{(\bar{Y}_1 - \bar{Y}_2) - (\mu_{Y_1} - \mu_{Y_2})}{\hat{\sigma}_{\bar{Y}_1 - \bar{Y}_2}}$$

In words, we compare the difference between the sample means $\bar{Y}_1$ and $\bar{Y}_2$ with the difference between the population means μ_{Y_1} and μ_{Y_2}, and then we divide this difference by the estimate of the standard error of the sampling distribution of the difference of means.

$$t = \frac{(\bar{Y}_1 - \bar{Y}_2) - (0)}{\sqrt{\dfrac{s_{Y_1}^2}{N_1 - 1} + \dfrac{s_{Y_2}^2}{N_2 - 1}}} = \frac{(91 - 85) - (0)}{\sqrt{\dfrac{64}{49} + \dfrac{100}{19}}}$$

$$t = \frac{6 - 0}{\sqrt{1.31 + 5.26}} = \frac{6}{\sqrt{6.57}} = \frac{6}{2.56} = \underline{2.34}$$

5. Deciding About Rejection of the Null Hypothesis

Since $t = 2.34$, we fail to reject the H_0 at the 0.01 level of significance. Thus we conclude that at the 0.01 level there is not a statistically significant difference on past academic performance between the means of the populations for the two types of majors.

Making a decision in the above problem is not too difficult because the t score is less than the required rejection score of $t = 2.66$. However, if we had set the 0.05 significance level, in which case the rejection level is $t = 2.00$, we would reject the H_0 at the 0.05 level. By expanding the critical region we increase the probability of a type I error. At the same time, however, we reduce the probability of a type II error.

Since others may disagree with our choice of a level of significance, and for greater analytical clarity, it is advisable to report the probability of a t test as falling between specified levels of significance. For example, in this instance we would say that "a two-tailed t test indicates that the probability of finding a difference of 6 percent in two such groups is $0.02 \leq \text{pr} \leq 0.05$."[4] Put differently, a t score of 2.34 represents the probability of finding a 6 percent difference between two samples of this size. Since we want to know just how likely this finding is, we may use the probability (pr) for comparison.

To measure differences between statistics, we determine the *difference between* the two statistics by *subtraction*, and then we find the *ratio of* that difference to the standard error of the sampling distribution of the statistic in question. Technically, the standard error also measures differences between two values. Thus, we can think of both the Z and the t scores as being ratios between the differences between the sample statistic itself in the numerator and the standard error of its sampling distribution in the denominator.

15.4 ANALYZING TWO PROPORTIONS

We can also test differences between sample proportions when N is large enough. The criteria regarding the difference between the means of two samples pertain to the differences between the proportions of two samples, and the fundamental steps in testing the difference between two sample proportions are the same as in testing the difference between two sample means.

In general, our interest is in analyzing nominal or ordinal variables. We hypothesize about the proportion possessing one category of a variable in one sample versus the proportion of respondents possessing the same category in a second sample. We want to test whether the difference of a certain number of percentage points is not statistically significant, inferring that it is due to random sampling error, or whether the difference is statistically significant, inferring that it is due to a true difference between the population proportions.

For example, suppose that a study of the civilian population between the ages of three and thirty-four reports that even though the proportion of males and females enrolled in school is nearly identical through age seventeen, for those students beyond the age of seventeen it is increasingly likely for a greater proportion of males than females to remain enrolled in school. This finding suggests that when age is not controlled and the entire civilian population between the ages of three and thirty-four is considered, it is likely that there will be a greater proportion of males than females enrolled in school.

In order to analyze the difference between proportions, let us first define the terms with which we will be working:

[4]For discussions of this problem in social statistics see Blalock, *Social Statistics*, p. 192; James K. Skipper, Jr., Anthony L. Guenther, and Gilbert Nass, "The Sacredness of .05: A Note Concerning the Uses of Statistical Levels of Significance in Social Science," *The American Sociologist*, 2 (February 1967), pp. 16–18.

p_{p_1} = proportion enrolled in the 1st population

q_{p_1} = proportion not enrolled in the 1st population

p_{p_2} = proportion enrolled in the 2nd population

q_{p_2} = proportion not enrolled in the 2nd population

p_{s_1} = proportion enrolled in the 1st sample

q_{s_1} = proportion not enrolled in the 1st sample

p_{s_2} = proportion enrolled in the 2nd sample

q_{s_2} = proportion not enrolled in the 2nd sample

As N becomes large, the sampling distribution of the difference of proportions, $p_{s_1} - p_{s_2}$, is approximately normal with

a mean $\qquad \mu_D = p_{p_1} - p_{p_2} = 0$

a variance $\qquad \sigma_D^2 = \dfrac{\sigma_1^2}{N_1} + \dfrac{\sigma_2^2}{N_2} = \dfrac{p_{p_1} q_{p_1}}{N_1} + \dfrac{p_{p_2} q_{p_2}}{N_2}$

and

a standard error $\quad \sigma_D = \sqrt{\dfrac{\sigma_1^2}{N_1} + \dfrac{\sigma_2^2}{N_2}} = \sqrt{\dfrac{p_{p_1} q_{p_1}}{N_1} + \dfrac{p_{p_2} q_{p_2}}{N_2}}$

Since the sampling distribution of the difference between proportions when the N's in both samples are fairly large is normally distributed, we can use the Z statistic. In order to use Z, we must calculate an estimate for the standard error of the sampling distribution of the difference between proportions. First, however, we need to compute an estimate for the values of the proportions in the population. This estimate for the population value, $\hat{p}_p$, is calculated from information we have in the samples.

To compute the estimate, we calculate the *overall* proportion of one category versus the *overall* proportion of the other. We are interested in the percent enrolled versus the percent not enrolled. The weighted estimate, $\hat{p}_p$, is the percent enrolled not considering sex, and it is based on the information we have about the percent enrolled for both sexes combined. This process is analogous to treating the two samples as if they were one and calculating the total overall values for $\hat{p}_p$ and $\hat{q}_p$. This can be expressed symbolically

$$\hat{p}_p = \frac{N_1 p_{s_1} + N_2 p_{s_2}}{N_1 + N_2}$$

$$\hat{q}_p = 1.00 - \hat{p}_p$$

We can now estimate the standard error of the sampling distribution of the difference between proportions. We take the square root of the proportions from the two samples combined times the ratio between the sum of the sample size to the product of the sample size. It can be expressed symbolically

$$\hat{\sigma}_{D_{p_s}} = \sqrt{\hat{p}_p \hat{q}_p \left(\frac{N_1 + N_2}{N_1 N_2}\right)}$$

This enables us to express the Z statistic as involving the difference between the two sample values for the proportion in one sample minus the proportion in the second. Since we are hypothesizing that $p_{p_1} = p_{p_2}$, under the null hypothesis it is equal to 0. We can express the Z statistic symbolically

$$Z = \frac{(p_{s_1} - p_{s_2}) - (p_{p_1} - p_{p_2})}{\hat{\sigma}_{D_{p_s}}} = \frac{(p_{s_1} - p_{s_2}) - 0}{\hat{\sigma}_{D_{p_s}}}$$

$$Z = \frac{D}{\hat{\sigma}_{D_{p_s}}}$$

With this background let us examine the difference between the proportion of males and females enrolled in school, setting the proportion of males enrolled equal to p_{s_1}, the proportion of females enrolled equal to p_{s_2}, the proportion not enrolled as q_{s_1} for males, and q_{s_2} for females. Table 15.4.1 presents two hypothetical samples of males and females.

Table 15.4.1. Percent enrolled in school by sex from population of civilians 3-34 years of age*

	Male	Female
Percent Enrolled	57.8 percent	52.0 percent
Percent Not Enrolled	42.2	48.0
Total	100.0 percent	100.0 percent
N	(542)	(553)

*This table represents two hypothetical samples of males and females. It is based on the information given in Table 166 in the *Statistical Abstract of the United States, 1973*, 94th ed. (Washington, D.C.: U.S. Bureau of the Census, 1973), p. 111.

1. Listing Our Assumptions

Level of Measurement: Sex is a dichotomous variable.

Model: We assume that we have two simple random samples.

Null Hypothesis (H_0): We hypothesize that the population proportions are equal. Thus, $H_0: p_{p_1} - p_{p_2} = 0$.

Alternative Hypothesis (H_1): We hypothesize that the population proportion of males enrolled in school is greater than the proportion of females. Thus, $H_1: p_{p_1} > p_{p_2}$.

2. Determining the Appropriate Sampling Distribution

We can utilize the Z statistic because we have large samples and because the sampling distribution of differences of proportion is normal with a mean $p_{p_1} - p_{p_2} = 0$ and a standard error

$$\sigma_{D_{p_*}} = \sqrt{\frac{p_{p_1} q_{p_1}}{N_1} + \frac{p_{p_2} q_{p_2}}{N_2}}$$

3. Choosing the Level of Significance and Specifying the Critical Region

We decide to select the 0.05 significance level. Since the alternative hypothesis H_1 predicts that p_{p_1} will be greater than p_{p_2}, we use a one-tailed directional test for the Z statistic. Using the Z table, we find that we will need a Z score that would reflect 95 percent of the sampling distribution to be in the acceptable region for the null hypothesis and 5 percent of the distribution to be in the critical region, as shown in Fig. 15.4.1. We need a Z score equal to 1.65 to reject the null hypothesis of no difference between proportions. Thus for a one-tailed test we need to find a Z score equal to or larger than 1.65 to reject the null hypothesis and to accept the alternative hypothesis.

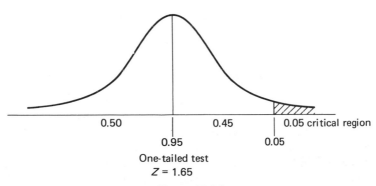

0.50   0.45   0.05 critical region

0.95   0.05

One-tailed test
$Z = 1.65$

Figure 15.4.1

4. Computing the Test Statistic

We use the information provided in Table 15.4.1 to compute the necessary values for $\hat{p}_p$, $\hat{q}_p$, and $\hat{\sigma}_{D_{p_*}}$. This enables us to use the Z statistic to measure the differences between the two sample proportions and the estimated standard error. The formula for Z is

$$Z = \frac{D}{\hat{\sigma}_{D_{p_s}}}$$

We can compute the necessary estimates as follows:

$$\hat{p}_p = \frac{N_1 p_{s_1} + N_2 p_{s_2}}{N_1 + N_2} = \frac{(542)(0.578) + (553)(0.520)}{542 + 553}$$

$$\hat{p}_p = \frac{313.3 + 287.6}{1095} = \frac{601}{1095} = 0.549$$

and

$$\hat{q}_p = 1 - \hat{p}_p = 0.451$$

Thus,

$$\hat{\sigma}_{D_{p_s}} = \sqrt{p_p q_p \left(\frac{N_1 + N_2}{N_1 N_2}\right)} = \sqrt{(0.549)(0.451)\frac{542 + 553}{(542)(553)}}$$

$$\hat{\sigma}_{D_{p_s}} = \sqrt{0.248 \frac{1095}{299,726}} = \sqrt{(0.248)(0.004)} = \sqrt{0.00099}$$

$$\hat{\sigma}_{D_{p_s}} = 0.0315$$

Since $D = 5.8$ percent, the proportion $= 0.058$, that is, $0.578 - 0.520 = 0.058$. Then for Z we find

$$Z = \frac{D}{\hat{\sigma}_{D_{p_s}}} = \frac{0.058}{0.0315} = \underline{\underline{1.84}}$$

5. Deciding About Rejection of the Null Hypothesis

Since $Z = 1.84$, we reject the H_0 at the 0.05 level of significance. Thus we conclude that there is a statistically significant difference between the two sexes in the proportion of civilians enrolled in school. Furthermore, we accept the alternative hypothesis H_1 that there is a greater proportion of males than females enrolled in school when all ages three to thirty-four are considered.

An important cautionary remark made in our discussion of the difference between two means is equally important at this point. Even though we have rejected H_0 at the 0.05 level and concluded that H_1 is acceptable and that there is a greater proportion of males than females enrolled in school, this, too, may be an error. Specifically, the Z score of 1.84 tells us that we have $0.5000 + 0.4671 = 0.9671$ of the normal sampling distribution included below the difference. This leaves 0.0329 of the population falling in what would be the critical region. This can be interpreted as indicating that approximately 33 times out of 1000, we could get a 0.058 difference between proportions as a result of random sampling fluctuations in our sampling procedure. Put differently, we cannot be completely sure that our sample is truly representative, and, as usual, our faith rests in the procedures of probability sampling.

15.5 ANALYZING THREE PROPORTIONS

A limitation of the Z test for the difference between two proportions is that it is limited to two samples and a dichotomous variable. When we are interested in three or more sample proportions, the chi-square test of significance is used. Earlier we learned that the chi-square technique enables us to measure whether the values we observe empirically are significantly different from those we would expect if our theoretical assumptions were correct.

First we formulate the null hypothesis that the population distributions are homogeneous; that is, that there are no differences between them. We then test the probability of obtaining larger-than-expected differences among the observed frequencies, and chi-square enables us to compute a general index that measures the observed distributions. Put differently, chi-square enables us to test whether or not there is a significant difference between the observed frequencies in each category compared to the expected frequencies in each category given the null hypothesis. Thus chi-square is a test to measure the independence of two variables and the goodness of fit of the data to the theoretical distribution.

Chi-square is computed by finding the sum of the squared differences between the frequencies observed (f_o) and the frequencies expected (f_e), divided by the frequencies expected. The general formula for chi-square may be expressed symbolically

$$\chi^2 = \sum \frac{(f_o - f_e)^2}{f_e}$$

The formula for chi-square tells us that the value for chi-square increases as the differences between the observed and expected frequencies increase. Chi-square will be zero when the observed and expected frequencies are the same.

Suppose that we are interested in comparing the proportions of three samples. Extending our analysis of the proportion enrolled in school, we will examine the proportion of the sexes enrolled in school for ages eighteen through thirty-four, grouped into three categories. Even though age is an interval variable, we can treat the three categories as if they were nominal, as shown in Table 15.5.1. There are strong arguments for using statistics that require the level of measurement at hand; however, there are many instances when chi-square is a useful measure even though only nominal assumptions are needed.

Table 15.5.1.   Percent of sexes enrolled in school from ages 18–34
from population of civilians 3–34 years of age

| | Age | | | |
	18–24	25–29	30–34	Total
Males	57.3%	68.0%	60.0%	59.0%
Females	42.7	32.0	40.0	41.0
Total	100.0%	100.0%	100.0%	100.0%
N	(715)	(126)	(57)	(898)

Source: *Statistical Abstract of the United States, 1973*, 94th ed. (Washington, D.C.: U.S. Bureau of the Census, 1973), Table 166, page 111.

Chi-square is computed directly from the actual observed frequencies rather than from their percentages. Thus, we first convert the percentage table to one that shows the actual frequencies in each of the categories, as shown in Table 15.5.2.

Table 15.5.2.   Observed frequencies (f_o) of sexes
enrolled in school, ages 18–34

| | Age | | | |
	18–24	25–29	30–34	Total
Males	410	86	34	530
Females	305	40	23	368
Total	715	126	57	898

Then we compute a table that shows the expected frequencies for each cell. The method for computing the expected frequencies is the familiar one of multiplying the relevant marginal frequencies by each other and dividing the product by the number in the sample. Even though it may result in some small rounding errors, we can calculate the expected frequencies by multiplying the marginal frequency of a category on one variable by each proportion of the categories on the other variable. For example, since 59.02 percent of the 898 are males, that is, 530/898 = 0.5902, we multiply 715 by 0.5902 and we find that we would "expect" (0.5902)(715) = 421.99 = 422 males in the 18–24-year-old group if the variables were statistically independent. All the expected values are shown in Table 15.5.3.

**Table 15.5.3. Expected frequencies (f_e) of sexes
enrolled in school, ages 18–34**

	Age			
	18–24	25–29	30–34	Total
Males	422	74	34	530
Females	293	52	23	368
Total	715	126	57	898

1. Listing Our Assumptions

Level of Measurement: Sex is a dichotomous variable. Age is a grouped interval variable that will be treated as a trichotomous nominal variable.

Model: We assume that the three categories represent simple random samples.

Null Hypothesis (H_0): We hypothesize that there are no differences between the proportions of males enrolled among the populations of age groups.

Alternative Hypothesis (H_1): We hypothesize that there is a statistically significant difference between the proportion of males enrolled among the age groups.

2. Determining the Appropriate Sampling Distribution

The sampling distribution for the chi-square statistic is presented in Table G in the Appendix. Chi-square depends upon the number of degrees of freedom, and we use the general computation for degrees of freedom for rows (r) and columns (c) in a cross-tabulated table, that is, $(r - 1)(c - 1)$. In this instance,

$$df = (r - 1)(c - 1) = (1)(2) = 2.$$

3. Choosing the Level of Significance and Specifying the Critical Region

We are not predicting the direction of the difference, but rather that there is a relationship (statistical dependence). However, it may appear that we are interested in performing a one-tailed test since we are seeking to test whether or not chi-square is larger than we would expect if only chance (random sampling error) were operating. We decide to select the 0.05 significance level. Consulting the chi-square table, we find that for 2 degrees of freedom at the 0.05 level, we need a chi-square score equal to or greater than 5.991.

4. Computing the Test Statistic

From the percentage table we computed the values for the observed frequency table. Then, using the marginals, we computed the expected frequency table. Now we can construct a computational table for chi-square as shown in Table 15.5.4. The chi-square computation table for our 2 by 3 table shows the cell value, the observed frequency, the expected frequency, their difference, the difference squared, and the ratio between the squared difference and the frequency expected. $\chi^2 = 5.56$ with two degrees of freedom, as shown in Table 15.5.4.

Table 15.5.4. Chi-square computations for 2 × 3

Cell	f_o	f_e	$(f_o - f_e)$	$(f_o - f_e)^2$	$(f_o - f_e)^2/f_e$
a	410	422	−12	144	0.34
b	86	74	12	144	1.95
c	34	34	0	0	0.00
d	305	293	12	144	0.49
e	40	52	−12	144	2.78
f	23	23	0	0	0.00
				$\chi^2 =$	5.56

$$df = (r - 1)(c - 1)$$
$$df = (1)(2) = 2$$

5. Deciding About Rejection of the Null Hypothesis

The chi-square value does not fall into the critical region, so we do not reject the null hypothesis. Thus we conclude that the null hypothesis cannot be rejected at the 0.05 level, and that the difference between the proportion of males to females for the three age groups is not statistically significant. To be more specific, however, the value for chi-square can be expressed as follows: A chi-square test indicates that the probability of finding a difference between the proportions of males to females as demonstrated by the age and sex table is $0.10 \le \chi^2 \le 0.05$.

EXERCISES

1. In your own words define, describe, or discuss the following terms and give an hypothetical example of each:

   σ_D

   $\hat{\sigma}_{D_{p_s}}$

   difference between statistics

sample statistic
sampling distribution of differences
independence between samples
two-tailed *t* test
chi-square

2. Give an hypothetical example of a research project in which you would be comparing statistics from two samples.

3. Explain why the *Z* test is limited to analyzing two proportions.

4. Using the figures in the following table.
   A. Determine the difference in geographic mobility between countries.
   B. Interpret your findings for a nonstatistical friend.

Percentage of population, age 5, Residentially Mobil during Five Years Prior to the Census, United States (1960), Canada (1960), England and Wales (1961), and Scotland (1961)

	U.S.	Canada	Eng. & Wales	Scotland
All ages 5 +	50.1	45.4	36.4	36.6

Source: Adapted form Larry H. Long. "On Measuring Geographic Mobility," in William Petersen (ed.) *Readings in Population* (New York: Macmillan Company) pp. 215–222.

5. Explain why chi-square is a suitable statistic when comparing two samples. Why is it better than some other statistics?

6. Using the data presented in Chapter 7, exercise 10, compare the two samples testing the difference between the means.

7. Using the data presented in Chapter 8, exercise 9, compare the following combinations and test the difference between the two proportions for the following combinations of variables:
   A. Group 1 vs. group 2
   B. Group 1 vs. group 3
   C. Group 1 vs. group 4
   D. Group 2 vs. group 3
   E. Group 2 vs. group 4
   F. Group 3 vs. group 4

8. Using the data presented in Chapter 8, exercise 9, analyze the difference between the proportions of the following combinations of 3 variables:
   A. 1, 2, and 3
   B. 1, 2, and 4
   C. 1, 3, and 4
   D. 2, 3, and 4

16

Analysis and Theory Construction

One of the major strengths of statistical methods is their use as analytical tools in the examination of social behavior. By definition, the word "analyze" means to separate an entity, event, or phenomenon into its constituent parts. From the common-sense point of view, we often speak of "analyzing the situation described in the newspapers," or "analyzing the prevailing mood on campus," or "analyzing our chances for success or failure," and so forth. Statistical analysis is more formal, but just as sensible; and we may speak of "analyzing the data," or "analyzing the statistical characteristics of the variables," or "analyzing the causal relationships between the variables," and so forth.

An essential aim of sociologists when they use statistics is to allow themselves to be more specific in the articulation of sociological theories. Central to this concern is the issue of causation. As is the case with most social scientists, the sociologist hopes to be able to clarify the relationship between variables and to show which independent variables contribute to the variation that occurs in the dependent variable. There are many ways this issue can be examined. One of them, however, fits in easily with what we already know about sociology and statistics: It involves the analytical use of a basic model.

This chapter discusses a *basic analytical model* and sets forth the assump-

341

tions and principles involved in it. This step is fundamentally statistical. However, the model must be applied with sound theoretical underpinnings for use in sociology. Thus the basic model is discussed as a tool for examining empirical evidence and as a step in examining the causal process within a system of variables.

One of the most important points we will learn in this chapter is that statistical models that may be applicable to empirical evidence cannot be used unless they are logically and appropriately handled in light of existing or developing social theories. Consequently, the last part of this chapter discusses the use of theory in developing analytical models that can be applied to empirical data.

16.1 EMPIRICAL EVIDENCE AND CAUSAL ANALYSIS

Before sinking our teeth into causal analysis per se, let us explore some of its underlying theoretical ideas. The general approach to causation grows out of the ideas behind classical experimental research. In experimental research, causal relationships can be examined because the researcher "controls" which subjects are exposed to experimental treatments.[1] That is, the researcher is able to manipulate the *experimental setting* in which the subjects are exposed to some kind of experimental treatment. We have learned that in sociological research such experimental studies are not conducted often. Instead, the sociological researcher "controls the variables" and the relationships between them. Thus the researcher literally holds constant or, more commonly, manipulates variable values, treating the variables in terms of exposure or treatments in the form of levels or categories of variables. In nonexperimental research, variables are most commonly controlled either by adjustment or by subdivision. Thus we manipulate the *variables* in our study rather than *the setting*.

It is important to distinguish between the observation of events in their natural setting and the laboratory setting in which the experimenter sets the stage in order to observe the possibly differential effects of two or more treatments. In the laboratory setting, the units of observation or units of experimentation are assigned at random to groups which are commonly called the "experimental group" and the "control group." In classical experimental design the researcher either manipulates the variables by applying varying levels of them or studies the results of varying levels which already exist. One of the basic ideas of experimental design is that the experimenter randomly

---

[1] For an easy-to-understand elementary treatment of this subject see Hubert M. Blalock, Jr., *An Introduction To Social Research* (Englewood Cliffs, New Jersey: Prentice-Hall, Inc., 1970); also see Donald T. Campbell, Lawrence Ross, and Glen V. Glass, *Experimental Methods* (Englewood Cliffs, New Jersey: Prentice-Hall, Inc. [forthcoming]).

assigns subjects to different treatment groups. In one group (experimental), the subjects are exposed to the presumed causal variable, and in the other group (control), the subjects are not exposed to it. The general idea is that the causal variable is manipulated because the experimental setting is different for the two groups.

Clearly, when it is possible to conduct an experiment, it can be an effective and efficient method for testing the causal hypothesis that variable X influences variable Y. However, there are many social situations in which it is not possible for the experimenter to manipulate the independent variable or to assign subjects at random to different treatment groups. If we were studying the effects of different methods of child rearing, for example, it would be difficult to randomly assign children to different families with different child rearing practices. Instead, the researcher probably would study randomly sampled children controlling for the way they had been reared.

In order to study variation, then, the sociological researcher generally does not directly manipulate the variable. Instead, the cases sampled are divided according to whether or not the variable is present or what level of it is possessed. Ideally, we should be able to eliminate alternative explanations that a given effect was actually caused by some other variable. In sociology, one complication that arises because of studying the natural situation is that many factors may be relevant while at the same time they may be difficult or impossible to control. Even so, the principles apply and the experimental design is used as a prototype for quantitative sociological research, and we *are* interested in *causal* relationships between variables. It is this interest that gives rise to the general idea of causal analysis through the use of empirical evidence and the constuction of causal models.

It is common for sociologists to deal with causality by using hypothetical, idealized "models" of the real world.[2] To construct such models we decide upon a limited number of variables that we treat as a *system*. This system has a network of causal relationships within it. In general, a *causal model* may be thought of as an attempt to portray the assumed (inferred) causal relationship within a system of variables. More specifically, in sociological research causal models are assertions or hypotheses used to describe, explain, or predict (1) the presence or existence of relationships, (2) the form or nature of the relationships (whether they are positive or negative), and (3) the direction of the causal influences.

Our concern is with which independent variables within a given system of variables influence or "cause" the dependent variables in the system and how

[2]This and much of the following discussion about causal models integrate some of the ideas of Hubert M. Blalock, Jr., *Causal Inferences in Nonexperimental Research* (Chapel Hill: The University of North Carolina Press, 1961); Otis Dudley Duncan, "Path Analysis: Sociological Example," in *The American Journal of Sociology*, 72 (July 1966), pp. 1–16; and James A. Davis, *Elementary Survey Analysis* (Englewood Cliffs, New Jersey: Prentice-Hall, Inc., 1971).

they do so. This means that we must consider *all* the pairs of relationships in the system we are studying. By pairs of relationships, we mean *two*-variable (a pair) relationships. For example, if there are three variables, we are interested in *all three* of the pair relationships between them. Thus if we are going to study *Y*, *X*, and *T*, we are concerned with the *YX* relationship, the *YT* relationship, and the *XT* relationship.

Causal models are often represented by *causal diagrams*, which graphically portray the causal relationships in a system of variables. Customarily, we show a nonnegligible relationship between two variables with a solid straight line labeled with a + or − sign and a negligible relationship with no line. For example, suppose we hypothesize that there is a direct positive relationship between *X* and *Y*, a direct negative relationship between *X* and *T*, and no direct relationship between *T* and *Y*. We can put these relationships into a diagram as shown in Figure 16.1.1.

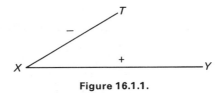

Figure 16.1.1.

This diagram does not tell us anything about the *direction* of the effects. We can easily take care of that by representing the direction of influence between the pair by putting an arrowhead on the end of each line. For instance, if we *theorize* (1) that *T* influences (causes) *X* negatively, that is, the possession of high levels of *X* leads to the possession of low levels of *T*, and (2) that *X* influences (causes) *Y* positively, that is, the possession of high levels of *X* leads to the possession of high levels of *Y*, we can add arrowheads in our diagram, as shown in Figure 16.1.2.

When we use such diagrams, we are theorizing (1) that certain relationships are present, (2) that the relationships have a particular form or nature, i.e., are positive or negative, and (3) that the direction of influence, or cau-

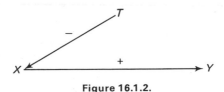

Figure 16.1.2.

sality, of the relationships is as drawn. Notice that the magnitudes of the relationships are not shown in either figure. They could be, however, and if they are, we usually indicate the magnitudes of the relationships by writing some coefficient value for the pair of variables above or beside the arrow connecting them.

Generally, when coefficients do not appear, it is because causal diagrams are constructed theoretically rather than empirically. We may start out with an existing theory or a previously constructed causal model. Then we may gather data and analyze them according to what they tell us about the model. However, it is not necessary to analyze data first in order to develop and draw a causal diagram.

16.2 THE USE OF THEORY IN DEVELOPING MODELS

So far we have talked about causal models in an abstract way, but we have not discussed how causal models can arise out of a particular theory or data. Earlier we learned that a theoretical interest is what usually initiates the research process. Since then we have seen that in all the studies cited there has been an underlying theory or theories from which the research grew. Let us go step by step through the process of stating a theory that then can be analyzed.

In the study *Socioeconomic Background and Achievement*, the schematic version of a basic model of occupational achievement, as shown in Figure 16.1.3, serves as the theoretical backdrop to their entire study.

Duncan, Featherman, and Duncan describe this figure as follows:

[The] Figure exhibits a diagrammatic arrangement of the variables discussed thus far as being implicated in a rudimentary model of the process of achievement. At the far right, as the ultimate outcome of the whole process, is the respondent's occupation. The letter Y stands for the variable, occupational socioeconomic status, as measured on the scale developed by Duncan (1961a). Four arrows lead to Y, representing the assumption that occupational status depends (directly) on educational attainment (measured by years of schooling, U), on family head's occupation (measured on socioeconomic scale X), on family head's education (years of schooling, V), and on unspecified residual factors summed up in variable B, which is taken to be uncorrelated with the other three determinants of Y. The second relation depicted by the diagram is the dependence of educational attainment (U) on family head's occupation (X) and family head's education (V) as well as unspecified residual factors, summed up in variable A, taken to be uncorrelated with X and V.[3]

[3]Otis Dudley Duncan, David L. Featherman and Beverly Duncan, *Socioeconomic Background and Achievement* (New York: Seminar Press, 1972), pp. 7–9.

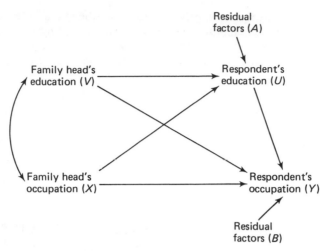

Figure 16.1.3. Schematic representation of the basic model of occupational achievement. (*From Otis Dudley Duncan, David L. Featherman, and Beverly Duncan, Socioeconomic Background and Achievement. New York: Seminar Press, 1972, p. 8.*)

Common sense tells us that time order is important and that we can say that past variables influence (cause) present variables. Put differently, when the present state of a variable depends upon the past state of another variable, then we can say that the latter "causes" the former. It would not make sense to suppose that the present could influence the past.

Duncan, Featherman, and Duncan assume that one's family head's occupation influences one's own occupation and not vice versa. That is, they view certain variables according to temporal occurrence. Put differently, the temporal order of events may enable us to theorize about the direction of influence. Thus we may say *theoretically* that the possession of a higher level of some past variable will "cause" the possession of a higher level of a present variable. This leads to an important theoretical perspective, which guides the present treatment of causal analysis.

We conceive of each individual's life cycle in terms of a particular temporal sequence. Ideally we ought to examine the variables at successive stages *over time*, and ideally we should carry out our study by actually observing each of the variables in a cohort of newborn infants at each stage of their lives. Such a procedure, however, is not practical. Therefore, for practical purposes, we generally use such instruments as a questionnaire that seeks retrospective answers about the life histories of the individual respondents as well as answers about their present circumstances. This enables us to examine a sample of individuals from an historical (over time) perspective.

A *causal relationship* is one in which a reordering of cases on one variable would be followed by a reordering of cases on the other variable. Thus if a reordering of cases on variable X is followed by a reordering of cases on variable Y, we infer that X influences (causes) Y.

Obviously, in sociological research the cases cannot be reordered because the subjects actually possess certain levels of past variables and certain levels of present variables as well as possessing certain levels of other nontemporal variables. In terms of *theoretical* causal relationships, however, we could say, for example, that "higher family head's occupation leads to higher respondent's occupation." In a sense, what we mean is that *if* we could somehow reorder things and give a person, say, a higher level of a past variable, then that person would possess a higher level of a present variable. This is what we mean when we say that a causal relationship is one in which a reordering on one variable would be followed by a reordering on the other.

16.3 A BASIC ANALYTICAL MODEL FOR EXPLANATION AND PREDICTION

We are interested in a research technique that enables us (1) to describe the relationships between two variables and (2) to determine the contribution of an independent variable to the behavior of a dependent variable.

The *basic linear model* is such a technique, and it provides us with an approach to measurement and prediction that takes into account the contributions of an unlimited number of independent variables on a given dependent variable.[4]

The basic linear model may be thought of as an equation in which the value of the dependent variable is shown to be made up of a linear combination of any number of independent variables. As such it is an effective way to examine multivariate influences upon the behavior of a dependent variable. From a familiar perspective, the basic linear model may be seen as a system of variables in which our concern is with finding the appropriate net contribution that each independent variable makes to the value of the dependent variable. It enables us to construct a composite equation that includes the partial weighted effects of the independent variables, the addition of which gives the value of the dependent variable. Thus a linear weighted composite is a way of showing the various components of the dependent variable that are due to the effects of the independent variables.

[4]In the present discussion we use a simplified version of the general linear model. This version combines the following treatments: James Fennessey, "The General Linear Model: A New Perspective on Some Familiar Topics," *American Journal of Sociology*, 74 (July, 1968), 1–27; Hubert M. Blalock, Jr., *Social Statistics*, 2nd ed. (New York: McGraw-Hill Book Company, 1972); and Robert A. Bottenberg and Joe H. Ward, Jr., *Applied Multiple Linear Regression*, 6570th Personnel Research Laboratory, Technical Document Report PRL-TDR-63-6 (Lackland Air Force Base, Texas, March 1963).

Without introducing complex mathematical properties, here is a set of definitions and an equation that enable us to express the basic model.

Y_i = the actual value of the dependent variable for the ith individual.

(Y) = a constant overall value that summarizes the general elevation of the dependent variable given all the individual observations.

$X_i, T_i, V_i, \ldots, K_i$ = the values of the K independent variables for the ith individual.

w_K = the partial weighted effects of the Kth independent variable.

u_i = the unique random effect of residual variables not included in the equation or the "error term" of the ith observation.

$$Y_i = (w_Y(Y)) + (w_X X_i) + (w_T T_i) + (w_V V_i) + \cdots + (w_K K_i) + (u_i)$$

This abstract version of the basic linear model can be stated in terms of linear regression. Let us begin with two interval variables, an independent variable X and a dependent variable Y.[5] We can state the two-variable version of the basic linear model with the following equation:

$$Y_i = (w_Y(Y)) + (w_X X_i) + (u_i)$$

Thus,

$$Y_i = a_{Y \cdot x} + (b_{YX} \cdot X_i) + e_{Y \cdot x}$$

We can expand the model to include any number of independent variables, with each new variable theoretically adding to and improving our predictive

[5]The connection between the two-variable linear regression model and the general linear model is clearest in what at first glance appears to be a complex equation in which the contributions and values of the dependent variable and the independent variable are considered in terms of their definitional formulas. The first term takes into account the overall value for Y minus a term that corrects for the overall value of X times the regression coefficient for Y given X. The second term uses the same regression coefficient times the individual score on variable X. Recalling what we learned earlier, these first two terms constitute the predicted value of Y given X. The last term in the formula enables us to compare the actual observed value Y_i with the predicted value $\hat{Y}$. Thus,

$$Y_i = \left[\frac{\sum Y}{N} - \frac{\dfrac{\sum \left(X - \frac{\sum X}{N} \right)\left(Y - \frac{\sum Y}{N} \right)}{N}}{\dfrac{\sum \left(X - \frac{\sum X}{N} \right)^2}{N}} \cdot \frac{\sum X}{N} \right]$$

$$+ \left[\frac{\dfrac{\sum \left(X - \frac{\sum X}{N} \right)\left(Y - \frac{\sum Y}{N} \right)}{N}}{\dfrac{\sum \left(X - \frac{\sum X}{N} \right)^2}{N}} \cdot X_i \right] + [Y_i - \hat{Y}]$$

efforts. Thus the basic linear model is a useful analytical tool for *multivariate* analysis. It is a straightforward extension to expand the equation for the basic linear model to take into account any number of independent variables. An equation that provides a multivariate version of the general linear model can be expressed symbolically:

$$Y_i = a_{Y \cdot XTS...K} + (b_{YX \cdot TS...K} \cdot X_i) + (b_{YT \cdot XS...K} \cdot T_i)$$
$$+ (b_{TS \cdot XT...K} \cdot S_i) + \cdots + (b_{YK \cdot XTS...} \cdot K_i) + e_{Y \cdot XTS...K}$$

When we make essentially the same simplifying assumptions that we did for two-variable linear regression, we can express the multiple-linear-regression equation symbolically:

$$\hat{Y} = a_{Y \cdot XTS...K} + (b_{YX \cdot TS...K} \cdot X_i) + (b_{YT \cdot XS...K} \cdot T_i)$$
$$+ (b_{YS \cdot XT...K} \cdot S_i) + \cdots + (b_{YK \cdot XTS...} \cdot K_i)$$

This equation states that we predict the value of Y from knowledge we have of the independent variables $X, T, S, \ldots, K$. Thus we estimate the actual value of Y_i with the predicted value, $\hat{Y}$, based on the simultaneous possession of known scores on all the variables in the multiple linear equation.

The error in estimating a value for Y given $X, T, S, \ldots, K$ follows the logic we developed above. Specifically, we compare each observed score (Y_i) with each estimated score ($\hat{Y}$) by subtracting $\hat{Y}$ from Y_i to measure how different each individual is from the linear regression estimates, that is, ($Y_i - \hat{Y}$). If we square each difference and sum the squares, we can measure the error in estimating actual individual scores with an appropriate linear estimate. Therefore, the variation we derive from the multiple-linear-regression predictions is merely an extension of the least-squares equation we used earlier, in which the sum of squares is a minimum.

Part of our interest is in the way variables vary as measured by the total sum of squares, or the variance. We are also interested in the ways in which the variables covary with each other as measured by the sum of the cross products, or the covariance. Finally, we are interested in how *combinations* of variables covary, as measured by examining the relationships between variables and the explained variance.

Sets of simplifying assumptions enable us to use various subtypes of the basic linear model when analyzing empirical data. These assumptions are straightforward and certainly helpful because they greatly simplify things for us. We learned about some of these assumptions when we studied linear regression, and correlation. We can also extend the basic linear model using the assumptions of multiple linear regression.

16.4 THE DEVELOPMENT OF THEORETICAL CAUSAL MODELS

Sociological analysis grows out of substantive and theoretical issues. For example, Blau and Duncan develop a theoretical model of the *American Occupational Structure*. They put it this way: "We seek to place our research findings into a theoretical framework and suggest theoretical interpretations for them."[6] In many ways theirs is a classical effort in that it serves as the basic framework for much of the current interest in path analytical methods and causal models.

Starting with the theoretical stance that intergenerational social mobility and mobility from occupational beginnings to occupational destinations reflect the dynamics of the occupational structure, Blau and Duncan analyze "the patterns of these occupational movements, the conditions that affect them, and some of their consequences."[7] Their aim is to explain aspects of the dynamics of social stratification in the United States.

Even though there are many conceptions of social class, Blau and Duncan view it from a specific perspective, explaining that:

> Occupational position does not encompass all aspects of the concept of class, but it is probably the best single indicator of it, . . . [and] conceptually, there is a closer relationship between economic class and occupational position than there is between occupational position and prestige status.[8]

Blau and Duncan tie mobility research, substantive problems, and sociological theory together, saying:

> A substantive problem of central concern to us . . . is how the observed patterns of occupational mobility are affected by various factors, such as a man's color, whether he has migrated, or the number of his siblings and his position among them. . . . The process of occupational mobility refers to the social metabolism that governs this allocation of manpower and hence underlies the dynamics of the occupational structure. The specification of the factors that affect the occupational achievements of individuals seeks to account for this dynamic process. In sum, the conventional mobility matrix represents the structure of occupational allocations to be explained, and the analysis of the conditions that determine the process of mobility is designed to furnish the required explanation.[9]

[6] Peter M. Blau and Otis Dudley Duncan, *The American Occupational Structure* (New York: John Wiley & Sons, Inc., 1967), p. 2.

[7] *Ibid.*, p. 1.

[8] *Ibid.*, p. 6.

[9] *Ibid.*, p. 9–10.

To do this, a basic model is proposed in which Blau and Duncan examine a system of five variables. They designate them by the following arbitrary letters:[10]

V: Father's educational attainment
X: Father's occupational status
U: Respondent's educational attainment
W: Status of respondent's first job
Y: Status of respondent's occupation

When the theory is located in the framework of multiple linear regression, Blau and Duncan explain:

> A basic assumption in our interpretation of regression statistics . . . has to do with the causal or temporal ordering of these variables. . . . We take the somewhat idealized assumption of temporal order to represent an order of priority in a causal or processual sequence, which may be stated diagrammatically as follows:
>
> $$(V, X) - (U) - (W) - (Y)$$
>
> In proposing this sequence we do not overlook the possibility of . . . "delayed effects," meaning that an early variable may affect a later one not only via intervening variables but also directly (or perhaps through variables not measured in the study).[11]

The gross effects including both causal effects and spurious noneffects of one variable on another are measured by the zero-order product-moment correlation coefficients for the five status variables. These values can be seen in Table 16.4.1.

Table 16.4.1.   Zero-order correlations for five status variables

	Y	W	U	X	V
Y: Occupational status	. . .	0.541	0.596	0.405	0.322
W: First job status			0.538	0.417	0.332
U: Education				0.438	0.453
X: Father's occupational status					0.516
V: Father's education					. . .

Source: Peter M. Blau and Otis Dudley Duncan, *The American Occupational Structure* (New York: John Wiley & Sons, Inc., 1967), p. 169, Table 5.1.

[10] *Ibid.*, p. 165.
[11] *Ibid.*, pp. 166–168.

With the exception of $r_{VX} = 0.516$ between father's education and father's occupation, the numbers in Figure 16.4.1 are path coefficients. Blau and Duncan offer a detailed explanation of how they arrived at these numbers and how the diagram was constructed, closing their discussion with the following remark:

> The technique of path analysis is not a method for discovering causal laws but a procedure for giving a quantitative interpretation to the manifestations of a known or assumed causal system as it operates in a particular population.[12]

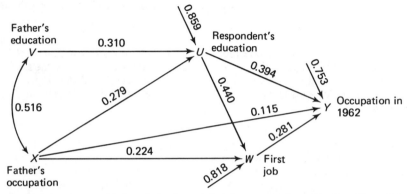

Figure 16.4.1.   Path coefficients in basic model of the process of stratification. (*From Peter M. Blau and Otis Dudley Duncan,* The American Occupational Structure. *New York: John Wiley & Sons, Inc., 1967, p. 170.*)

Common sense suggests the possibility that there are delayed or indirect effects between variables as well as direct ones. This means we suspect that a variable occurring early in the sequence may influence a later one, not just through the intervening variables as shown in their proposed sequence, but also directly. This suspicion raises the issue that becomes a major reason for our analytic efforts. How we get at the direct and indirect effects is one of the purposes of path analysis.

EXERCISES

1. In your own words, define, describe, or discuss the following terms and give an hypothetical example of each.
   experimental setting
   causal model

[12] *Ibid.,* p. 177.

inferred relationship
basic linear model
variables
direct relationship
theory
multivariate analysis
empirical data
operational terms

2. Give an hypothetical example of a research project in which the researcher would manipulate the variables in the study rather than the setting.

3. Using a research report from a recent issue of the *American Sociological Review* or the *American Journal of Sociology*, verbally describe and construct a causal model of the hypothesis.

4. Explain in words how the basic model equation can be expanded to take into account any number of independent variables. Discuss the value of doing this in an analysis.

5. Develop or adapt an existing sociological theory that is amenable to statistical analysis. Be explicit in stating the operational as well as verbal definitions of the variables.

6. Construct a causal model that graphically portrays the theory you presented for problem 5. Label the variables and indicate the direction and form of the relationships.

$$17$$

Path
Analysis

In this chapter, we will learn the specific techniques of a form of multiple linear regression called *path analysis*. The concepts and methods we will learn in this chapter can be extended to many other forms of multivariate analysis which you will meet as you study more areas of sociology.

Path analysis aims at determining the direct and indirect effects among a number of variables. Such effects are measured by partial-regression beta weights which are called *path coefficients*. Importantly, zero-order product-moment correlation coefficients and path coefficients may be interpreted identically. Thus, either a zero-order product-moment correlation coefficient or a path coefficient tell us the degree to which two interval variables are related.

17.1 THE TECHNIQUE OF PATH ANALYSIS

Path analysis focuses on the direct and indirect effects of each "independent" variable in a closed system of variables.[1] It is an analytical statistical tech-

[1] Path analysis has grown out of the early work of Sewell Wright, "The Method of Path Coefficients," *Annals of Mathematical Statistics*," V (September, 1934), pp. 161–215. The version treated here is based largely on the work of Otis Dudley Duncan, "Path Analysis:

nique that enables us to give a quantitative interpretation to the interrelationships within a known or an assumed causal system that exists in some specific population. In general, we must simplify reality by making three assumptions. We assume that the set of relationships between the variables in a causal system is linear, additive, and *recursive* (operates asymmetrically, i.e., in only one direction). Importantly, these assumptions are not about the specific individuals in the study but only about the set of relationships between the variables.

For our purposes, we assume that all of the variables *within* the system are measured. An unmeasured variable (or set of variables) outside the system is called a *residual variable*. When the variation in one variable within the system is not completely explained by the other variables within it, we assume that the unexplained variation is caused by a residual variable. We also assume that the residual variables are not correlated with any of the others within the system or with each other. Thus, we use the residual variable to account for the variation in a given variable that is *unexplained* by the measured variables within our system. Quite simply, the residual stands for the variation left over after the system has explained all it can.

We can describe our system further as (1) one in which the unmeasured variables are not examined in our analysis, (2) one in which the residual factors are not correlated with each other, and (3) one in which each of the variables is directly related to all of the variables preceding it according to our assumed causal sequence.

To look at the relationships within such a system as "effects," we must translate them to path coefficients. For our purposes, we are dealing with a "completely recursive" system in which *all* possible partial coefficients are estimated, as opposed to a system in which only specified partials are estimated and then a check on goodness of fit between model and data is made. The latter are called *over-identified recursive systems*, and are treated in more advanced statistics books.

Earlier we learned how to calculate the generalized case of partial regression coefficients. Now all we must do is extend our knowledge somewhat, for in a completely recursive system, path coefficients are nothing more than a specialized case of the partial b coefficients. Path coefficients are the standardized partial b coefficients of all preceding "independent" or "test" variables on each of the successive "dependent" variables in a closed system. Path coefficients are expressed symbolically as p_{YX}, where the *first* subscript Y indicates the dependent variable and the *second* subscript X indicates the independent variable whose causal influence is being considered.

Sociological Examples," *The American Journal of Sociology*, 72 (July, 1966), pp. 1–16; and Kenneth C. Land, "Principles of Path Analysis," in *Sociological Methodology 1969*, ed. Edgar F. Borgatta (San Francisco: Jossey-Bass, Inc., 1969), pp. 3–17.

Path analysis actually ought to be easy to understand because path coefficients and zero-order product-moment correlation coefficients may be interpreted similarly. For instance, you may recall that if we find a zero-order r_{YX} of 0.57, we can say that a change of one standard deviation in X produces a (*gross effect*) change of 0.57 of a standard deviation in Y. A p_{YX} of 0.57 tells us that a change of one standard deviation in X produces a (*net effect*) change of 0.57 of a standard deviation in Y. Thus, a zero-order product-moment correlation coefficient tells us the *direct* and *indirect* degree to which two interval variables are related, and a path coefficient tells us the *direct* degree to which two interval variables are related. Therefore, path analysis may be thought of as a set of procedures that enables us to interpret data by decomposing correlation coefficients into direct and indirect effects, and path coefficients indicate the extent to which the variance in a dependent variable is determined by the variances of the independent variables.

Let us examine several formulas that prove that the zero-order correlation coefficient is equal to the sum of a set of standardized b coefficients, also called beta weights or path coefficients. The basic theorem of path analysis states that the zero-order correlation between any two variables is equal to the sum of the products of the paths and correlations between all the variables in the system. It may be expressed symbolically as:

$$r_{ij} = \sum_k p_{ik} r_{kj}$$

An example helps clarify this theorem. Let us assume that variables Y, X, T, and S are correlated at the zero-order level. Let us also assume that the independent variables X, T, and S are interrelated. We can state the zero-order correlation between X and Y in terms of the paths from and correlations with the other variables.

$$r_{YX} = p_{YX} r_{XX} + p_{YT} r_{TX} + p_{YS} r_{SX}$$

The correlation of X with itself, r_{XX}, is, of course, 1.00; thus,

$$r_{YX} = p_{YX} + p_{YT} r_{TX} + p_{YS} r_{SX}$$

We can say that the gross effect of X on Y is equal to the direct effect of X on Y (p_{YX}), plus the indirect effect of X on Y that operates via T ($p_{YT} r_{TX}$) plus the indirect effect of X on Y that operates via S ($p_{YS} r_{SX}$).

Thus far, we cannot measure the effect of variables outside our system. This effect is based on the error term and indicates the influence of residual variables not measured in the analysis. Since any variable's correlation with itself is 1.00, we can extend the equation as follows:

$$r_{YY} = 1.00 = p_{YX} r_{XY} + p_{YT} r_{TX} + p_{YS} r_{SX} + p_{YE} r_{EX}$$

where E = the error term or the path from the aggregate of residual variables.

As we learned earlier, the zero-order correlation coefficient also may be thought of as a path if we do not have any control variables. Since

$$r_{EY} = p_{YE}$$

we can multiply the path coefficient times the correlation coefficient and use the product of the two coefficients as a squared value that also is the proportion of variance explained

$$p_{YE}r_{EY} = r_{EY}^2 = p_{YE}^2$$

In light of these steps, we can now state the formula for the multiple correlation coefficient squared in terms of paths and correlations. Using another general theorem of path analysis for the multiple correlation coefficient, we can see that it, too, can be stated in terms of paths and correlations.

Let us set up the formula for the special case of the correlation of the dependent variable with itself. The basic theorem would state

$$r_{ii} = 1.00 = \sum_k p_{ik}r_{ki}$$

We can expand this by setting the range of k to include all variables (both those measured and those unmeasured) equal to k'. Thus,

$$r_{ii} = 1.00 = \sum_k p_{ik}^2 + 2\sum_{}^{k,k'} p_{ik}r_{kk'}p_{ik'}$$

Thus, in terms of the basic theorem we can state the multiple correlation coefficient for i given j and k:

$$R_{i\cdot jk}^2 = \sum_{l=1}^{k} p_{ik}^2 + 2\sum p_{ik}r_{kj}p_{ij}$$

The formula in our example of Y, X, T, and S now can be shown as the squared path to the dependent variable Y from the residual variables E, and it is derived by subtracting the sum of the products of the path and correlation coefficients.

$$p_{YE}^2 = 1.00 - (p_{YX}r_{XY} + p_{YT}r_{TY} + p_{YS}r_{SY})$$

$$p_{YE} = \sqrt{1.00^2 - [p_{YX}r_{XY} + p_{YT}r_{TY} + p_{YS}r_{SY}]^2}$$

$$p_{YE} = \sqrt{1.00 - R_{Y\cdot XTS}^2}$$

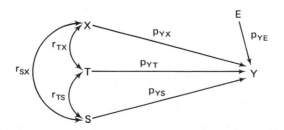

17.2 CALCULATING PATH COEFFICIENTS

Knowing how to conceptualize path coefficients is essential but not too difficult, as we just learned. Calculating them is also essential and is often considered difficult. However, some of the material we have already learned serves as the basis for the necessary calculations. How that happens grows out of a question neither posed nor answered earlier. Namely, how can b coefficients be interpreted verbally when we are analyzing variables measured in different units?

For example, suppose that we measure academic performance in terms of grade point average, academic effort in terms of hours spent studying each week, past performance in terms of high school average, and father's occupation in terms of Blau and Duncan's 17-level occupational status scale.[2] These variables could be analyzed with regular b coefficients, but since the units they are measuring are not equivalent, it would be like trying to compare apples, oranges, bananas, and grapes. The sticky statistical problem is that we need some means to correct the different scales and adjust them to one standardized scale.

As with most sticky problems, there is a statistical procedure to do just this, and it is really quite simple. We merely multiply the b coefficient by the ratio of the standard deviation in the independent variable to the standard deviation of the dependent variable, thereby standardizing it. In this way, we can convert regular regression coefficients into equivalent units from one variable to the other. Thus we can measure changes in one variable in terms of changes in standard deviation units in the other. We call these coefficients *standardized b coefficients* or *beta weights* and represent them with the Greek letter β.[3] Zero-order total beta weights can be expressed symbolically

$$\beta_{YX} = b_{YX} \cdot \frac{s_X}{s_Y}$$

or

$$\beta_{XY} = b_{XY} \cdot \frac{s_Y}{s_X}$$

The formulas simply tell us to multiply the zero-order total b coefficient by the ratio of the standard deviation of the independent variable X to the standard deviation of the dependent variable Y and vice versa. Zero-order total beta weights have a neat verbal interpretation: "β_{YX} measures how much

[2] Peter M. Blau and Otis Dudley Duncan, *The American Occupational Structure* (New York: John Wiley & Sons, Inc., 1967), pp. 165–171.
[3] For a complete discussion see Hubert M. Blalock, Jr., *Social Statistics*, 2nd edition (New York: McGraw-Hill Book Company, 1972), pp. 450–53.

change in the dependent variable (Y) is produced by a change of one standard deviation in the independent variable (X)."

Let us consider the beta weight equation with some known values for variables for Y and X. Given the following values for the standard deviations and the zero-order b coefficients for the variables, present academic performance (Y) and present academic effort (X),

$$s_Y = 1.14 \qquad b_{YX} = 0.284$$
$$s_X = 2.29 \qquad b_{XY} = 1.15$$

and using the equation for beta weights, we find:

$$\beta_{YX} = b_{YX} \cdot \frac{s_X}{s_Y} = 0.284 \cdot \frac{2.29}{1.14} = 0.284 \cdot 2.01 = \underline{\underline{0.571}}$$

and

$$\beta_{XY} = b_{XY} \cdot \frac{s_Y}{s_X} = 1.15 \cdot \frac{1.14}{2.29} = 1.15 \cdot 0.497 = \underline{\underline{0.571}}$$

These identical beta weight values may be interpreted as indicating that a change of one standard deviation in hours spent studying (X) produces a change of 0.57 of a standard deviation in grade point average (Y) and vice versa. In other words, if we could "give" someone 2.29 (the standard deviation of X) more hours spent studying, it would produce a change of 57 percent of 1.14 (the standard deviation of Y) or $0.57 \cdot 1.14 = 0.65$, thereby "causing" an increase of 0.65 grade point average units.

A potentially useful equivalence in terms is apparent if we compare the square of the beta weight to the product of the two total b coefficients. Since the two total beta weights are equal, that is, $\beta_{YX} = \beta_{XY}$, if we square the beta weight, we find that the square is equal to the product of the two total b coefficients. Thus,

$$\beta_{YX}^2 = \beta_{XY}^2 = b_{YX} \cdot b_{XY}$$
$$(0.57)^2 = 0.284 \cdot 1.149$$
$$0.33 = 0.33$$

This equivalence is even more meaningful when we recall that the product of the two zero-order b coefficients is equal to the square of the product-moment correlation coefficient, that is, $b_{YX} \cdot b_{XY} = r_{YX}^2$. Since, $b_{YX} \cdot b_{XY} = \beta_{YX}^2$, and since $b_{YX} \cdot b_{XY} = r_{YX}^2$, then $\beta_{YX} = r_{YX}$. The trap now closes and we see that the zero-order beta weight is equal to the zero-order product-moment correlation coefficient. Thus, in the zero-order instance, we can use either the beta weight or the product-moment correlation coefficient and make the same interpretive statements we just learned.

We can extend the zero-order beta weight procedure to the partial by multiplying the partial b coefficient by the ratio of the standard deviation of the independent variable to the standard deviation of the dependent variable. In this way, we can adjust partial b coefficients so that they are equivalent in terms of changes in standard deviation units. We call these standardized partial b coefficients *partial beta weights*. The first-order beta weight of Y on X controlling for T can be expressed symbolically:

$$\beta_{YX \cdot T} = b_{YX \cdot T} \cdot \frac{s_X}{s_Y}$$

This formula tells us to multiply the first-order partial b coefficient by the ratio of the standard deviation of the independent variable X to the standard deviation of the dependent variable Y. Notice that we do not "control for T" when calculating the beta weight. This is because we are "holding T constant," thus it is not varying, and we are interested in the XY relationship at each level of T distinct from the others.

The first-order partial beta weights can be interpreted verbally as follows: "$\beta_{YX \cdot T}$ measures how much of a change in the dependent variable Y is produced by a change of one standard deviation in the independent variable X holding test variable T constant."

The interpretive equivalence between β_{YX} and r_{YX} can be expanded to the level of partial beta weights. However, at the higher order partial level, the influence of the independent and control variables must be considered. In the first place, $\beta_{YX \cdot T}$ does not equal $\beta_{XY \cdot T}$. Neither the partial b coefficients nor the ratio of the standard deviations are equal. That is,

$$\beta_{YX \cdot T} = b_{YX \cdot T} \cdot \frac{s_Y}{s_X}$$

$$\beta_{XY \cdot T} = b_{XY \cdot T} \cdot \frac{s_X}{s_Y}$$

$$\beta_{YX \cdot T} \neq \beta_{XY \cdot T}$$

The partial correlation coefficient, however, is the same for both. That is, $r_{YX \cdot T} = r_{XY \cdot T}$. Importantly, the square of the partial correlation coefficient is equal to the product of the two beta weights. Thus,

$$\beta_{YX \cdot T} \cdot \beta_{XY \cdot T} = r^2_{YX \cdot T}$$

This means that the beta weights for two variables controlling for a third are different even though the partial correlation coefficients for the same two variables are equal. Herein lies one of the advantages of beta weight analysis (path analysis) over partial correlation analysis. Namely, since it is possible to distinguish between $\beta_{YX \cdot T}$ and $\beta_{XY \cdot T}$, we can examine the *influence of X on*

Y in the former case and of Y *on* X in the latter case. In other words, we can study the influence of one variable on another according to the *assumed causal direction*.

The connections between partial correlation and beta weights are useful in calculating higher order beta weights. We learned that in regression analysis the partial correlation coefficient refers to that part of the relationship between two variables that is left over after controlling for independent and test variables. In order to calculate the partial coefficient, we started with a matrix of zero-order product-moment correlation coefficients for a set of variables. As we just saw, the coefficients in the zero-order matrix can be seen as showing the gross amount of change that the assumed independent variable produces in the assumed dependent variable for a specific combination of two variables. Partial correlation coefficients show the amount of variation in the dependent variable, which is explained by one independent variable after the other variables in the system of relationships have explained all they could. In this sense, the first-order partial correlation coefficient $r_{YX \cdot T}$ tells us the amount of variation in Y that is explained by T in addition to that which is explained by X alone. Similarly, $r_{YT \cdot X}$ tells us the amount of variation in Y that is explained by X in addition to that which is explained by T alone. Recall $r_{YX \cdot T}$ can be expressed symbolically:

$$r_{YX \cdot T} = \frac{(r_{YX}) - (r_{YT}) \cdot (r_{XT})}{\sqrt{1.00 - r_{YT}^2} \cdot \sqrt{1.00 - r_{XT}^2}}$$

Fortunately, the calculation of path coefficients is similar to the calculation of partial correlation coefficients just shown. Technically, in causal models such as ours which deal with one-way causation, path coefficients are sequentially-ordered partial regression coefficients in *standard form*, that is, they are standardized b coefficients. In other words, path coefficients are what we have been calling "beta weights." Beta weights, or path coefficients, may be interpreted as showing the net change in standard deviation units in the dependent variable that is produced by one standard deviation change in one independent variable controlling for the others. Beta weights can be expressed symbolically in terms of the matrix of zero-order product-moment correlations:

$$\beta_{YX \cdot T} = b_{YX \cdot T} \cdot \frac{s_X}{s_Y} = \frac{r_{YX} - (r_{YT}) \cdot (r_{XT})}{1.00 - r_{XT}^2}$$

and

$$\beta_{XY \cdot T} = b_{XY \cdot T} \cdot \frac{s_Y}{s_X} = \frac{r_{YX} - (r_{YT}) \cdot (r_{XT})}{1.00 - r_{YT}^2}$$

Now all that remains is to translate the beta weights to path coefficients, and that is an easy matter because, as we learned earlier, they are identical.

Thus,

$$\beta_{YX \cdot T} = p_{YX} = \text{the direct path of } X \text{ to } Y \text{ or the direct effect of } X \text{ on } Y \text{ controlling for } T, \text{ and}$$

$$\beta_{YS \cdot UTX} = p_{YS} = \text{the direct path of } S \text{ to } Y \text{ or the direct effect of } S \text{ on } Y \text{ controlling for } U, T, \text{ and } X$$

17.3 DEVELOPING A PATH ANALYSIS CAUSAL MODEL

Since it is a useful statistical technique, let us devote some detailed attention to developing our own path analysis causal model. The substantive theoretical issue with which we are concerned can be phrased as a question that specifies our interest and delineates the variables that comprise our system: How is present occupational status influenced by such factors as family background, educational background, and occupational background?

To investigate this issue, we use the ideas of path analysis and some of the skills we have learned in this book. To put these to use requires two important conditions. First, we must use *quantitative* measurements of these factors, enabling us to use statistical rather than qualitative analysis. Second, we must limit our analysis to a specified number of variables. In this case, let us suppose that we select father's education, father's occupational status, respondent's education, respondent's first job, and respondent's present occupational status. The choice of *five* variables conforms to Blau and Duncan's basic model of stratification; however, we could select more or fewer variables and still use path analysis. Actually, such a limitation is not a handicap because the five variables we have chosen describe a meaningful life pattern in the life cycle of a cohort of respondents. This enables us to develop a model of the determinants of occupational status by examining the relationships among a set of five variables. Furthermore, these variables are amenable to quantitative treatment and are suitable for multiple-linear-regression analysis.

Our study is concerned with the responses of a National Opinion Research Center sample of 10,557 men to questions about the American occupational structure.[4] These questions enable us to analyze the patterns of occupational mobility among American males.

---

[4]The samples from which the present data were gathered were prepared by Current Population Surveys and by the research units of the various branches of the military service in October 1964. Moreover, our universe is an unusually complete one, including the following: Civilian veterans and nonveterans, Army enlisted men and officers, and Marine Corps enlisted men and officers, as of October 1964. Our data are from a sample that includes 3045 veterans, 6548 nonveterans, and 964 military personnel, for a total of 10,557 male respondents between the ages of 16 and 34.

Since many of our ideas follow from those of Blau and Duncan,[5] we use the same five status variables. For convenience and consistency, our designation of these variables conforms identically to those of Blau and Duncan. These variables are:

V: Father's Education was determined by the respondent's answer to the question "What is the highest grade in school completed by your father or male head of the household?"

X: Father's Occupation was based on the respondent's designation as to the kind of work done by his father or the male head of the household at the time the respondent was 15 years old.[6]

U: Respondent's Education was operationalized by the respondent's answer to the question "What is the highest grade of regular school you have completed?"

W: Respondent's First Job was determined by the respondent's description of his first *full-time* civilian job.

Y: Respondent's Occupation was based on the respondent's answer to the question "What kind of work are you doing? What kind of business or industry is this?"

It is at this point that the assumed causal order of the variables becomes a substantive issue. In our theoretical system, we assume ideally that although father's education precedes father's occupation, we do not wish to analyze this relationship; thus, we handle the two variables at the zero-order level. Ideally, we theorize that this relationship precedes respondent's education, which precedes respondent's first job, which precedes respondent's present occupation. Thus, our variables can be ordered in the following linear sequence:

$$(V, X) \longrightarrow (U) \longrightarrow (W) \longrightarrow (Y)$$

This arrangement indicates that we theoretically predict direct positive relationships between sequential variables.

Putting the old and new knowledge together enables us to carry out a path analysis of the process of occupational mobility. Table 17.3.1 presents the zero-order product-moment correlations for the five status variables. In this table, we can examine the *gross* (direct plus indirect) *effects* of our findings.

One important finding stands out in terms of gross effects. Namely, there is a clear ordering of influences on the respondent's present occupational status. This may be seen by comparing the decreasing magnitude of the correlation coefficients (1) between first job and respondent's occupational status, $r_{YW} = 0.571$, (2) between respondent's education and respondent's occupational status, $r_{YU} = 0.524$, (3) between father's occupation and respondent's

---

[5]Blau and Duncan, *The American Occupational Structure*, esp. pp. 165–71.
[6]The data for occupational status were classified according to Blau and Duncan's seventeen major occupational categories.

Table 17.3.1. Matrix of zero-order product-moment correlation coefficients for five status variables

		Variable				
Variable		Y	W	U	X	V
Y: Respondent's Occupation		...	0.571	0.524	0.325	0.240
W: First Job			....	0.450	0.311	0.224
U: Respondent's Education				....	0.376	0.335
X: Father's Occupation					....	0.386
V: Father's Education						....

occupational status, $r_{YX} = 0.325$, and (4) between father's education and respondent's occupational status $r_{YV} = 0.240$. Symbolically, we can express this ordering of influences as

$$r_{YW} > r_{YU} > r_{YX} > r_{YV}$$

Verbally, we can say that respondent's occupational status is most strongly correlated with first job followed by respondent's education followed by father's occupation and then by father's education. The data also show a similar pattern regarding first job and respondent's education; thus,

$$r_{WU} > r_{WX} > r_{WV}, \quad \text{and} \quad r_{UX} > r_{UV}$$

We are beginning to set the stage for a theoretical finding of this path analysis. For the moment, however, suffice it to say that in terms of gross effects, the most recent variables have progressively more influence in determining present occupational status than those temporally most distant.

We are interested in the *direct* effect of one variable on another, and this is where we use the path coefficient. The complexity of the formulas makes hand calculation a tedious task, and virtually all path analyses utilize computer programs for the computations. The values for the desired path coefficients can be summarized in tabular form and placed in a matrix.

Table 17.3.2 shows a matrix of sequentially-ordered partial regression coefficients in standard form, called beta weights or path coefficients, above the diagonal. The amount of change between the zero-order correlations and the partial beta weights is shown below the diagonal. All of the *values of change* (differences between zero-order and partial) are statistically significant at the 0.05 level. The partial-regression beta weights appearing above the diagonal become the path coefficients that we use in a path diagram, and they show the direct effect of one variable on another, while the differences show the indirect effect between the same two variables.

Table 17.3.2.   Matrix of sequentially ordered beta weights (standardized partial regression coefficients) [above diagonal], the total indirect effects (difference between zero-order correlations and the beta weights) [below diagonal], and the coefficients of determination (R^2)

		Variable					
Variable		Y	W	U	X	V	R^2
Y:	Respondent's Occupation	....	0.405	0.307	0.077	0.017	0.421
W:	First Job	0.166	....	0.380	0.154	0.037	0.228
U:	Respondent's Education	0.217	0.070	....	0.290	0.223	0.184
X:	Father's Occupation	0.248	0.157	0.086	....	0.386	....
V:	Father's Education	0.223	0.187	0.112	0.000	....	....

17.4 CONSTRUCTING A PATH ANALYSIS CAUSAL DIAGRAM

Path causal diagrams provide a graphic representation of the relationships among a system of linear, additive, and recursive causally related variables. There are several conventions commonly used when drawing path diagrams.[7]

1. We represent the *assumed causal relations* or *paths* between variables within our system by *unidirectional* (one head) straight arrows that connect each independent variable to each variable dependent on it.
2. *Unanalyzed correlations*, that is, those between two independent variables within the system for which no attempt is made to analyze the causal relation, are represented by *bidirectional* (two heads) curvilinear arrows to distinguish them from the straight *causal* arrows.
3. The *paths from residual variables* are represented by unidirectional straight arrows from outside the system headed toward the system directly to the dependent variable in question.
4. The *path coefficients* of the assumed causal relationships are the values that appear along the arrows in the path diagram.
5. The labeled arrows of the causal diagram show the highest order partial coefficients at *that* point in the model; that is, the diagram displays the paths in their assumed causal sequence.

Since the relationship between father's education V and father's occupation X is not part of the problem at hand, we use convention number 2 and the path between V and X is depicted by a curved bidirectional arrow. This appears to contradict the original statement about the order of relationships; however, we treat the relationship between these two variables as not being part of our analysis as described in convention 2. Put differently, the regression slope of X on V or vice versa is not controlled for any of the other vari-

---

[7]These conventions are adapted from Land, "Principles of Path Analysis," pp. 6–7; and Duncan, "Path Analysis: Sociological Examples" p. 3.

ables in our system; thus, the zero-order correlation $r_{VX} = 0.386$ is used. It may be expressed diagrammatically as in Figure 17.4.1.

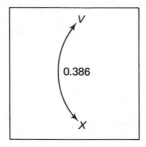

Figure 17.4.1.

The next variable in our causal system is respondent's education U. Our first interest is in the direct effect of father's education on respondent's education controlling for father's occupation, called the path to U from V. Thus, we calculate the standardized partial-linear-regression beta weight $\beta_{UV \cdot X}$. This path coefficient may be expressed symbolically as:

$$\beta_{UV \cdot X} = p_{UV} = 0.223$$

Now we examine the relationship between father's occupation and respondent's education. In this case, we are interested in the direct effect of father's occupation on respondent's education controlling for father's education, called the path to U from X. This means that we calculate the beta weight for the dependent variable U from the independent variable X controlling for the test variable V. It may be expressed symbolically:

$$\beta_{UX \cdot V} = p_{UX} = 0.290$$

We can enter these two paths as arrows in our diagram as shown in Figure 17.4.2.

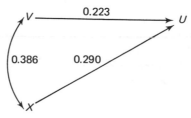

Figure 17.4.2.

The next sequential variable in our causal model is the measure of occupational background which is operationalized according to the respondent's first job W. Our first interest is in the most distant variable, father's education, and we are concerned with the path from V to W. Put verbally, we are interested in the direct influence of father's education on one's first job, controlling for father's occupation and respondent's education as indicated by the second-order partial beta weight $\beta_{WV \cdot XU}$.

It may be expressed symbolically:

$$\beta_{WV \cdot XU} = p_{WV} = 0.037$$

When we control for father's occupation and respondent's education, we find that the direct effect path from father's education to respondent's first job almost "disappears," because the beta weight $\beta_{WV \cdot XU} = 0.037$. This reflects that almost all of the influence of father's education on respondent's first job is indirect.[8]

Next we are interested in the direct influence of father's occupation on respondent's first job, controlling for father's education and respondent's education, that is, the path to W from X. Since $\beta_{WX \cdot VU} = 0.154$, then $p_{WX} = 0.154$.

Now our interest is in how much direct influence respondent's education has on respondent's first job. This is determined by the beta weight controlling for father's education and father's occupation. Since $\beta_{WU \cdot VX} = 0.380$, then $p_{WU} = 0.380$. We can enter these three paths in our diagram as shown in Figure 17.4.3.

The basic procedures are the same as those described above, no matter how many variables we use. The only difference is the order of the partial coefficients used in the formula. The third-order beta weight for Y given V,

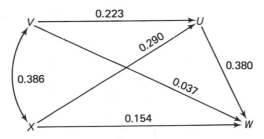

Figure 17.4.3.

---

[8]A convention in certain path analyses is not to enter paths less than 0.05. Thus, since $\beta_{WV \cdot XU} < 0.05$, we could treat the path as negligible, i.e., $p_{WV} = 0.00$. However, for our purposes, it seems clearer to enter all direct paths and their value.

controlling for X, U, and W, is the path from father's education to respondent's occupation. It may be expressed symbolically:

$$\beta_{YV \cdot XUW} = p_{YV} = 0.017$$

Since $\beta_{YX \cdot VUW} = 0.077$, $p_{YX} = 0.077$; since $\beta_{YU \cdot VXW} = 0.307$, $p_{YU} = 0.307$; and since $\beta_{YW \cdot VXU} = 0.405$, $p_{YW} = 0.405$. We can enter these paths in our causal diagram as shown in Figure 17.4.4.

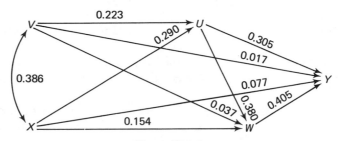

Figure 17.4.4.

Up until this point, all the values entered in the path diagram are taken directly from Table 17.3.2, which shows the sequentially-ordered partial beta weights just calculated. Now, we must calculate the residual paths to each of the variables that were considered to be "dependent" as we progressed through the causal sequence, that is, respondent's education, respondent's first job, and respondent's present occupational status, in that order.

To calculate these residual paths we use the three multiple correlation coefficients R^2, which also are shown in Table 17.3.4. Remember, R^2 measures the proportion of variation in the dependent variable that is *explained* by the multiple linear relationships with the independent variables in a set of variables. However, now we are interested in the proportion of the variation that is *not explained* by the multiple linear relationships. This difference may be shown by subtracting R^2 from 1.00. For example, since $R^2_{U \cdot VX}$ is the proportion of variation in U which is explained by its linear relationship with V and X, then the proportion of unexplained variation in U is $(1.00 - R^2_{U \cdot VX})$. We take the square root of this difference to calculate the residual path to variable U, with the question mark representing the unknown or residual variables. We can do this for all three variables:

Since $\quad R^2_{U \cdot VX} = 0.184$, then $p_{U?} = \sqrt{1.00 - 0.184} = \sqrt{0.816} = 0.903$

Since $\quad R^2_{W \cdot VXU} = 0.228$, then $p_{W?} = \sqrt{1.00 - 0.228} = \sqrt{0.772} = 0.879$

Since $\quad R^2_{Y \cdot VXUW} = 0.421$, then $p_{Y?} = \sqrt{1.00 - 0.421} = \sqrt{0.579} = 0.761$

These residual paths from variables outside our system to variables U, W, and Y may be entered in our diagram as shown in Figure 17.4.5.

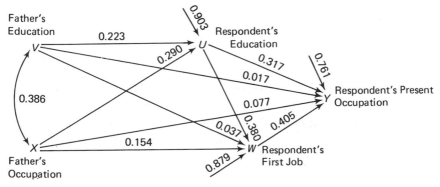

Figure 17.4.5.

17.5 INTERPRETING PATH ANALYSIS CAUSAL MODELS

There are many approaches to interpreting path diagrams and causal models, the most meaningful offering theoretical and substantive statements rather than merely descriptive or methodological ones. Let us start with a discussion of the causal effects.

Our path analysis enables us to compare the direct versus the indirect effects of independent variables on dependent variables. In order to determine the indirect effect of one variable on another, we can use the difference between the zero-order correlation coefficient and the path coefficient for the two variables. The reason we are able to use these two values is that the correlation coefficient r_{YX} can be interpreted in terms of changes in standard deviations in X, producing specific changes in standard deviations in Y in terms of gross effects. The path coefficient p_{YX} means essentially the same thing in terms of net effects.

For example, the zero-order correlation coefficient between respondent's present occupational status and respondent's education is $r_{YT} = 0.524$. This means that a change in one standard deviation in U produces a change of 0.524 of a standard deviation in Y in gross effects. The path coefficient between the same two variables is $p_{YU} = 0.307$. This tells us that nearly 60 percent, i.e., $(0.307/0.524) = 0.589$, of the gross effect results from the *direct* influence of U on Y. Actually, some or all of the "direct" effect may be mediated by another variable *not* included in the analysis.

In general, we may calculate the indirect effects by taking the difference between the zero-order correlation coefficient and the path coefficient that connects the two variables. In the example just seen, the indirect effect of education on occupation operates through the variable first job. In other words, the remainder of the effect of U on Y,

$$(r_{YU} - p_{YU}) = (0.524 - 0.307) = 0.217,$$

is indirect via W. The temporally most recent dependent variables have common antecedent causes; thus, the total indirect effect denotes the common causes of a two-variable relationship, which spuriously inflate the zero-order correlation between them. The total indirect effects are shown in Table 17.3.2 below the diagonal.

Notice that even respondent's first job W, the temporally closest variable in our causal system to Y, has some indirect effects, that is, common antecedent causes, that are between one-third and one-half as large as the direct effects. For example, $r_{YW} = 0.571$ and $p_{YW} = 0.405$; thus, the sum total of indirect effects is 0.166. In this instance the indirect effects are common determinants of Y and W that "spuriously inflate the correlation between them."[9]

We can interpret our path analysis causal model in terms of either variable abbreviations or words. Let us put it both ways. First, in abbreviated variables, we can say that in our model Y appears as being influenced directly by W, U, X, and V. The very small influence of V on Y does not suggest that V has no influence on Y. Clearly, V affects U, which affects Y directly as well as indirectly via W. V also is correlated with X, thereby sharing in the gross influence of X on Y. Furthermore, the influence of X on Y is partly direct and partly indirect. Thus, the zero-order correlation r_{YV}, which we said indicates the gross effect of V on Y, may be interpreted as being largely indirect and resulting primarily because of V's effect on intervening variables and its correlation with other causes of Y.

Since this abbreviated statement does not provide us with useful connections to the theory or the substance of the problem, let us put it into words. In our causal model, present occupational status is influenced directly by respondent's first job, respondent's education, father's occupation, and father's education. The very small influence of father's education on respondent's present occupational status does not suggest that father's education has no influence. Clearly, father's education affects respondent's education, which affects respondent's occupational status directly as well as indirectly by way of respondent's first job. Father's education is also correlated with father's occupation, thereby sharing in the gross influence of father's occupation on the respondent's present occupational status. Furthermore, the influence of father's occupation on present occupational status is partly direct and partly indirect. Thus, our path analysis suggests that most of the influence of

---

[9]Blau and Duncan, *The American Occupational Structure*, p. 176.

father's education on both respondent's first job and respondent's occupational status is largely indirect.

When we examine all the paths in our model working backwards, we find that occupational status is influenced more strongly by first job than by education. Similarly, first job is more strongly influenced by education than by father's occupation. Finally, respondent's education is more strongly influenced by father's occupation than by father's education. The data continue to support the early observation that the most recent variables are more influential than the most distant ones in determining occupational status. All of our findings consistently support this "recency hypothesis." This could be interpreted to mean that attention must be paid to the theoretical temporal ordering of such variables, because it appears that the most recent variables are the most influential.

One of the major benefits of a *replication study* such as the one we have just completed is that it enables direct comparison with an existing causal model. Let us compare and test the differences between our findings and those of Blau and Duncan.

Figure 17.5.1 is a graphic representation of the system of relationships among the five status variables that are proposed in the basic model of Blau

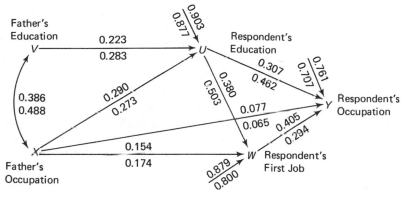

Figure 17.5.1. Path coefficients in basic model of the process of stratification, All-American sample males ages 20–34 (top values), and Blau and Duncan OCG sample, males ages 25–34 (bottom values).

and Duncan.[10] The values in the diagram above the arrows are the path coefficients we presented in Table 17.3.2 for the NORC All-American Sample. The values below the arrows are those of Blau and Duncan.[11] Ignoring

[10]*Ibid.*, p. 170.
[11]*Ibid.*, p. 178.

the value differences along each path, our model allows us to quote Blau and Duncan exactly:

> Y is shown here as being influenced directly by W, U, and X, but not by V ...
> But this does not imply that V has no influence on Y. V affects U, which does
> affect Y both directly and indirectly (via W). Moreover, V is correlated with
> X, and thus shares in the gross effect of X on Y, which is partly direct and
> partly indirect. Hence the gross effect of V on Y, previously described in
> terms of the correlation r_{YV}, is here interpreted as being entirely indirect, in
> consequence of V's effect on intervening variables and its correlation with
> another cause of Y.[12]

This means that path analyses of both samples produce causal schemes that suggest that all of the influence of father's education both on one's first job and on one's subsequent occupation is indirect; hence, no arrows connect these variables. This has been interpreted as indicating considerable inter-generational mobility. As stated above, if one ignores the values along the paths, the basic pattern of connecting arrows is identical with that of Blau and Duncan, indicating that the net effects are quite similar. In other words, although our coefficients are somewhat different from theirs, our causal scheme shows the same pattern.

However, when we compare the values of our path coefficients with theirs, there are some statistically significant differences at the 0.05 level. To begin with, working backwards, two notable differences are the paths between respondent's first job and respondent's occupation and between respondent's education and respondent's occupation. Blau and Duncan's diagram indicates that respondent's education is a more influential determinant of one's occupation.[13] They explain that "occupational status ... (Y) apparently is influenced more strongly by education than first job."[14] Our data suggest the opposite. The All-American Sample continues to support our earlier observation that the most recent status variables are more influential than the most distant ones in determining respondent's occupation.

Another interesting difference between the causal diagram of Blau and Duncan and ours is that in the All-American Sample father's occupation controlling for father's education is more influential in determining respondent's education than is father's education controlling for father's occupation.[15] Along with most of our findings, this supports our "recency" hypothesis, and each stage of our path analysis lends further credence to this notion.

[12] *Ibid.*, p. 171.
[13] Their $p_{YU} = 0.462$ while ours is 0.155 points lower, $p_{YU} = 0.307$, and their $p_{YW} = 0.294$ while ours is 0.143 points higher, $p_{YW} = 0.405$.
[14] Blau and Duncan, *The American Occupational Structure*, p. 170.
[15] *Ibid.*, p. 180.

From all this, we can summarize briefly by saying that our findings generally support the findings of Blau and Duncan concerning their basic model.
In addition, our findings indicate that the theoretical temporal ordering of the
status variables is important. Moreover, we hypothesize that the most recent
variables are the most influential in determining one's occupational status.

As a result of the large size of the sample, almost all the findings are "statistically significant." In addition to testing the significance of the differences
between path coefficients, it is possible to compute confidence intervals for
each coefficient. In this instance, the large sample helps create relatively small
intervals. This might suggest a greater degree of "confidence" than is warranted, given the possibility of sampling error and chance fluctuations. With
smaller samples, however, it is beneficial to compute such intervals.

EXERCISES

1. In your own words define, describe, or discuss the following terms and give
   an hypothetical example of each:
   indirect effects
   residual variable
   path coefficients
   beta weight
   unanalyzed correlation
   completely recursive system
   over-identified recursive system
   assumed causal direction
   direct influence

2. Consider the following matrix of correlations of characteristics of males
   age 25–64 in the U.S. and Great Britain. Correlations for the U.S. are given
   above the diagonal, and for Britain below the diagonal; means and standard
   deviations are given in the two right columns for the U.S. and the two bottom
   rows for Britain.

Variable	O_F	S	O	I	Mean	S.D.
O_F Prestige of father's occ.	—	0.300	0.266	0.150	41.1	11.0
S School leaving age	0.251	—	0.564	0.310	17.8	3.0
O Prestige of respondent's occ.	0.364	0.455	—	0.340	42.1	11.7
I Annual income ($)	0.298	0.448	0.595	—	7080	5670
Mean	35.5	14.5	38.9	2507		
Standard deviation	10.5	1.4	11.6	1234		

Source: Adapted from: Donald J. Treiman, *Occupational Prestige in Comparative
Perspective*. New York: Academic Press, 1975; and Donald J. Treiman and Kermit
Terrell, "Sex and the Process of Status Attainment: A Comparison of Working
Women and Men," in *American Sociological Review*, 40 No. 2, April, 1975, pages
174–200.

A. Calculate the values for the following model, for the U.S. and for Britain, utilizing path coefficients.

B. Write a brief paragraph describing and comparing the process of status attainment in the two countries implied by the models.

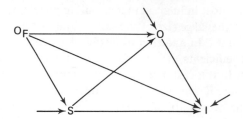

3. The following causal argument is presented by Duncan, Featherman, and Duncan in their study of intergenerational mobility:

(1) Educational attainment depends on three characteristics of the family of orientation, the respondent's number of siblings and the occupational level and educational attainment of his father (actually, the head of the family, in the event of the father's absence).

(2) Occupational status in 1962 depends on educational attainment and the foregoing three family background items.

(3) Income depends on occupational status, educational attainment, and the three characteristics of the family of orientation.[16]

A. Construct a causal model which corresponds to the above argument.

B. Discuss the model in light of their causal argument and general sociological theory.

4. Using the data presented below and the model developed in problem 3

A. Calculate the path coefficients, including residual effects.

B. What conclusions can you reach from the model?

C. Discuss the model in light of the findings in A and B.

Simple correlations between variables entering into the basic model, for non-negro men with nonfarm background, in experienced civilian labor force, by age group 35–44: March 1962

Age group and variable 35–44	Correlation with					Mean	Standard deviation
	X	T	U	Y	H		
V^*	0.5300	−0.2871	0.4048	0.3194	0.2332	8.55	3.72
X	—	−0.2476	0.4341	0.3899	0.2587	34.41	23.14
T	—	—	−0.3311	−0.2751	−0.1752	3.77	2.88
U	—	—	—	0.6426	0.3759	11.95	3.20
Y	—	—	—	—	0.4418	44.78	24.71
H	—	—	—	—	—	7.50	5.36

[16]Otis Dudley Duncan, David L. Featherman, and Beverly Duncan, *Socioeconomic Background and Achievement*. New York: Seminar Press, 1972, page 39.

Partial regression coefficients in standard form for recursive model
relating achieved statuses to family background factors, by age,
for non-negro men with nonfarm background, in experienced
civilian labor force: March 1962 (parentheses enclose each
coefficient less than its standard error in absolute value)

Age and dependent variable 35–44	Independent variables*					Coefficient of determination
	Y	U	T	X	V	
U	—	—	−.2053	.2780	.1985	.269
Y	—	.5668	−.0540	.1266	(.0073)	.431
H	.3247	.1193	−.0201	.0492	.0494	.216

*V = Father's education
X = Father's occupation
T = Number of siblings
U = Education
Y = Occupation
H = Income

Source: Adapted from Otis Dudley Duncan, David L. Featherman, and Beverly Duncan, *Socioeconomic Background and Achievement*. New York: Seminar Press, 1972, pages 38 and 40.

5. The following matrix shows the correlations among four variables taken from the Cole and Cole studies of the award system in American physics, from a sample of 120 American physicists, stratified by eminence.

Variable	A	P	D	Q
A Number of awards received	1.00	0.45	0.50	0.68
P Number of papers published		1.00	0.24	0.72
D Quality of department			1.00	0.33
Q Quality of work (measured by citation count)				1.00
Mean	1.61	34.40	3.62	202
Standard deviation	1.50	30.20	0.96	304

Source: Adapted from Stephen Cole and Jonathan R. Cole, "Visibility and the Structural Bases of Awareness of Scientific Research," in *American Sociological Review*, 33, No. 3, June 1968, pages 397–413.

One causal argument might be that (1) number of awards received depends on the number of papers published, and the quality of the work, and that (2) the quality of the department depends on the number of awards received, the number of papers published, and the quality of work.

Another argument might be that the quality of the department depends on (1) the number of papers published and the quality of the work, and that (2) the number of awards received depends on the quality of the department, the number of papers published, and the quality of work.

    A. Construct two different path models, corresponding to the two above arguments.

    B. Calculate the path coefficients for both of your models including residual effects.

    C. What conclusions do you draw from your results?

6. The following causal argument can be used in studying inter-generational mobility. Education depends on father's occupation and education. Occupation depends on education, father's occupation and father's education.

    A. Construct a path model based on the above argument.

    B. The data in the following table are from three different studies used by Duncan, Featherman, and Duncan in their study of intergenerational mobility. Using your model, compute path coefficients for all three sets of data.

    C. Compare your results for the three sets of data.

Simple correlations in OCG and DAS data sets, with alternative DAS results for different measures of occupational status.

Data set	Variable	Father's occupation $(X$ or $X')$	Respondent's education (U)	Respondent's current occupation $(Y$ or $Y')$	Mean	SD
OCG, non-Negro nonfarm background, age 25–64	Father's education (V)	0.506	0.393	0.306	8.63	3.67
	Father's occupation (X)	—	0.419	0.371	34.07	22.72
	Respondent's education (U)	—	—	0.610	11.70	3.30
	Respondent's occupation (Y)	—	—	—	43.47	24.58
DAS–1 occupations scored on socioeconomic index	V	0.481	0.322	0.271	8.80	3.20
	X	—	0.338	0.306	33.90	23.76
	U	—	—	0.599	12.00	3.20
	Y	—	—	—	45.84	24.03
DAS–2 occupations scored on prestige scale	V	0.424	0.322	0.256	8.80	3.20
	X'	—	0.240	0.211	39.22	12.78
	U	—	—	0.567	12.00	3.20
	Y'	—	—	—	42.74	13.44

Source: Otis Dudley Duncan, David L. Featherman, and Beverly Duncan, *Socioecomomic Background and Achievement.* New York: Seminar Press, 1972, page 46.

18

Survey Analysis

One of the common methodological techniques in sociology is the use of sample surveys, census materials, and cross-cultural data. Questionnaires or interview schedules are basic data-gathering methods, and it is important to have a clear understanding of how statistical survey data can be analyzed. This chapter provides one approach to survey analysis. We will discuss how to calculate, use, and interpret survey analysis coefficients for ordinal data.

Survey analysis is different in some important aspects from path analysis. Survey procedures include examination of joint effects, the use of differential coefficients, and ordinal measures of association. To assess the degree of association between dichotomous ordinal variables, we use Yule's Q. Earlier we learned that Q is a coefficient that has a meaningful upper and lower limit and that has an intrinsic meaning that is easy to understand. As you may remember, Yule's Q is the special dichotomous case of γ (gamma), a measure of association that tells us the degree to which ordinal variables are related. In order to carry out multivariate analysis, it is necessary to look at the two-variable associations controlling for one or more test variables.

18.1 SOME PRINCIPLES UNDERLYING SURVEY ANALYSIS

Partial coefficients for Yule's Q are analogous to partial correlation and regression coefficients; however, the calculations are different. Partial regression correlations and b coefficients are derived by controlling variables on the basis of the *zero-order* coefficients. On the other hand, partial coefficients for Q are derived by partitioning the *respondents* according to the possession of a number of variable attributes. For our purposes, the variables will be controlled in a temporally ordered sequential fashion so that the relationship between any two variables is determined by controlling only those test variables that are antecedent (prior) to the most recent one. In no case do we look at the relationship between two variables controlling for one that is consequent (following).

By putting together the partial and differential Q's and their appropriate weights, we can show that

$$Q_{XY} = [(Q_{XY \cdot PART\ T})(PW)] + [(Q_{XY \cdot DIFF\ T})(DW)]$$

From this formula for three dichotomous variables, we can say that Q_{XY} is a weighted average of the partial and differential Q values in which the weights for each are the proportions of cross products tied on T and differing on T.[1] The first-order partial Q values can be interpreted in relation to the zero-order Q value in the same way we interpret partial correlation coefficients.

The following four statements summarize the approach:[2]

A. With three variables analyzed simultaneously no neat classification of all possible outcomes exists. However, causal models serve to clarify a number of interesting possibilities.

B. Causal models boil down to three rules:
   1. Predicting partials from the signs in the models,
   2. Predicting $D - P$ from doglegs [the legs which connect two variables via a third variable] in the models, and
   3. Looking at one zero-order and two partials when one variable is assumed to be consequent.

C. Even in three-variable systems there are many different causal models. They may be classified by the number of causal assertions, 0, 1, 2, 3, ... From the viewpoint of sociological theory, three-assertion models containing a cycle of arrowheads are especially important because they provide operational definitions of "historical" and "functional" theories.

---

[1] James A. Davis, *Elementary Survey Analysis* (Englewood Cliffs, New Jersey: Prentice-Hall, Inc., 1971) uses $Q_{XY:\ TIED\ T}$ to express the partial coefficient.
[2] *Ibid.*, pp. 128–29.

D. It is easy to determine whether the data fit a particular model, but sometimes very difficult to tell which model the data fit best.

There are many possible applications and extensions of these ideas and
formulas for higher-order Q coefficients. One important step is to use these
analytical tools to develop *causal models*. Causal models for survey analysis
are hypotheses about the presence, sign, and direction of influence for the
relationships among all the pairs of variables in a causal system. The procedures, rules, conventions, and applications of four-variable causal models
may be summarized as follows:[3]

A. Analysis with two test variables involves elaboration of "the equation"
   into four terms whose weighted average equals Q_{XY}.
   1. The second-order partial $Q_{XY: PART\ S,\ T}$
   2. The partial differential for T or $Q_{XY: PART\ S,\ DIFF\ T}$
   3. The partial differential for S or $Q_{XY: DIFF\ S,\ PART\ T}$
   4. The second-order differential or $Q_{XY: DIFF\ S,\ T}$

B. These coefficients may be used to test causal models by adopting the following conventions:
   1. The second-order partial should have the sign of the causal assertion
      for XY unless there is a consequent variable.
   2. The partial differential effects should have the signs of the relevant
      dogleg products unless the fourth variable is consequent.

C. Applying these concepts to the problem of explanation, we conclude that:
   1. Two test variables may or may not provide a better explanation of XY
      than either taken alone in three-variable tabulations.
   2. Successful four-variable explanations usually have relatively small
      joint effects, or joint effects opposite in sign to XY.

It is possible to extend these ideas and to use them in the construction of
causal diagrams.

18.2 CALCULATING MULTIVARIATE
SURVEY ANALYSIS COEFFICIENTS

Let us examine survey analysis and the way it can be used in the construction of a causal diagram. In this case we will use an unusually complete
sample which includes all possible categories of American males. Our data
are from the same All-American Sample of veterans, nonveterans, and
military personnel that we used in our discussion of path analysis. There are
3773 male respondents between the ages of twenty and thirty-four who

[3] *Ibid.*, pp. 164–65.

answered all the questions for the five status variables.[4] We are concerned
with the responses of the 3773 men to questions about the American occupational structure. Since our analysis follows from that of Blau and Duncan we
use the same five status variables, and our designation of these variables
conforms identically to theirs.[5] These variables are:

  V: *Father's Education* was determined by the respondent's answer to the
     question "What is the highest grade in school completed by your father
     or male head of the household?" It is dichotomized according to
     whether or not the father had graduated from high school, — = 2748,
     + = 1025.

  X: *Father's Occupation* was based on the respondent's answer about the
     kind of work done by his father or the male head of the household at
     the time the respondent was 15 years old. It is dichotomized into
     — = 2900, those whose fathers were blue-collar workers or farmers,
     and + = 873, those whose fathers were white-collar workers,
     managers, proprietors, officials, and professionals.

  U: *Respondent's Education* was based on the response to the question
     "What is the highest grade of regular school you have completed?"
     It is dichotomized into those who had graduated from high school or
     less, — = 2453, and those who had had some college or more,
     + = 1320.

  W: *Respondent's First Job* was determined by the respondent's description
     of his first *full-time* civilian job. It is dichotomized into — = 2515,
     those whose first job was as a blue-collar worker or farmer, and
     + = 1,258, those whose first job was as a white-collar worker, man-
     ager, proprietor, official, or professional.

  Y: *Respondent's Occupation* was based on the respondent's answer
     to the question "What kind of work are you doing? What kind of
     business or industry is this?" It is dichotomized into those whose work
     was blue-collar or farm, — = 2304, and those whose work was white-
     collar, managerial, proprietorial, official, or professional, + = 1469.

Like Blau and Duncan, we assume that the causal sequencing of the five
status variables over time follows a temporal ordering and that although
father's education is presumed to have preceded father's occupation, their
relationship will remain unanalyzed; that respondent's education followed
father's occupation and father's education; and that respondent's education

---

[4]Briefly summarized, the all-American sample consists of 10,557 males from the following groups: Civilian veterans and nonveterans, Army enlisted men and officers, Navy
enlisted men and officers, Air Force enlisted men and officers, and Marine Corps enlisted
men and officers. The samples from which the present data were gathered were prepared by
Current Population Surveys and by the research units of the various branches of the military service in October, 1964.

[5]Peter M. Blau and Otis Dudley Duncan, *The American Occupational Structure* (New
York: John Wiley & Sons, Inc., 1967), pp. 165-71.

was followed by respondent's first job, which was followed in turn by respondent's occupation.

In order to carry out survey analysis, we look at the two-variable associations and then control by cross-tabulation one or more test variables. These are introduced to *test* the properties of the original Q_{XY} by recalculating the Q coefficient among subgroups of cases that are similar and dissimilar in their category of the test variable. These Q values may then be compared with the zero-order coefficient to assess the amount of change or explanatory quality of the test variable(s). The variables are controlled in a temporally ordered sequential fashion so that the relationship between any two variables is determined by controlling only those test variables that are antecedent to the most recent one. In no case do we look at the relationship between two variables controlling for a consequent one. Thus we operate essentially as we did in path analysis, and, beyond the zero-order relationship between two variables, we are interested only in the relationship as it is influenced by all the antecedent variables in our model. Put differently, we are concerned with the *net* effects between variables.

Table 18.2.1 presents the raw data for the subsequent calculations for the cross-tabulated frequencies of the five status variables. Table 18.2.2 presents

Table 18.2.1. Raw data for the five status variables cross-tabulated in temporal order

Father's Education (V)	Father's Occupation (X)	Respondent's Education (U)	Respondent's First Job (W)	Respondent's Occupation (Y) −	+
+	+	+	+	20	283
+	+	+	−	33	42
+	+	−	+	14	22
+	+	−	−	56	18
+	−	+	+	11	124
+	−	+	−	79	33
+	−	−	+	21	38
+	−	−	−	194	37
−	+	+	+	12	128
−	+	+	−	22	33
−	+	−	+	16	34
−	+	−	−	104	36
−	−	+	+	38	232
−	−	+	−	136	94
−	−	−	+	136	129
−	−	−	−	1412	186

$N = 3773$

Table 18.2.2. Matrix of zero-order Q coefficients for five status variables,
All-American sample, males ages 20–34 (N = 3773)

Variable	Variable				
	Y	W	U	X	V
Y: Respondent's Occupation	...	0.880	0.830	0.666	0.500
W: Respondent's First Job		...	0.799	0.642	0.503
U: Respondent's Education			...	0.693	0.644
X: Father's Occupation				...	0.696
V: Father's Education					...

the zero-order Q coefficients for all pairs of the five status variables used in the analysis.

The pattern of associations between the variables reveals that at the 95 percent confidence level, there are very strong positive associations between respondent's first job and respondent's occupation ($Q_{WY} = 0.880$ with confidence limits of 0.830 and 0.930), and between respondent's education and respondent's occupation ($Q_{UY} = 0.830$ with confidence limits of 0.793 and 0.867), a substantial positive association between father's occupation and respondent's occupation ($Q_{XY} = 0.666$ with confidence limits of 0.621 and 0.711), and a moderate positive association between father's education and respondent's occupation ($Q_{VY} = 0.500$ with confidence limits of 0.445 and 0.555).

In terms of gross effects, the most influential determinants of one's occupation are those variables that are temporally closest in the causal scheme. That is, as one moves from right to left across the columns in Table 18.2.2, there is a direct relationship between the recency of variable influence and higher positive correlations.

The fact that there is a greater degree of association between respondent's education and respondent's occupation than there is between father's education and father's occupation ($Q_{VX} = 0.696$ with confidence limits of 0.646 and 0.746) confirms a similar finding by Blau and Duncan[6]. We can conclude that in terms of "uncontrolled" effects, our sample indicates that one's father's occupation is a more important determinant of one's subsequent status characteristics than is one's father's education. This can be seen by comparing variables X and V in Table 18.2.2.

Table 18.2.3 presents a matrix of sequentially-ordered partial Q coefficients (above the diagonal), and the difference (amount of change) between the zero-

[6]*Ibid*,. p. 169.

Table 18.2.3. Matrix of sequentially ordered partial Q coefficients*
(below diagonal) and the difference between zero-orders
and partials (above diagonal)

Variable		Variable				
		Y	W	U	X	V
Y:	Respondent's Occupation	....	−0.106	−0.166	−0.275	−0.379
W:	Respondent's First Job	0.774	....	−0.089	−0.242	−0.358
U:	Respondent's Education	0.664	0.710	....	−0.087	−0.138
X:	Father's Occupation	0.391	0.400	0.606	....	0.000
V:	Father's Education	0.121	0.145	0.506	0.696	....
Partial Coefficient Order		3	2	1	0	

*The formulas for partial Q's depend upon the number of control variables. In general, the partial amounts to dividing the difference between consistent pairs and inconsistent pairs tied on the test variable(s) by the sum of the consistent and inconsistent pairs tied on the test variable(s).

order and partial coefficients (below the diagonal). All the differences are statistically significant at the 0.05 level of significance. The order may be seen at the bottom of the table, moving from left to right, 3, 2, 1, and 0. The scores in column Y are third-order partials, those below the diagonal in column W are second-order partials, those below the diagonal in column U are first-order partials, and the one below the diagonal in column X is the zero-order coefficient from the original matrix.

The differences between the zero-orders and the partials may be read in the top right-hand half of the matrix (above the diagonal). For example, reading up column V, we see that there is no change in the association between variables X and V, a reduction of 0.138 Q points between variables U and V, a reduction of 0.358 Q points between variables W and V, and a reduction of 0.379 Q points between variables Y and V. Similar interpretations may be made for columns X, U, and W.

18.3 CONSTRUCTING A SURVEY ANALYSIS CAUSAL DIAGRAM

Instead of presenting a completed causal diagram, we will develop one step by step. To construct a temporally-ordered causal diagram, we must introduce test variables in the order of their occurrence. In our example, we begin with the zero-order association between father's education and father's occupation. Inasmuch as no variables in the system could have "worked backwards" to influence this relationship, we do not "control" it in subsequent analysis. This zero-order association will appear in the causal diagram as a bidirectional arrow indicating that there is a positive unanalyzed relationship between the variables father's occupation and father's education. This relationship is shown in Figure 18.3.1.

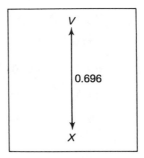

Figure 18.3.1.

The first-order partial ($Q_{VU \cdot PART\ X} = 0.506$) association between father's education and respondent's education controlling for father's occupation is still a substantial but somewhat reduced positive association, and the arrow in the causal model between V and U is designated by this partial Q. The first-order partial between father's occupation and respondent's education controlling for father's education is reduced, but it too is still substantial ($Q_{XU \cdot PART\ V} = 0.606$). Put another way, the associations between father's education and respondent's education and between father's occupation and respondent's education are somewhat accounted for by each other, but there are still substantial associations between them. Thus the causal diagram expands to show three variables, as shown in Figure 18.3.2.

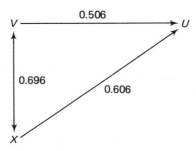

Figure 18.3.2.

The next variable in the model is respondent's first job. It is important to note that the question about one's first job asked what work was done "*after* you stopped going to school." Since our model locates first job as occurring after respondent's education, the arrows leading to first job in our causal diagram will be second-order partial Q coefficients. These values result from calculations using a sixteenfold table showing the relationship between respondent's first job and one of the other variables controlling for *both* of the other variables.

The association between father's education and respondent's first job controlling for father's occupation and respondent's education is considerably reduced ($Q_{VW \cdot PART\ X, U} = 0.145$, reduced from $Q_{VW} = 0.503$). We conclude that the relationship between father's education and respondent's first job cannot be completely accounted for by father's occupation and son's education, but that they do go a long way toward explaining the original moderate-positive-zero-order association of $Q_{VW} = 0.503$. Our data indicate that respondent's first job is to some extent influenced by the level of father's education when measured through survey analysis.

The association between father's occupation and respondent's first job controlling for father's education and respondent's education is also markedly reduced, but not explained ($Q_{XW \cdot PART\ V, U} = 0.400$, reduced from $Q_{XW} = 0.642$). Apparently, there is less intergenerational influence between the father's occupation and the son's first job than between the father's education and the son's first job, for the former second-order partial association is 2.5 times larger than the latter. This may be interpreted as indicating that the sort of career beginnings a man has will depend to a proportionately greater extent upon his father's occupation during his childhood than it will upon his father's level of education.

The second-order partial coefficient between respondent's education and first job is only slightly reduced and still a substantial positive association ($Q_{UW \cdot PART\ V, X} = 0.710$ reduced from $Q_{UW} = 0.799$). This means that the level of respondent's education is the single most important determinant of first job within our system of status variables.

By entering these three second-order partial coefficients in our model, we continue its expansion in the causal diagram, as shown in Figure 18.3.3 on page 386.

The final variable in the model is respondent's occupation. Figure 18.3.4 shows that all four independent status variables are associated with respondent's occupation controlling for all the others. That is, the third-order partial Q coefficients are all nonnegligible, although all are to some extent reduced. Our five-variable causal model of the process of stratification shows the zero-, first-, second-, and third-order Q coefficients between the variables in temporally-ordered causal sequence as shown in Figure 18.3.4 on page 386.

The association between father's education and respondent's occupation

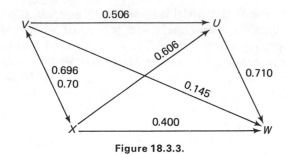

Figure 18.3.3.

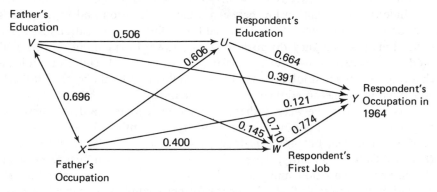

Figure 18.3.4. Five-variable causal model of the process of stratification: variables in temporally arranged causal sequence, zero-, first-, second-, and third-order.

has been reduced 0.379 Q points to a low positive relationship ($Q_{VY \cdot PART \, X, U, W}$ = 0.121). This may be interpreted as indicating that even when all the other status variables are controlled, there is still an association between one's father's education and one's occupation. Similarly, father's occupation and respondent's occupation are not as strongly correlated, as indicated by the zero-order coefficient. The substantial Q_{XY} of 0.666 is reduced to a moderate one of $Q_{XY \cdot PART \, V, U, W}$ = 0.391, or a change of 0.275 Q points. This too is in the direction of explanation, but not enough to suggest that one's father's occupation does not influence one's occupation directly.

18.4 INTERPRETING A SURVEY-ANALYSIS CAUSAL MODEL

Our replication of the Blau and Duncan path analysis approach to occupational mobility produced similar and supporting results. They are careful to note that the "numerical estimates" of their model are valid only for the particular population under study and that the values are unlikely to be the same for other populations or even subpopulations within the United States. This leads them to explain that:

> The technique of path analysis is not a method for discovering causal laws but a procedure for giving a quantitative interpretation to the manifestations of a known or assumed causal system as it operates in a particular population. When the same interpretive structure is appropriate for two or more populations there is something to be learned by comparing their respective path coefficients and correlation patterns. We have not yet reached the stage at which such comparative study of stratification systems is feasible.[7]

We have attempted to make such a quantitative interpretation using a sample from a different subpopulation of the United States. Moreover, by comparing the respective path coefficients and correlation patterns, we have carried out the suggested comparative study of stratification and occupational mobility between the two samples.

The causal model that emerged from our path analysis is essentially like theirs, except for the differences between our coefficient values and theirs. Our causal diagram exactly resembles theirs in regard to the pattern of arrow connections between the five status variables. Our path analysis causal model indicates that the influence of father's education on respondent's first job and subsequent occupation is completely indirect and operates through the variables father's occupation and respondent's education. Blau and Duncan interpret this as indicative of intergenerational mobility.

In our survey analysis, in terms of gross effects the most influential determinants of one's occupation are those variables that are temporally closest in the causal scheme. That is, as one moves from right to left across the columns in Table 18.2.2 which shows the zero-order Q coefficients, there is a direct relationship between the recency of influence and higher positive associations. This finding confirms the same one mentioned in our earlier path analysis regarding the zero-order correlation coefficients and recency.

The fact that there is a greater degree of association between respondent's education and respondent's occupation ($Q_{UY} = 0.830$) than there is between father's education and father's occupation ($Q_{VX} = 0.696$) confirms a similar finding by Blau and Duncan, who interpret this difference as suggesting "a heightening of the effect of education on occupational status between fathers' and sons' generations."[8]

In the present findings, the association between father's occupation and respondent's education is higher than the association between father's

---

[7] *Ibid.*, p. 177.
[8] *Ibid.*, p. 169.

education and respondent's education, whether examined by path analysis or survey analysis. Blau and Duncan found just the reverse.[9] In terms of "uncontrolled" effects, then, the all-American sample indicates that one's father's occupation is consistently a more important determinant of one's subsequent status characteristics than is one's father's education. This can be seen quite clearly by comparing the coefficient values for variables X and V.

Father's occupation and respondent's education go a long way toward explaining the original moderate-positive-zero-order relationship between father's education and first job. To see this change, we compare the zero-order Q_{WV} of 0.503 with the second-order partial $Q_{WV \cdot TIED\ XU} = 0.145$, obtaining a difference of 0.358. This indicates that the variables father's occupation and respondent's education explain $0.358/0.503 = 71$ percent of the gross influence of father's education on respondent's first job as measured in Q units.

Blau and Duncan found that this relationship is completely explained by the same two test variables in terms of path analysis, i.e., that the influence is all indirect.[10] Notably, partial Q values are generally higher than path coefficients computed on the same data. Also, when the present data were assessed using path coefficients, there was a negligible correlation between father's education and respondent's first job ($p_{WV} < 0.05$). Therefore, our data indicate that respondent's first job *is* to some extent influenced by the level of father's education when measured in terms of survey analysis (Yule's Q partialed), but *is not* so influenced when measured in terms of path analysis.

The association $Q_{XY} = 0.666$ may be interpreted as meaning that we can do about 67 percent better than chance in predicting order on one's occupation knowing order on one's father's occupation, not controlling for any other variables. Now the partial Q coefficient $Q_{XY \cdot TIED\ V, U, W} = 0.391$ means that we can still do about 39 percent better than chance in predicting order on one's occupation knowing order on one's father's occupation, even after controlling for father's education, respondent's education, and respondent's first job.

This finding is in accord with the Blau and Duncan model, but the magnitude of the third-order partial Q is much greater than that of the corresponding path coefficient. Thus, our survey analysis model implies considerably less intergenerational occupational mobility than does our path analysis model or Blau and Duncan's. The original Blau and Duncan model shows that $p_{YX} = 0.115$, and ours that $p_{YX} = 0.092$, compared to our partial

---

[9] In terms of structural analysis, we find $Q_{XU} = 0.709$ with confidence limits of 0.691 and 0.727, and $Q_{VU} = 0.630$ with confidence limits of 0.609 and 0.650. In terms of path analysis, we find $r_{XU} = 0.376$, $r_{VU} = 0.335$, while they found the opposite, $r_{VU} = 0.453$, and $r_{XU} = 0.438$.

[10] Blau and Duncan, *op. cit.* p. 170.

Q of 0.391. Clearly, when this relationship is examined in terms of path analysis, there is a low positive association, but in terms of survey analysis, the same association is a moderate positive one. We can say, therefore, that there may be different theoretical implications between a path analysis causal model and a survey analysis causal model.

Our formulations suggest some additional differences, and some of these differences are important in terms of theories of occupational mobility.[11] For example, their model of stratification indicates that respondent's education is more influential on respondent's occupation than is respondent's first job. Our data suggest otherwise, both in our survey analysis model *and* when we use path analysis. The all-American sample, no matter how analyzed, appears to have just the opposite pattern. This finding continues to support our notion that the recency of occurrence of each of the status variables is an important aspect of occupational mobility. Some of the implications of this finding are apparent. One ought to remain in school as long as possible and attain the maximum amount of schooling in order to ensure a "better" career beginning.

Other implications are not so apparent but are equally important. For instance, our path analysis produces a pattern similar to our survey analysis in that the relationships between sequential variables controlling for all antecedent variables and ignoring all other connecting arrows produce a causal model with substantial and very strong relationships. That is, as we progress from father's education to father's occupation to respondent's education to respondent's first job to respondent's occupation, we find less step-by-step mobility than is implied by the Blau and Duncan model, which indicates a combination of high and low associations. Our models suggest either that our respondents actually possess different patterns of status characteristics than theirs, or that a different theory is warranted, or both. Whatever the reason may be, however, the implication is just as clear. One ought to have parents with a higher education to begin with in order to be assured of a better occupation. But how many of us can so choose?

It is important that our path analysis of the data shows that the All-American Sample possessed some characteristics different from those samples used by Blau and Duncan; however, they also possessed some very similar ones in terms of the causal schemes. For instance, one of our important findings is that the recency of occurrence of the status variables seems to be a powerful determining factor at any and all points in the process of stratification. Thus, our statistical analysis suggests that it is important to consider the theoretical temporal ordering of events in life in an investigation of occupational mobility.

Our survey analysis consistently supported our recency-of-occurrence hypothesis. In addition, it indicated that when respondents are classified

---

[11] All of the differences are statistically significant at the 0.05 level.

according to their locations among several status variables controlled simultaneously, i.e., when the variables are cross-tabulated, a causal model that continues to show association between any pair of variables is produced. This configuration leads us to a different formulation than path analysis does.

As a case in point, let us look at the comparative magnitude of the association between father's occupation and respondent's occupation. Path analysis leads us to say that almost all the influence on respondent's occupation from father's occupation is indirect and operates through two other variables. Survey analysis, on the other hand, indicates that much of the influence remains and that we can do considerably better than chance in predicting order on respondent's occupational status given order on father's occupation, even after controlling the other two variables.

One of the unique features of the present study is that it is a detailed example of the way different statistical analyses of the same data can result in different implications and suggestions. This leads to a cautionary remark of theoretical and methodological concern to fledgling social statisticians. We must pay careful attention to the theoretical assumptions underlying our analyses. If we are to use causal inference meaningfully, we must determine what it is we are trying to investigate rather than blindly trying to apply sophisticated statistical formulas, deriving complicated quantitative measures, plugging them into a high-speed electronic computer, and producing a causal diagram. The point is that we must be careful to be neither theoretically lazy nor dependent upon the computer for our theories. Sophisticated statistical methods are no substitute for clear thinking.

EXERCISES

1. In your own words, define, describe, or discuss the following terms and give an hypothetical example of each:
   consistent cross products
   inconsistent cross products
   zero-order association
   differential weight
   partial Q coefficients
   sixteen-fold table
   gross effects

2. Explain Table 18.2.2 to your roommate who has not studied sociology or statistics. Explain the computation for Q and interpret the results.

3. Give three statistical definitions of a zero-order Q coefficient.

4. In a short essay, compare the model established in Figure 18.4.3 with the path model described in Chapter 17.

5. We have pointed out some weaknesses of path analysis. Discuss these

weaknesses in terms of survey analysis and explain how they can be overcome.

6. Point out the major differences between survey analysis and path analysis.
   A. Which technique is better?
   B. For which types of analysis is one better than the other?
   C. Why is one better than the other?

7. Here is a model of the variables sex, education, occupation, and earnings.[12]
   1. Sex and education are unrelated.
   2. Sex affects occupation negatively.
   3. Sex affects earnings positively.
   4. Education affects occupation positively.
   5. Education affects income positively.
   6. Occupation affects income positively.

   Using the data presented below, answer the following questions:
   A. State a plausible sociological theory about the interrelationships between the variables sex, education, occupation, and income.
   B. Construct a causal model and diagram of the findings.
   C. Interpret the diagram.

Raw data for sex, education, occupation, and income (figures are from 1960 U.S. census tables, rounded to the nearest thousand)

S Sex	T Education	X Occupation	Y Income −	Y Income +	S Sex	T Education	X Occupation	Y Income −	Y Income +
+	+	+	3451	7294	−	+	+	6982	1150
+	+	−	4188	3817	−	+	−	2005	76
+	−	+	2397	1986	−	−	+	2538	168
+	−	−	13303	5871	−	−	−	6063	107

$N = 61396$

Four-variable results for data in Table 6.15

Q_{ST}

	Y Differs	Tied	3 Variable		
Tied	+.184	−.126	+.019	Second-order partial	−.126
X Differs	−.047	−.521	−.305	X effect	−.395
3 Variable	+.042	−.373	−.182	Y effect	+.310
				Joint effect	+.164

Q_{SX}

	Y Differs	Tied	3 Variable		
Tied	−.283	−.510	−.406	Second-order partial	−.510
T Differs	−.155	−.612	−.404	T effect	−.102
3 Variable	−.206	−.572	−.405	Y effect	+.227
				Joint effect	+.230

[12]Text and tables adapted from James A. Davis, *Elementary Survey Analysis* (Englewood Cliffs, N. J.: Prentice-Hall, 1971), pp. 154–157.

Q_{SY}

		X			Second-order partial	+.884
		Differs	Tied	3 Variable	T effect	−.017
	Tied	+.800	+.884	+.849	X effect	−.084
T	Differs	+.713	+.867	+.772	Joint effect	−.070
	3 Variable	+.748	+.877	+.810		

Q_{TX}

		Y			Second-order partial	+.705
		Differs	Tied	3 Variable	S effect	+.059
	Tied	+.748	+.705	+.726	Y effect	+.043
S	Differs	+.748	+.764	+.757	Joint effect	−.059
	3 Variable	+.748	+.733	+.740		

Q_{TY}

		X			Second-order partial	+.372
		Differs	Tied	3 Variable	S effect	−.112
	Tied	+.585	+.372	+.495	X effect	+.213
S	Differs	+.202	+.260	+.225	Joint effect	−.271
	3 Variable	+.414	+.326	+.378		

Q_{XY}

		T			Second-order partial	+.365
		Differs	Tied	3 Variable	S effect	−.475
	Tied	+.576	+.365	+.495	T effect	+.211
S	Differs	+.112	−.110	+.024	Joint effect	+.011
	3 Variable	+.369	+.146	+.282		

8. Set up an alternative theory and carry out steps A, B, and C from question 7.

9. Here is a model of the variables proximity, interaction, sentiment, and norms:

> "Integrated housing will make whites more favorable to [blacks] through the intervening variables of interaction (whites who live near [blacks] will tend to interact with them and interaction leads to liking, liking to interaction) and also norms (integrated housing developments will tend to develop social norms favoring positive interracial sentiments and people tend to adopt the sentiments favored by local norms). In addition, we assume that norms and interaction have a mutual positive relationship.[13]

Using the data below, answer the following questions:

A. State a plausible sociological theory about the interrelationships between the variables proximity, interaction, sentiment, and norms.
B. Construct a causal model and diagram of the findings.
C. Interpret the diagram.

Raw data for variables in Wilner study

S	T	X	Y −	Y +	S	T	X	Y −	Y +
+	+	+	33	76	−	+	+	20	43
+	+	−	31	25	−	+	−	37	36
+	−	+	19	14	−	−	+	36	27
+	−	−	27	15	−	−	−	118	41
								N =	598

S = proximity   T = interaction   X = norms   Y = sentiment

[13] *Ibid.*, p. 148.

Four-variable results for data in Table 6.12

Q_{ST}

		Y				
		Differs	Tied	3 Variable	Second-order partial	+.517
	Tied	+.499	+.517	+.508	X effect	+.056
X	Differs	+.638	+.573	+.610	Y effect	−.018
	3 Variable	+.580	+.546	+.564 = Zero order	Joint effect	+.083

Q_{SX}

		Y				
		Differs	Tied	3 Variable	Second-order partial	+.376
	Tied	+.349	+.376	+.363	T effect	+.092
T	Differs	+.566	+.468	+.524	Y effect	−.027
	3 Variable	+.480	+.425	+.455 = Zero order	Joint effect	+.125

Q_{SY}

		X				
		Differs	Tied	3 Variable	Second-order partial	+.064
	Tied	+.155	+.064	+.111	T effect	+.181
T	Differs	+.452	+.245	+.366	X effect	+.091
	3 Variable	+.335	+.162	+.258 = Zero order	Joint effect	+.116

Q_{TX}

		Y				
		Differs	Tied	3 Variable	Second-order partial	+.310
	Tied	+.450	+.310	+.387	S effect	+.154
S	Differs	+.601	+.464	+.542	Y effect	+.140
	3 Variable	+.534	+.394	+.473 = Zero order	Joint effect	−.003

Q_{TY}

		X				
		Differs	Tied	3 Variable	Second-order partial	+.452
	Tied	+.528	+.452	+.491	Effect of S	−.024
S	Differs	+.591	+.428	+.523	Effect of X	+.076
	3 Variable	+.563	+.440	+.508 = Zero order	Joint effect	+.087

Q_{XY}

		T				
		Differs	Tied	3 Variable	Second-order partial	+.389
	Tied	+.470	+.389	+.431	Effect of S	−.020
S	Differs	+.560	+.369	+.485	Effect of T	+.081
	3 Variable	+.520	+.380	+.459 = Zero order	Joint effect	+.110

Tables adapted from James A. Davis, *Elementary Survey Analysis* (Englewood Cliffs, N.J.: Prentice-Hall, 1971), pp. 149 and 151.

10. Set up an alternative theory and carry out steps A, B, and C. Discuss the two theories in terms of the findings.

APPENDIX

Statistical Tables

INSTRUCTIONS FOR TABLE A

How to Use the Table of Square Roots

To find the square root of a number rounded to two places you must select either the $\sqrt{N}$ column or the $\sqrt{10N}$ column. If the number is between 1.0 and 9.9, use the $\sqrt{N}$ column. If the number is between 10.0 and 99.0, use the $\sqrt{10N}$ column.

If the number is 100.0 or more, it must be converted to one between 1.0 and 99.0 inclusive. To do this, move the decimal point an even number of places to the *left* until you get a number between 1.0 and 99.0 inclusive. Then, use the same two rules. Once you select the correct column, find the square root entry in the body of the table. Since the decimal point was moved an even number of places to the left to get a number between 1.0 and 99.0, inclusive, you now must locate the decimal point of the square root in the table. To do so, move the decimal point of the square root entry *half* as many places to the right.

If the number if less than 1.0, that is, a fraction, move the decimal point an even number of places to the *right* to convert the number to one between 1.0 and 99.0. Once again, if the number is between 1.0 and 9.9, use the $\sqrt{N}$

Table A Table of square roots

1.0	1.00000	3.16228	5.5	2.34521	7.41620
1.1	1.04881	3.31662	5.6	2.36643	7.48331
1.2	1.09545	3.46410	5.7	2.38747	7.54983
1.3	1.14018	3.60555	5.8	2.40832	7.61577
1.4	1.18322	3.74166	5.9	2.42899	7.68115
			6.0	2.44949	7.74597
1.5	1.22474	3.87298	6.1	2.46982	7.81025
1.6	1.26491	4.00000	6.2	2.48998	7.87401
1.7	1.30384	4.12311	6.3	2.50998	7.93725
1.8	1.34164	4.24264	6.4	2.52982	8.00000
1.9	1.37840	4.35890			
			6.5	2.54951	8.06226
2.0	1.41421	4.47214	6.6	2.56905	8.12404
2.1	1.44914	4.58258	6.7	2.58844	8.18535
2.2	1.48324	4.69042	6.8	2.60768	8.24621
2.3	1.51658	4.79583	6.9	2.62679	8.30662
2.4	1.54919	4.89898			
			7.0	2.64575	8.36660
2.5	1.58114	5.00000	7.1	2.66458	8.42615
2.6	1.61245	5.09902	7.2	2.68328	8.48528
2.7	1.64317	5.19615	7.3	2.70185	8.54400
2.8	1.67332	5.29150	7.4	2.72029	8.60233
2.9	1.70294	5.38516			
			7.5	2.73861	8.66025
3.0	1.73205	5.47723	7.6	2.75681	8.71780
3.1	1.76068	5.56776	7.7	2.77489	8.77496
3.2	1.78885	5.65685	7.8	2.79285	8.83176
3.3	1.81659	5.74456	7.9	2.81069	8.88819
3.4	1.84391	5.83095			
3.5	1.87083	5.91608	8.0	2.82843	8.94427
3.6	1.89737	6.00000	8.1	2.84605	9.00000
3.7	1.92354	6.08276	8.2	2.86356	9.05539
3.8	1.94936	6.16441	8.3	2.88097	9.11043
3.9	1.97484	6.24500	8.4	2.89828	9.16515
4.0	2.00000	6.32456	8.5	2.91548	9.21954
4.1	2.02485	6.40312	8.6	2.93258	9.27362
4.2	2.04939	6.48074	8.7	2.94958	9.32738
4.3	2.07364	6.55744	8.8	2.96648	9.38083
4.4	2.09762	6.63325	8.9	2.98329	9.43398
4.5	2.12132	6.70820	9.0	3.00000	9.48683
4.6	2.14476	6.78233	9.1	3.01662	9.53939
4.7	2.16795	6.85565	9.2	3.03315	9.59166
4.8	2.19089	6.92820	9.3	3.04959	9.64365
4.9	2.21359	7.00000	9.4	3.06594	9.69536
5.0	2.23607	7.07107	9.5	3.08221	9.74679
5.1	2.25832	7.14143	9.6	3.09839	9.79796
5.2	2.28035	7.21110	9.7	3.11448	9.84886
5.3	2.30217	7.28011	9.8	3.13050	9.89949
5.4	2.32379	7.34847	9.9	3.14643	9.94987

Source: Adapted from John E. Freund, Statistics: A First Course, *1st ed.* (*Englewood Cliffs, N.J.: Prentice-Hall, 1976*), *pp. 317–18.*

column. And, if the number is between 10.0 and 99.0, use the $\sqrt{10N}$ column. The decimal point of the square root entry in the table must then be moved half as many places to the *left* in order to locate it in the right place.

Some examples help make this clearer. Let us take the whole numbers 2.3, 23, 230, and 2,300. The first number, 2.3, is easy to locate in the table in the N column, and its square root may be found in the $\sqrt{N}$ column, i.e., $\sqrt{2.3} = 1.51658$. Since the second number, 23, is between 10.0 and 99.0, you change the 2.3 in the N column to 23, that is $(10 \times 2.3) = 10N$. Thus, you use the $\sqrt{10N}$ column, i.e., $\sqrt{23.0} = 4.79583$. The third number, 230, must be converted. Follow the rule and move the decimal point an *even* number of places to the *left*, 2.30, getting the number 2.3. Thus, you use the $\sqrt{N}$ column, i.e., $\sqrt{230} = 15.1658$. The fourth number, 2,300, also must be converted. Move the decimal point an even number of places to the left, 23.00, getting 23. Thus, you use the $\sqrt{10N}$ column, i.e., $\sqrt{2,300} = 47.9583$.

Now, let us take the fractions .76, .076, .0076 and .00076. The first number, .76, must be converted. Move the decimal an *even* number of places to the *right*, .76, getting a number between 10.0 and 99.0. Thus, you use the $\sqrt{10N}$ column, i.e., $\sqrt{.76} = .871780$. The second number, .076, is converted by moving the decimal an even number of places to the right, 07.6. Thus, you use the $\sqrt{N}$ column, i.e., $\sqrt{.076} = .275681$. The third number, .0076, is converted to .0076. Thus, you use the $\sqrt{10N}$ column, i.e., $\sqrt{.0076} = .087178$. The fourth number, .00076, is converted to .0007.6. Thus, you use the $\sqrt{N}$ column, i.e., $\sqrt{.00076} = .027568$.

To use Table B we transform the raw score of a variable having the standard normal distribution into a z-score. The entries in Table B are the proportions of area in the standard normal curve with a mean of 0, a standard deviation of 1.00, and a total area also equal to 1.00. The standard normal distribution is symmetrical; thus Table B which shows the areas in positive z-scores may also be used for negative z-scores. They would appear to the left of 0 in the diagram.

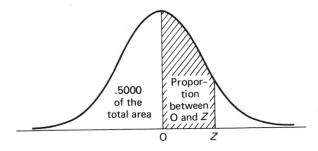

Table B The standard normal distribution

z	.00	.01	.02	.03	.04	.05	.06	.07	.08	.09
0.0	.0000	.0040	.0080	.0120	.0160	.0199	.0239	.0279	.0319	.0359
0.1	.0398	.0438	.0478	.0517	.0557	.0596	.0636	.0675	.0714	.0753
0.2	.0793	.0832	.0871	.0910	.0948	.0987	.1026	.1064	.1103	.1141
0.3	.1179	.1217	.1255	.1293	.1331	.1368	.1406	.1443	.1480	.1517
0.4	.1554	.1591	.1628	.1664	.1700	.1736	.1772	.1808	.1844	.1879
0.5	.1915	.1950	.1985	.2019	.2054	.2088	.2123	.2157	.2190	.2224
0.6	.2257	.2291	.2324	.2357	.2389	.2422	.2454	.2486	.2517	.2549
0.7	.2580	.2611	.2642	.2673	.2704	.2734	.2764	.2794	.2823	.2852
0.8	.2881	.2910	.2939	.2967	.2995	.3023	.3051	.3078	.3106	.3133
0.9	.3159	.3186	.3212	.3238	.3264	.3289	.3315	.3340	.3365	.3389
1.0	.3413	.3438	.3461	.3485	.3508	.3531	.3554	.3577	.3599	.3621
1.1	.3643	.3665	.3686	.3708	.3729	.3749	.3770	.3790	.3810	.3830
1.2	.3849	.3869	.3888	.3907	.3925	.3944	.3962	.3980	.3997	.4015
1.3	.4032	.4049	.4066	.4082	.4099	.4115	.4131	.4147	.4162	.4177
1.4	.4192	.4207	.4222	.4236	.4251	.4265	.4279	.4292	.4306	.4319
1.5	.4332	.4345	.4357	.4370	.4382	.4394	.4406	.4418	.4429	.4441
1.6	.4452	.4463	.4474	.4484	.4495	.4505	.4515	.4525	.4535	.4545
1.7	.4554	.4564	.4573	.4582	.4591	.4599	.4608	.4616	.4625	.4633
1.8	.4641	.4649	.4656	.4664	.4671	.4678	.4686	.4693	.4699	.4706
1.9	.4713	.4719	.4726	.4732	.4738	.4744	.4750	.4756	.4761	.4767
2.0	.4772	.4778	.4783	.4788	.4793	.4798	.4803	.4808	.4812	.4817
2.1	.4821	.4826	.4830	.4834	.4838	.4842	.4846	.4850	.4854	.4857
2.2	.4861	.4864	.4868	.4871	.4875	.4878	.4881	.4884	.4887	.4890
2.3	.4893	.4896	.4898	.4901	.4904	.4906	.4909	.4911	.4913	.4916
2.4	.4918	.4920	.4922	.4925	.4927	.4929	.4931	.4932	.4934	.4936
2.5	.4938	.4940	.4941	.4943	.4945	.4946	.4948	.4949	.4951	.4952
2.6	.4953	.4955	.4956	.4957	.4959	.4960	.4961	.4962	.4963	.4964
2.7	.4965	.4966	.4967	.4968	.4969	.4970	.4971	.4972	.4973	.4974
2.8	.4974	.4975	.4976	.4977	.4977	.4978	.4979	.4979	.4980	.4981
2.9	.4981	.4982	.4982	.4983	.4984	.4984	.4985	.4985	.4986	.4986
3.0	.4987	.4987	.4987	.4988	.4988	.4989	.4989	.4989	.4990	.4990

Source: Adapted from John E. Freund, Statistics: A First Course, *2nd ed.* (*Englewood Cliffs, N.J.: Prentice-Hall, 1976*), p. 335.

Table C   Random numbers

24418	23508	91507	76455	54941	72711	39406
57404	73678	08272	62941	02349	71389	45605
77644	98489	86268	73652	98210	44546	27174
68366	65614	01443	07607	11826	91326	29664
64472	72294	95432	53555	96810	17100	35066
88205	37913	98633	81009	81060	33449	68055
98455	78685	71250	10329	56135	80647	51404
48977	36794	56054	59243	57361	65304	93258
93077	72941	92779	23581	24548	56415	61927
84533	26564	91583	83411	66504	02036	02922
11338	12903	14514	27585	45068	05520	56321
23853	68500	92274	87026	99717	01542	72090
94096	74920	25822	98026	05394	61840	83089
83160	82362	00350	98536	38155	42661	02363
97425	47335	69709	01386	74319	04318	99387
83951	11954	24317	20345	18134	90062	10761
93085	35203	05740	03206	92012	42710	34650
33762	83193	58045	89880	78101	44392	53767
49665	85397	85137	30496	23469	42846	94810
37541	82627	80051	72521	35342	56119	97190
22145	85304	35348	82854	55846	18076	12415
27153	08662	61078	52433	22184	33998	87436
00301	49425	66682	25442	83668	66236	79655
43815	43272	73778	63469	50083	70696	13558
14689	86482	74157	46012	97765	27552	49617
16680	55936	82453	19532	49988	13176	94219
86938	60429	01137	86168	78257	86249	46134
33944	29219	73161	46061	30946	22210	79302
16045	67736	18608	18198	19468	76358	69203
37044	52523	25627	63107	30806	80857	84383
61471	45322	35340	35132	42163	60332	98851
47422	21296	16785	66393	39249	51463	95963
24133	39719	14484	58613	88717	20980	77900
67253	67064	10748	16006	16767	57345	42285
62382	76941	01635	35829	77516	98468	51686
98011	16503	09201	03523	87192	66483	55649
37366	24386	20654	85117	74078	64120	04643
73587	83993	54176	05221	94119	20108	78101
33583	68291	50547	96085	62180	27453	18567
02878	33223	39109	49536	56199	05993	71201
91498	41673	17195	33175	04994	09879	70337
91127	19815	30219	55591	21725	43827	78862
12997	55013	18662	81724	24305	37661	18956
96098	13651	15393	69995	14762	69734	89150
97627	17837	10472	18983	28387	99781	52977
40064	47981	31484	76603	54088	91095	00010
16239	68743	71374	55863	22672	91609	51514
58354	24913	20435	30965	17453	65623	93058
52567	65085	60220	84641	18273	49604	47418
06236	29052	91392	07551	83532	68130	56070

Source: Adapted from John E. Freund, Statistics: A First Course, *2nd ed. (Englewood Cliffs, N.J.: Prentice-Hall, 1976), pp. 357–58.*

Table C   Random numbers (continued)

48611	62866	33963	14045	79451	04934	45576
78812	03509	78673	73181	29973	18664	04555
19472	63971	37271	31445	49019	49405	46925
51266	11569	08697	91120	64156	40365	74297
55806	96275	26130	47949	14877	69594	83041
77527	81360	18180	97421	55541	90275	18213
77680	58788	33016	61173	93049	04694	43534
15404	96554	88265	34537	38526	67924	40474
14045	22917	60718	66487	46346	30949	03173
68376	43918	77653	04127	69930	43283	35766
93385	13421	67957	20384	58731	53396	59723
09858	52104	32014	53115	03727	98624	84616
93307	34116	49516	42148	57740	31198	70336
04794	01534	92058	03157	91758	80611	45357
86265	49096	97021	92582	61422	75890	86442
65943	79232	45702	67055	39024	57383	44424
90038	94209	04055	27393	61517	23002	96560
97283	95943	78363	36498	40662	94188	18202
21913	72958	75637	99936	58715	07943	23748
41161	37341	81838	19389	80336	46346	91895
23777	98392	31417	98547	92058	02277	50315
59973	08144	61070	73094	27059	69181	55623
82690	74099	77885	23813	10054	11900	44653
83854	24715	48866	65745	31131	47636	45137
61980	34997	41825	11623	07320	15003	56774
99915	45821	97702	87125	44488	77613	56823
48293	86847	43186	42951	37804	85129	28993
33225	31280	41232	34750	91097	60752	69783
06846	32828	24425	30249	78801	26977	92074
32671	45587	79620	84831	38156	74211	82752
82096	21913	75544	55228	89796	05694	91552
51666	10433	10945	55306	78562	89630	41230
54044	67942	24145	42294	27427	84875	37022
66738	60184	75679	38120	17640	36242	99357
55064	17427	89180	74018	44865	53197	74810
69599	60264	84549	78007	88450	06488	72274
64756	87759	92354	78694	63638	80939	98644
80817	74533	68407	55862	32476	19326	95558
39847	96884	84657	33697	39578	90197	80532
90401	41700	95510	61166	33757	23279	85523
78227	90110	81378	96659	37008	04050	04228
87240	52716	87697	79433	16336	52862	69149
08486	10951	26832	39763	02485	71688	90936
39338	32169	03713	93510	61244	73774	01245
21188	01850	69689	49426	49128	14660	14143
13287	82531	04388	64693	11934	35051	68576
53609	04001	19648	14053	49623	10840	31915
87900	36194	31567	53506	34304	39910	79630
81641	00496	36058	75899	46620	70024	88753
19512	50277	71508	20116	79520	06269	74173

Table D   Factorial values, 1–19

N	$N!$
0	1
1	1
2	2
3	6
4	24
5	120
6	720
7	5,040
8	40,320
9	362,880
10	3,628,800
11	39,916,800
12	479,001,600
13	6,227,020,800
14	87,178,291,200
15	1,307,674,368,000
16	20,922,789,888,000
17	355,687,428,096,000
18	6,402,373,705,728,000
19	121,645,100,408,832,000

Table E   Binomial coefficients

n	$\binom{n}{0}$	$\binom{n}{1}$	$\binom{n}{2}$	$\binom{n}{3}$	$\binom{n}{4}$	$\binom{n}{5}$	$\binom{n}{6}$	$\binom{n}{7}$	$\binom{n}{8}$	$\binom{n}{9}$	$\binom{n}{10}$
0	1										
1	1	1									
2	1	2	1								
3	1	3	3	1							
4	1	4	6	4	1						
5	1	5	10	10	5	1					
6	1	6	15	20	15	6	1				
7	1	7	21	35	35	21	7	1			
8	1	8	28	56	70	56	28	8	1		
9	1	9	36	84	126	126	84	36	9	1	
10	1	10	45	120	210	252	210	120	45	10	1
11	1	11	55	165	330	462	462	330	165	55	11
12	1	12	66	220	495	792	924	792	495	220	66
13	1	13	78	286	715	1287	1716	1716	1287	715	286
14	1	14	91	364	1001	2002	3003	3432	3003	2002	1001
15	1	15	105	455	1365	3003	5005	6435	6435	5005	3003
16	1	16	120	560	1820	4368	8008	11440	12870	11440	8008
17	1	17	136	680	2380	6188	12376	19448	24310	24310	19448
18	1	18	153	816	3060	8568	18564	31824	43758	48620	43758
19	1	19	171	969	3876	11628	27132	50388	75582	92378	92378
20	1	20	190	1140	4845	15504	38760	77520	125970	167960	184756

If necessary, use the identity $\binom{n}{k} = \binom{n}{n-k}$.

Source: Adapted from John E. Freund, Statistics: A First Course, *2nd ed. (Englewood Cliffs, N.J.: Prentice-Hall, 1976), p. 344.*

Table F   Student's *t* distribution (*t* statistic)

df	Level of significance for one-tailed test					
	.10	.05	.025	.01	.005	.0005
	Level of significance for two-tailed test					
	.20	.10	.05	.02	.01	.001
1	3.078	6.314	12.706	31.821	63.657	636.619
2	1.886	2.920	4.303	6.965	9.925	31.598
3	1.638	2.353	3.182	4.541	5.841	12.941
4	1.533	2.132	2.776	3.747	4.604	8.610
5	1.476	2.015	2.571	3.365	4.032	6.859
6	1.440	1.943	2.447	3.143	3.707	5.959
7	1.415	1.895	2.365	2.998	3.499	5.405
8	1.397	1.860	2.306	2.896	3.355	5.041
9	1.383	1.833	2.262	2.821	3.250	4.781
10	1.372	1.812	2.228	2.764	3.169	4.587
11	1.363	1.796	2.201	2.718	3.106	4.437
12	1.356	1.782	2.179	2.681	3.055	4.318
13	1.350	1.771	2.160	2.650	3.012	4.221
14	1.345	1.761	2.145	2.624	2.977	4.140
15	1.341	1.753	2.131	2.602	2.947	4.073
16	1.337	1.746	2.120	2.583	2.921	4.015
17	1.333	1.740	2.110	2.567	2.898	3.965
18	1.330	1.734	2.101	2.552	2.878	3.922
19	1.328	1.729	2.093	2.539	2.861	3.883
20	1.325	1.725	2.086	2.528	2.845	3.850
21	1.323	1.721	2.080	2.518	2.831	3.819
22	1.321	1.717	2.074	2.508	2.819	3.792
23	1.319	1.714	2.069	2.500	2.807	3.767
24	1.318	1.711	2.064	2.492	2.797	3.745
25	1.316	1.708	2.060	2.485	2.787	3.725
26	1.315	1.706	2.056	2.479	2.779	3.707
27	1.314	1.703	2.052	2.473	2.771	3.690
28	1.313	1.701	2.048	2.467	2.763	3.674
29	1.311	1.699	2.045	2.462	2.756	3.659
30	1.310	1.697	2.042	2.457	2.750	3.646
40	1.303	1.684	2.021	2.423	2.704	3.551
60	1.296	1.671	2.000	2.390	2.660	3.460
120	1.289	1.658	1.980	2.358	2.617	3.373
∞	1.282	1.645	1.960	2.326	2.576	3.291

Source: Adapted from R. A. Fisher and F. Yates, Statistical Tables for Biological, Agri-cultural and Medical Research (*Edinburgh and London: Oliver & Boyd, Ltd., 1948*), *by permission.*

Table G   The chi-square distribution (X^2 statistic)

Probability

df	.99	.98	.95	.90	.80	.70	.50	.30	.20	.10	.05	.02	.01	.001
1	$.0^3157$	$.0^3628$	.00393	.0158	.0642	.148	.455	1.074	1.642	2.706	3.841	5.412	6.635	10.827
2	.0201	.0404	.103	.211	.446	.713	1.386	2.408	3.219	4.605	5.991	7.824	9.210	13.815
3	.115	.185	.352	.584	1.005	1.424	2.366	3.665	4.642	6.251	7.815	9.837	11.341	16.268
4	.297	.429	.711	1.064	1.649	2.195	3.357	4.878	5.989	7.779	9.488	11.668	13.277	18.465
5	.554	.752	1.145	1.610	2.343	3.000	4.351	6.064	7.289	9.236	11.070	13.388	15.086	20.517
6	.872	1.134	1.635	2.204	3.070	3.828	5.348	7.231	8.558	10.645	12.592	15.033	16.812	22.457
7	1.239	1.564	2.167	2.833	3.822	4.671	6.346	8.383	9.803	12.017	14.067	16.622	18.475	24.322
8	1.646	2.032	2.733	3.490	4.594	5.527	7.344	9.524	11.030	13.362	15.507	18.168	20.090	26.125
9	2.088	2.532	3.325	4.168	5.380	6.393	8.343	10.656	12.242	14.684	16.919	19.679	21.666	27.877
10	2.558	3.059	3.940	4.865	6.179	7.267	9.342	11.781	13.442	15.987	18.307	21.161	23.209	29.588
11	3.053	3.609	4.575	5.578	6.989	8.148	10.341	12.899	14.631	17.275	19.675	22.618	24.725	31.264
12	3.571	4.178	5.226	6.304	7.807	9.034	11.340	14.011	15.812	18.549	21.026	24.054	26.217	32.909
13	4.107	4.765	5.892	7.042	8.634	9.926	12.340	15.119	16.985	19.812	22.362	25.472	27.688	34.528
14	4.660	5.368	6.571	7.790	9.467	10.821	13.339	16.222	18.151	21.064	23.685	26.873	29.141	36.123
15	5.229	5.985	7.261	8.547	10.307	11.721	14.339	17.322	19.311	22.307	24.996	28.259	30.578	37.697
16	5.812	6.614	7.962	9.312	11.152	12.624	15.338	18.418	20.465	23.542	26.296	29.633	32.000	39.252
17	6.408	7.255	8.672	10.085	12.002	13.531	16.338	19.511	21.615	24.769	27.587	30.995	33.409	40.790
18	7.015	7.906	9.390	10.865	12.857	14.440	17.338	20.601	22.760	25.989	28.869	32.346	34.805	42.312
19	7.633	8.567	10.117	11.651	13.716	15.352	18.338	21.689	23.900	27.204	30.144	33.687	36.191	43.820
20	8.260	9.237	10.851	12.443	14.578	16.266	19.337	22.775	25.038	28.412	31.410	35.020	37.566	45.315
21	8.897	9.915	11.591	13.240	15.445	17.182	20.337	23.858	26.171	29.615	32.671	36.343	38.932	46.797
22	9.542	10.600	12.338	14.041	16.314	18.101	21.337	24.939	27.301	30.813	33.924	37.659	40.289	48.268
23	10.196	11.293	13.091	14.848	17.187	19.021	22.337	26.018	28.429	32.007	35.172	38.968	41.638	49.728
24	10.856	11.992	13.848	15.659	18.062	19.943	23.337	27.096	29.553	33.196	36.415	40.270	42.980	51.179
25	11.524	12.697	14.611	16.473	18.940	20.867	24.337	28.172	30.675	34.382	37.652	41.566	44.314	52.620
26	12.198	13.409	15.379	17.292	19.820	21.792	25.336	29.246	31.795	35.563	38.885	42.856	45.642	54.052
27	12.879	14.125	16.151	18.114	20.703	22.719	26.336	30.319	32.912	36.741	40.113	44.140	46.963	55.476
28	13.565	14.847	16.928	18.939	21.588	23.647	27.336	31.391	34.027	37.916	41.337	45.419	48.278	56.893
29	14.256	15.574	17.708	19.768	22.475	24.577	28.336	32.461	35.139	39.087	42.557	46.693	49.588	58.302
30	14.953	16.306	18.493	20.599	23.364	25.508	29.336	33.530	36.250	40.256	43.773	47.962	50.892	59.703

For larger values of df, the expression $\sqrt{2\chi^2} - \sqrt{2df - 1}$ may be used as a normal deviate with unit variance, remembering that the probability for χ^2 corresponds with that of a single tail of the normal curve.

Source: Adapted from R. A. Fisher and F. Yates, Statistical Tables for Biological, Agricultural and Medical Research (*Edinburgh and London: Oliver & Boyd, Ltd., 1948*), *by permission.*

Index